Soundscape Ecology

Almo Farina

Soundscape Ecology

Principles, Patterns, Methods and Applications

Springer

Almo Farina
Department of Basic Sciences
and Foundations
Urbino University
Urbino, Pesaro-Urbino
Italy

ISBN 978-94-007-7373-8 ISBN 978-94-007-7374-5 (eBook)
DOI 10.1007/978-94-007-7374-5
Springer Dordrecht Heidelberg New York London

Printed on acid-free paper

Springer is part of Springer Science+Business Media (www.springer.com)

Preface

The world is full of sounds that are shared by humans and many other organisms, but little is known about the role and importance of the acoustic context to guide evolutionary processes at different aggregation scales.

Recent advances in sound technology offer a friendly use of a broad collection of different devices able to store large digital sound files and to enter into contact with worldwide acoustic libraries from different biogeographic areas.

The style of the modern life is characterized by an increasing exposure to technological noise produced by cars, airplanes, trains, ships, and industrial activity, and such exposure, named sonic pollution, is responsible in animal populations and in human societies as well for profound changes in their behavior and in many life traits, creating a growing risk of extinction, especially for marine mammals. In addition, annoyance and relevant pathologies are effects that humans suffer when exposed to massive doses of noise.

The relationships between sounds and organisms have become a matter of interest for ecology and fully justify the creation of a distinct branch of ecological research: soundscape ecology.

This terms is not new. Truax used such a term in defining soundscape ecology as the "The study of the systematic relationships between humans and sonic environments" [Truax B (ed) (1999) *Handbook for acoustic ecology*. Cambridge Street Publishing, Toronto]. From that time other scientists and scholars have utilized sound as a proxy to study several processes in animal communication, in community aggregation, in long-term monitoring, etc. We are sure we shall not disturb anyone if we consider soundscape ecology in a broader sense as an ecological discipline that investigates the effects of sounds on the structure and dynamics of animal species at the level of individual populations and community, humans included. The soundscape is an important theme to be developed, especially in urban societies, with reverberations in still-natural environments in remote regions. At the same time, soundscape ecology can approach the effects of landscape structures and biological dynamics on sounds.

The aim of this book is to create a learning tool primarily to teach soundscape ecology in advanced courses of ecology at universities, and at the same time to offer

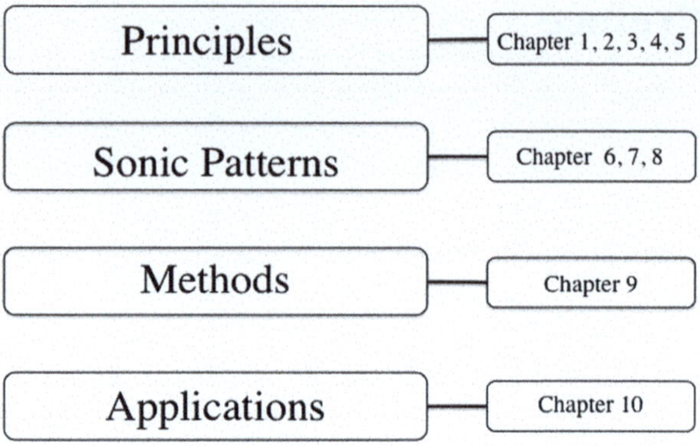

Fig. 1 Schematic organization of the content

a guide to consultants and practitioners and finally updated reading for experts and amateurs.

Despite being aware of the limitations of this book, which may appear fragmented for the numerous topics discussed, our aim is to create an educational path able to introduce and guide students to the discovery of the soundscape domain and its complexity. To do this, in the text is included all the epistemic and technological information required to achieve the best results.

Definitively, this book is not a review of the state of the art of bioacoustics, ecological acoustics, or psychoacoustics, but a first annotated guide to explore some emergent aspects of the different approaches to sonic phenomena, introducing the "world of the sounds" with the clear and hopeful purpose to integrate the separate aspects of knowledge into a new emergent common disciplinary framework.

Up to the present, the epistemological context of soundscape ecology has not been well delineated, and probably this represents the major difficulty: to prime a new ecological discipline able to face the challenge to consider the acoustics of natural environments and the anthropogenic domain as components of the same phenomenological domain.

The book is organized in four main sections: Principles (Chaps. 1, 2, 3, 4, and 5), Sonic Patterns (Chaps. 6, 7, and 8), Methods (9), and Applications (10) (Fig. 1). In the first section are introduced the main sonic characters of the environment, the theories that actually have been developed to explain acoustic patterns and dynamics in natural and human-influenced landscapes, and all the other epistemic tools necessary for the best understanding of acoustic phenomenon. For this, elements of biosemiotics, cognitive ecology, and landscape ecology are briefly described (Chap. 1, Soundscape and Landscape Ecologies). The relationship between the geographic landscape and the soundscape is discussed, presenting models and conjectures. In particular, the principles that guide landscape ecology are compared with the principles adopted in soundscape ecology, trying to find points of contact and integrations that make sense across appropriate spatial and temporal scales.

The sonic character of the landscape under the constraint of the environmental proxies such as soil morphology, vegetation, and weather conditions is described (Chap. 2, Sonic Characters of Landscapes).

Theories, hypotheses, and models that explain the relationship between animals and soundscape are described and discussed according to the most advanced evolutionary and adaptive visions in terms of bioacoustics theories (Chap. 3, Bioacoustics Theories) and of communication theories (Chap. 4, Communication Theories).

Particular attention has been paid to the description of the acoustic dimension of humans (Chap. 5, Human Dimension of the Soundscape: From Individuals to Society).

In the second section, some sonic patterns such as noise (Chap. 6, Sonic Patterns I: The Noise), choruses (Chap. 7, Sonic Patterns II: The Choruses), and sound and vibrations from the soil (Chap. 8, Sonic Patterns III: Sounds and Vibrations from Soil) are described.

In the third section, acoustic fundamentals, parameters, and methodologies offer a short guide how to handle the sounds, the most updated list of devices, software and metrics used to capture, store, and process sound files (Chap. 9, Methods). Special emphasis is reserved to the metrics that are available to extract information on processes and patterns emerging from the sonic environment.

In the fourth section (Chap. 10, Applications), applications of soundscape ecology procedures are described in detail, with special emphasis on the protection of the sonic context in fragile and valuable landscapes such as natural parks and protected areas. Climatic change monitoring schemes are also considered, as well as all the problems related to acoustic pollution, that in the marine environment affects the large pelagic fishes and marine mammals so seriously, and in terrestrial landscapes the human societies that live in urban contexts.

At the beginning of each chapter a short abstract summarizes the main arguments considered, and full references of the authors cited in the text are also provided.

Urbino, Italy Almo Farina

Contents

Chapter 1
Soundscape and Landscape Ecology

The soundscape is defined as the entire sonic energy produced by a landscape and is the result of the overlap of three distinct sonic sources: geophonies, biophonies, and anthrophonies.

Geophonies are the result of sonic energy produced by nonbiological natural agents such as winds, volcanoes, sea waves, running water, rain, thunderstorms, lightning, avalanches, earthquakes, and flooding.

Biophonies are the results of animal vocalizations (song, contact and alarm calls, voices).

Anthrophonies are the result of all the sounds produced by technical devices (engines, blades, wheel revolutions, industries, etc.).

The spatial overlap of geophonic, biophonic, and anthrophonic patterns creates the sonotopes. Each sonotope for the effect of behavioral constraints can be further subdivided into soundtopes. In the zone of contact between different soundtopes, a tension zone of acoustic uncertainty/overlap is called a sonotone.

The sonic environment, important for most organisms that perceive, in a species-specific way, acoustic information coming from different components of the ecosystems, represents a promising field of ecological research: "soundscape ecology." This new ecological discipline was presented to a landscape ecology congress in 2009 with a symposium titled "Soundscape Ecology: Merging Bioacoustics and Landscapes," where a primary role was recognized for different components such as acoustic ecology, landscape ecology, bioacoustics, urban and environmental acoustics, behavioral ecology, and biosemiotics.

Soundscape ecology finds important applications in the assessment of the environmental quality of parks and protected areas, in urban planning and design, in ethology and anthropology, and finally in long-term monitoring of the effects of climatic changes.

A. Farina, *Soundscape Ecology: Principles, Patterns, Methods and Applications*,
DOI 10.1007/978-94-007-7374-5_1,

1.1 Introduction

Access of all sensory perceptions represents a necessary step toward sustainability, expressed not only in terms of ecosystem carrying capacity but also by the growth and maintenance of the quality of human life. The aesthetic dimension of sustainability is largely represented by the quality of the sonic environment in urban, rural, and wild contexts.

Although sonic cues contribute to landscape shaping and functioning, little is known about the relationships between the sonic environment and landscapes. As well expressed by Hedfors and Berg (2003) "*the sparseness of literature on auditory aspects of landscapes reflects the hegemony of vision*" when compared with the sonic component of the environment.

Moreover, an informed design could contribute to accomplish this neglected perspective and open a way toward a more healthy environment.

It is our aim in this first chapter to introduce the theoretical bases necessary to fully understand the importance and the function of the sonic environment for so many organisms, humans included, and to describe the mechanisms that are at the basis of active and passive sonic communication. We try to describe the highest number of problematics in which acoustics emissions are embedded, such as overexposure, sound degradation, and sound masking.

Sound is an important component of the majority of terrestrial, freshwater, and marine ecosystems and becomes an essential context with which organisms daily interact performing their vital functions. Sound is transmitted in air and in water as well, but soil and rock are permeable to vibration to some extent. Thus, the transmission of sounds that occurs in solids, in liquids, and in a gas medium has different and distinct physical modalities (such as speed, impedance, excess of attenuation). For this reason, we can expect different mechanisms of perception and strategies of communication used by organisms living in these different contexts, that is, we expect different "sonic ecologies."

We can say that every phase of matter and every type of material reacts differently to sound waves (vibration), and this can be useful from remote sensing procedures to monitoring the dynamics of environmental context and an organism's behavior.

1.2 A Semantic Caveat: Sound, Sonic, Acoustic

Although there are not precise rules about the more appropriate use of some terms (Brown et al. 2011), in this book we have used words such as sound, sonic, and acoustic with precise meanings and for specific purposes:

Sound refers to the informative energy that participates in the physical phenomenon of pressure waves of a vibrating object.

Sonic is the adjective that refers to every physical or conceptual object created by the sound (e.g., sonic environment, sonic ambience, sonic information, sonic patterns).

Acoustic is the adjective that refers to the organismic perception, interpretation, and reaction to a sound (e.g., acoustic adaptive hypothesis, acoustic alarm, acoustic behavior).

1.3 The Definition of Soundscape

As for the landscape, there are many definitions of the soundscape. The soundscape can be simply defined as an acoustical composition that results from the voluntary or involuntary overlap of different sounds of physical or biological origin.

Payne et al. (2009) has defined the soundscape thus: "*Soundscapes are the totality of all sounds within a location with an emphasis in the relationship between individual's or society's perception of, understanding of and interaction with the sonic environment.*"

The Oxford Dictionary defines the soundscape as "*the sounds heard in a particular location, considered as a whole.*" In this last definition we can distinguish two key words: sounds and location. Later, we will connect this "location" with the landscape and related concepts. Another definition of soundscape, "*An atmosphere or environment created by or with sound,*" attributes to sound the important role of a specific dimension that embraces organisms. Not least, soundscape can be defined as an emotive environment.

According to a more linguistically rigorous perspective, "sound" + "scape" results a semantic paradox because sound can be heard but not seen. It would be more correct to define the soundscape as the distribution of sounds across a landscape when the landscape is considered a geographic entity and not a complex cognitive agency (Farina 2006). But despite this semantic remark, in this book we use the term "soundscape" to indicate the geographic distribution of multisource sound perceptions as usually considered in a broad scientific and educational context.

The soundscape is the result of the overlapping of the sounds from geophonic (wind, flowing water, sea waves, eruptions), biophonic (vocalizations, contact and alarm calls, songs), and anthrophonic (industrial and urban activities; road, marine, and air traffic) sources, that in turn are strictly related to the structure and functioning of the geographic landscapes (Pijanowski et al. 2011a).

The soundscape is the result of sounds that are produced by abiotic or biotic agents and that are first perceived and successively interpreted by organisms: this means that a cognitive component must be added or at least considered in the soundscape process.

Different terms are used to create a semantic regarding this complex subject. In particular, different terms are used as synonym of soundscape:

1. The acoustic environment
2. The sonic environment

3. The sound environment
4. The environment of sound
5. The aural space
6. The natural acoustic environmental sounds
7. The ambient conditions
8. Quiet areas (areas where environmental noise quality is good)
9. City soundscape
10. The total ambient acoustic environment
11. The total soundscape
12. The acoustic soundscape

And according to a human context, Jennings and Cain (2013) argue: "*The perception of the soundscape is inherently personal and affected by what a listener, each with unique set of experiences and preferences, brings to the listening situation. This framework is therefore underpinned by the proposition that a person's perception of a soundscape depends most strongly on the activity they are doing at the time, and consequently their corresponding state of listening.*"

Environmental sound often is considered simply a noise, but reducing the noise levels often does not provide improvement of the negative sensations that such a sound produces on the listeners. Noise mitigation in many cases is not sufficient to create a pleasant sonic environment.

The soundscape approach can solve in part the difficulties connected with the sonic environment evaluation and management.

According to an anthropogenic perspective, the term soundscape emphasizes the way the acoustic environment is perceived, understood, and interpreted by different human aggregations (individuals, groups, and society). The soundscape, in this case, is a perceptual construct and a physical phenomenon too. When the soundscape is considered the noise of a place, apparently a different approach is requested to address the problem, but later it will be possible to see how the two aspects are strictly connected.

The soundscape can be considered also as a type of language uttered by a landscape, the emergent sonic information that creates a living context in which sonic information and acoustic communication represents biosemiotics tools that are used by vocal organisms in active and passive cognitive processes.

Interest in the soundscape, at its first beginning considered a peculiarity only of musicians, pertains to many categories of scholars, practitioners, and decision makers ranging from psychologists, engineers, physicians, architects, designers, and sociologists to biologists and ecologists.

The soundscape is an acoustic context but also a cultural domain that greatly contributes to defining the characteristics of a region, the culture of the people, and more in general the cultural heritage (Scarre and Lawson 2006). The strict and continuous contacts between sound and environment emphasizes the importance of this sonic context in shaping use and traditions and reinforcing the sense of the place and every other issue related to human culture and heritage.

It is relevant to consider that the globalization we find in the economic domain has been extended progressively to other domains such as the psychological aspects, degrading some identitarian feelings. The soundscape could work to reconcile the local to the global dimension of our perceptional Umwelt because it maintains also, involuntarily, the "sense of place" in people.

Any landscape is characterized by a sonic environment, and a sonic environment is composed of a plethora of messages from a transmitter to volunteer or non-volunteer receivers. Often sonic ambience is associated with noise pollution, and the sonic environment, when not appropriately identified, does not receive the necessary attention. However, sounds such as bird songs or a waterfall create a sense of well-being and relaxation in humans. The soundscape becomes a dimension in which the integration of acoustic and visual stimuli creates a new perceptive dimension in which cognition represents a fundamental synthesis. The soundscape can receive much attention from the field of nature conservation and therapeutic evaluation, and collection of information on the soundscape represents a valuable asset in the fields of nature conservation, behavioral sciences, and human social processes.

1.4 A Brief History of Soundscape Ecology: Origins and Perspectives

The soundscape ecology represents a new transdisciplinary field of research strictly connected to landscape ecology that has recently been posed to the attention of scholars and students (Pijanowski et al. 2011a; Truax and Barrett 2011). We utilize the term transdisciplinary with the same meaning used by Zev Naveh to describe the role and function of landscape ecology at the beginning of its scientific foundation Naveh and Lieberman (1984). The origin of soundscape ecology has been the result of the intentions of small and isolated groups of landscape ecologists who had recognized the necessity to investigate the patterns and processes created by sounds at all levels of biological and ecological complexity.

Despite a long tradition in soundscape studies focused on the human dimension (Schafer 1977), the idea of a "Soundscape Ecology" was presented for the first time by Bryan Pijanowski and Almo Farina at the 24th annual meeting of US-IALE Branch in Snow Bird, Utah (USA), April 2009, in a symposium titled "Soundscape Ecology: Merging Bioacoustics and Landscapes." The genuine interest in this new topic has gained the interest of the scientific community with its presentation in the journal *BioScience* (AIBS) (Pijanowski et al. 2011a) and successively with seminal contributions that were collected in a special issue of the journal *Landscape Ecology* (Springer) (Pijanowski and Farina 2011). Soon, the soundscape ideas attracted other scientific competencies in the applied field of acoustics (Davies 2013). Since that time, interest in soundscape ecology has grown very quickly, actually chronicled by the organization of symposia and seminars on this topic. For example, one of the most recent, organized by Almo Farina in Florence (14 June 2012) under the aegis of Uniscape, has produced the Careggi Landscape Declaration on Soundscapes (June 2012) (see Chap. 10).

Definitively, soundscape ecology recognizes the integration of other scientific disciplines such as acoustic ecology, landscape ecology, bioacoustics, urban and environmental acoustics, behavioral ecology, and biosemiotics and appears to be a promising field to investigate the role of the sonic ambience in maintaining vital functions in animals and humans (Farina et al. 2011a).

The soundscape, which means the sonic information that we collect in a place, is today easier to record and process than in the recent past. Similar to images that can be recorded with a digital compact camera, sound can be easily handled, stored, reproduced, and processed outside its context.

Advanced digital technologies of field recording at low cost offer new tools to investigate frequency and temporal patterns (such as Zoom H4, Zoom, JP; Songmeter SM2, Wildlife Acoustics, MA, USA).

New metrics such as the acoustic entropy index (H) (Sueur et al. 2008) and the acoustic complexity index (ACI) (Pieretti et al. 2011; Farina et al. 2011b, 2012), supported by new remote sensing technologies (Bouzid et al. 2013), are constantly being tested (see Chap. 9).

The complexity that emerges from the information which is contained in a soundscape can be analyzed and interpreted by the application of functional models such as the soundtope hypothesis (Farina et al. 2011a), the acoustic niche hypothesis (Krause 1993), or the acoustic adaptation hypothesis (Morton 1975). Many other theories and hypotheses can also find room in the soundscape ecology framework, and in Chap. 3 such theories are presented and discussed in detail.

Soundscape ecology can find several environmental applications such as assessment of the environmental quality of parks and protected areas, urban planning and design, ethology and anthropology, and finally in the long-term monitoring of the effects of climatic changes in geophysical (the sonic prospection of vanishing glaciers) and biological contexts (sonic dynamics of reproductive phenology). The soundscape ecology coupled to a bioacoustics approach can explain the effects of some landscape patterns such as edges, and size and shape of patches, on vocalizing animals.

1.5 The Disciplinary Contribution to the Soundscape Ecology

The study of the soundscape in terms of relationship with the abiotic and biotic environment in the broadest perspective is a matter of ecology. This fact should not be a surprise. Often ecology is used to investigate complex aspects of our environment because this science has robust and rigorous epistemic tools to face such a challenge.

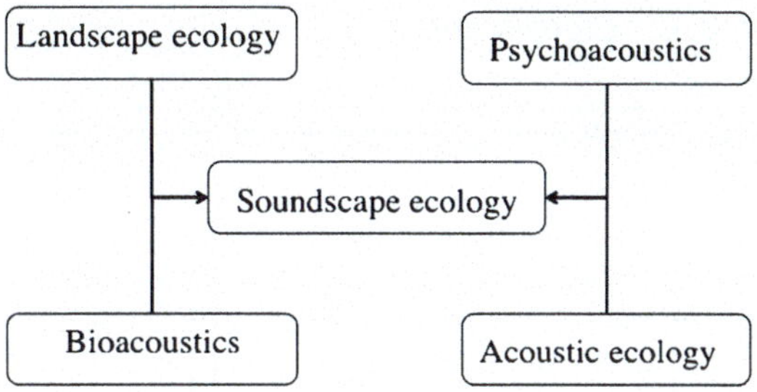

Fig. 1.1 Landscape ecology, bioacoustics, psychoacoustics, biosemiotics, and acoustic ecology concur to create the transdisciplinary character of soundscape ecology

However, the sonic environment has been for a long time a focus of several different disciplines, ranging from animal behavior studies to urban engineering (Fig. 1.1).

The relationship between the landscape and the soundscape is very strict; in fact, the soundscape is an extension of the concept of landscape. The (geographic) landscape is the physical spatial dimension in which every population, community and ecosystem process occurs in an explicit (multidimensional) way. The landscape dimension is indispensable to fully understand the patterns and the processes that concern a soundscape. We describe later the different models that try to connect the soundscape with the landscape, but most of the patterns that we observe at soundscape level are the result of the character of a landscape matrix. Nevertheless, new models that integrate soundscape with landscape must be elaborated in the future to better connect these two dimensions.

Definitively, soundscape ecology requires several approaches and competencies for its transdisciplinary character where its goal may be focused:

- Analysis of spatiotemporal acoustic patterns borne by the interaction between geophonic, biophonic, and anthrophonic sonic components.
- Analysis of the relationship between patterns and processes of the landscape and patterns and processes of the soundscape.
- Conservation of soundscapes as important components of the landscape.
- Short- and long-term monitoring of ecosystem dynamics.

1.6 The Three Sources of the Sonic Environment

The soundscape is the result of the mixture of different and concurring sounds. According to the source of such sounds it is possible to distinguish a geophonic component, a biophonic component, and an anthrophonic component. We refer to a physical dimension of the soundscape as a first descriptive approach, and this represents the more simplified and objective classification of a soundscape.

This distinction is particularly useful especially when we intend to explore the relationship between soundscape patterns (sonic patches or sonotopes) and landscape configurations.

1.6.1 Geophonies

The geophonies are represented by all the sounds produced by nonbiological natural agents such as winds, volcanoes, sea waves, running water, rain, thunderstorms, lightning, avalanches, earthquakes, and floods and represent the sonic background with which other sounds can overlap, mix, or mask. The geophonies are strongly affected by the geomorphic trait of a region (steepness, exposition, etc.) and by climatic conditions and local weather.

The morphology of a region, the presence of valleys, canyons, ridges, and cliffs, determines a different propagation of sounds strongly affecting the sonic degradation. Breeze regime, air humidity, and temperature affect the propagation patterns of sound waves. In the water environment, depth, salinity, and temperature are important drivers of the sonic processes.

1.6.2 Biophonies

The biophonies are defined as the emerging nonhuman sound produced by living organisms in a given biome (Krause 2012, p. 68). Every biome, and inside the biomes every landscape patch (ecotopes), have distinct biophonies.

In temperate terrestrial biomes, the majority of the biophonies are produced by birds, frogs, insects, and mammals.

In tropical regions, insects and frogs dominate, followed by birds and mammals. In the freshwater environment, insects and fishes are important components. In shallow seawater, crustaceans are the dominant "musicians," but in deep-sea waters whales, dolphins, and killer whales (orcas) are the most frequent biophonic sources.

Although not included by Krause, human voices must be considered to pertain to the biophonies. The complexity and plasticity of such acoustic expression known as voice is a human peculiarity.

Biophonies have different patterns according to latitude, season, and hours of day. In fact, vocal animals are more active during the day in particular hours (e.g., birds and frogs perform choruses at dawn and at dusk), varying with the seasons (e.g., birds are acoustically active in springtime), and according to latitude (e.g., in intertropical regions birds are acoustically active at all times, but are silent outside the breeding season at high latitudes) (Fig. 1.2). In subtropical regions, seasonal invariance reduces the seasonal effect, whereas the effects of seasonal changes are maximal in boreal and polar regions. This reasoning can be applied to terrestrial biomes: in marine biomes, other factors are concurrent such as currents, light penetration, sea depth, and water transparency.

A long-term fluctuation of vocal activity can be expected according to the changes in phenological cycles under climatic changes.

The acoustic activity of animals is finalized to communicate actively or passively with other intra- and interspecific individuals for several reasons and functions.

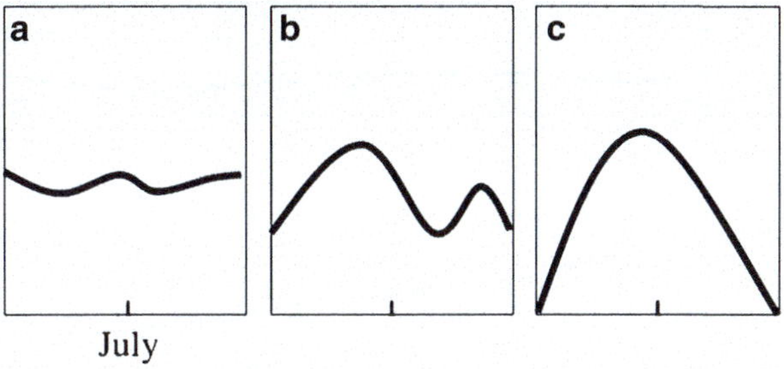

Fig. 1.2 Biophonies have different patterns during the year according to latitude. (**a**) At intertropical latitudes, animals are acoustically active all year around. (**b**) In temperate regions, biophonies have a major peak in spring and a second peak in early autumn. (**c**) At the highest latitudes, there is a peak only in the middle of summer, with a dramatic reduction of biophonic sounds during the rest of the year

This phenomenon activates communicative channels that concur to the acoustic complexity that we can hear around us. The apparent complexity of the acoustic communication largely depends on factors such as species-specific acoustic performance, population density, number of species composing the local aggregation (community), time of the day, and season, and finally by the typology of the vegetational cover (biomes and ecoregions). All these factors are strongly affected by geographic positioning along the parallels.

With the use of digital technology, the expansion of magnetic memory and highly sophisticated microphones and amplifiers assembled into programmable devices for long-time autonomous (in terms of power supply) operation, the recording of soundscapes has made a leap forward during the last decennia. Despite these tremendous improvements, however, until now there have not been processes able to identify automatically every type of sound. This limitation poses problems in the identification of the components of a soundscape and largely reduces the possibility of efficient evaluation of the potentialities expressed in distinct sites.

At the moment it is not possible to identify all the species that compose a soundscape, but nevertheless new approaches are available.

Acoustic diversity, expressed in terms of frequencies, became an important matter of investigation and of estimation of the complexity that we can expect in a specific site or habitat. Acoustic diversity cannot be correlated with biological diversity, at least in birds (Gasc et al. 2013), but nevertheless a high acoustic diversity is associated with a specific assemblage of species. In fact, species such as birds that have a special organ, the syrinx, to produce sounds, are able to produce a great variety of sounds.

Biophonies present different dynamics according to the temporal scale of resolution.

Observations of a breeding colony of European starlings at the University Farm, Wytham (Oxfordshire, UK) by Wright and Cotton (1994) have demonstrated the alternation of singing momentum with feeding activity at the ground. This alternation occurred twice per hour, but as egg-laying time approached, the majority of birds spent their time close to the breeding colony.

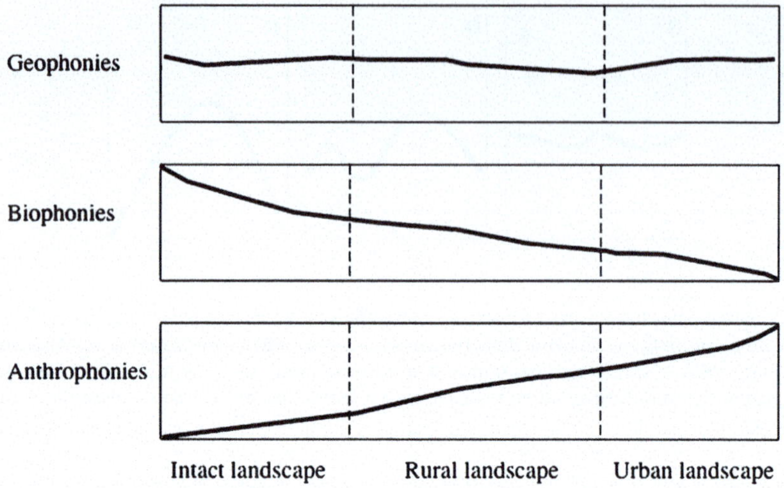

Fig. 1.3 Along a gradient of increased human intrusion (from intact natural landscape, through rural landscape, and to urban landscape), geophonies seem to be not affected, biophonies show a clear decrease, and finally anthrophonies increase

This behavior decreased in sharpness and intensity later in the season when breeding behavior was more demanding: the switch between episodes of feeding and singing was less definite as the breeding season advanced. Singing time at the breeding stage represents a way to affirm social cohesion and to transmit a focused message to the females.

1.6.3 Anthrophonies

The anthrophonies are the results of the movements of artificial devices such as cars, train, airplanes, industrial machinery, and bells. This component of the soundscape is becoming intrusive and dominant in large parts of the Earth, associated with urban development and globalized trade. Antrophonies are the major cause of noise pollution, a phenomenon that has revealed dangerous consequences to all organisms and to human health also, producing relevant modifications of behavior in both human and nonhuman animals. In humans, high levels of noise can cause annoyance and aggression, hypertension, hearing loss, additional stress, tinnitus, hearing loss, sleep disturbance, etc. Long-term exposure to high levels of noise can contribute to cardiovascular pathology and to an increment in blood pressure. In Chaps. 5 and 6 are presented more details about the characteristics and consequences of exposure to noise.

Anthrophonies increase in importance moving toward urban areas and close to industrial and transport infrastructures (Fig. 1.3) such as highways, airports, and air corridors. Anthrophonies change according to the different structure and function of cities and the social and economic context. A sonic gradient across a city is intercepted by the economic value of houses, fixing their price.

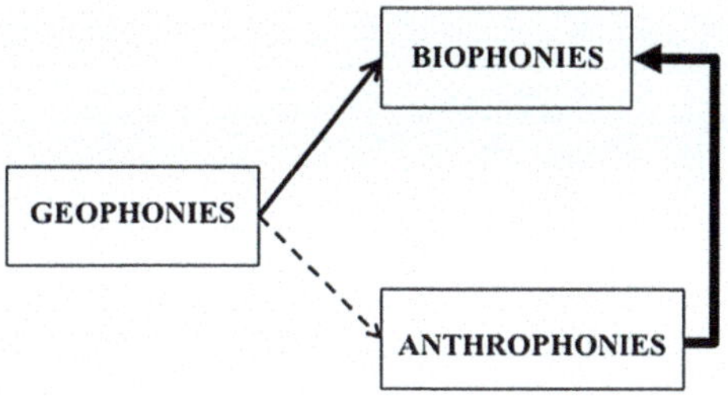

Fig. 1.4 In the soundscape, geophonies are the sonic sources that have a direct influence on biophonies and secondarily on anthrophonies. Anthrophonies can strongly impact on biophonies

1.6.4 The Relatioships Between Geophonies, Biophonies and Anthrophonies

The interaction between geophonies, biophonies, and anthrophonies determines peculiar patterns in the sonic environment. The geophonies represent the independent variable when compared with the other two sources; its influence is the main constraint especially affecting the biophonies (Fig. 1.4).

It is demonstrated that high levels of geophonies or anthrophonies depress the biophonies or modify the way in which organisms emit sounds (songs, calls, alarms, vocalizations).

For instance, a windy day depresses the song of most birds, and proximity to a heavily traveled highway prevents acoustic communication among the majority of vocal species living in the neighborhoods. Figure 1.5 shows a map of a soundscape created plotting recording sessions with sampling stations.

The depression in soundscape activity that is observable on the right side of the picture is coincident with the presence of a small river, whose noise masks or depresses the biophonies (mainly produced by birds).

Climatic events are important constraint for biophonies. The anthrophonies can produce severe changes in biophonies when these are of strong intensity, as close to an airport. Anthrophonies (ship engines, drilling platforms) have dramatic effects in both shallow and deep waters, affecting the survival of whales and other cetaceans.

In urban areas, the presence of vegetation in parks and along the avenues reduces the effect of traffic noise and noise in general, as well documented by Voichita Bucur in the 2006 book *Urban Forest Acoustic* (Bucur 2006).

1.7 Foreground and Background Sonic Environment

Sonic events have qualitative (e.g., frequency categories) and quantitative (e.g., pressure level) characteristics. The spatial and temporal variations of these characteristics create a dynamic sonic gradient across the landscapes, providing additional information not otherwise available.

In fact, the addition of sonic variables to a landscape analysis allows us to consider the temporal variability of the landscape.

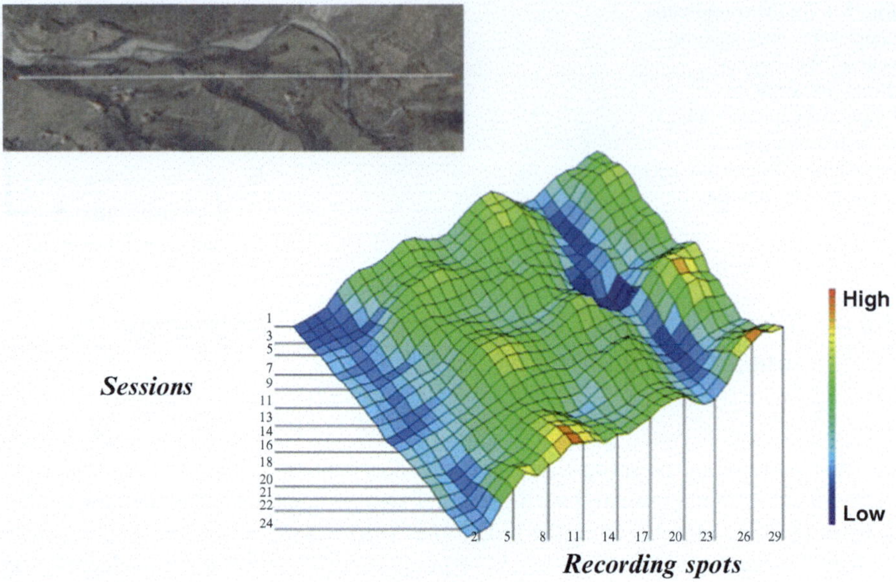

Fig. 1.5 Sonic map created plotting recording sessions and sampling station along a sonic transect, 900 m long, of a rural landscape (Northern Italy, Fivizzano Commune). The evident depression observable on the *right* part of the maps is coincident with the presence of a noisy stream (From Farina 2010, unpublished data)

Sonic ambience becomes an important source of information that acts at two scales: a local scale dominated by foreground sounds (unpredictable in space and time) and by cyclic background that becomes a landmark for some organism functions.

Foreground and background sounds refer to the position of a sound source from a listener. The foreground is an acoustic event close to the observer that can create an immediate reaction in the listener and represents a signal of urgent information. Background sounds, which can be used as landmarks, refer to an indistinct amount of low-frequency sounds that results from the blend of several individual sound sources degraded by distance.

The relationship between soundscape and landscape can be tentatively discriminated according the background/foreground duality. The unpredictability and the temporal variability of foreground sounds are less connected with landscape configuration than the background. The foreground sound moving across the landscape creates temporary sonic configurations, the sonotope, that when further shaped by vocal organisms have been considered soundtopes.

Background noise is more constant and is more influenced by landscape configuration, which could depend on the fact that background sounds are the result of coarser geophysical and anthropogenic processes. Landscape features such as forest edges or cliffs can act as sonic barriers or acoustic amplifiers. Some background sounds, such as ocean waves or winds and breezes, are constant and persistent in time, characterizing the sonic ambience. Often the intensity of the background noise can be masked by foreground sounds, but they represent the sonic landmarks that can be actively discriminated by organisms to fix their home range without using visual cues or integrating them. For instance, the background sound from coral reefs can orient pelagic fishes, in the same manner that the traffic of a highway can discourage birds and mammals from crossing the area, reducing the risk of collisions with vehicles.

The landscape approach, which presently is so popular for investigating the role of land use policies to preserve the biodiversity or to evaluate the effects of human intrusion by logistic infrastructures and urban development, often cannot allow detection of most of the processes that occur at a short time scale of hours or days.

For instance, close to human settlements, as in urban areas, some hours of the day are full of background noise and high intense foreground sound (sirens, horns, jackhammers, loud human voices, music) in the middle of the morning and during the evening.

For this reason, integration with a soundscape investigation approach allows exploring short-term dynamics that have a great impact on the ecological and behavioral processes that occur at the population and community scale.

It is recognized that visual and acoustic stimuli are strictly connected during the perception of the environment. In their investigation, Carles et al. (1999) have found that in urban green spaces and in cultural landscapes the sound environment is essential to complete environmental evaluation. The information from a landscape often does not provide sufficient cues for animals such as humans, and the acoustic component become increasingly important with the use of cognitive sensing. The soundscape can be considered an additional layer of spatial information across a landscape.

Foreground sound strongly concurs with animal behavior, to animal displacement in time and space, and this effect has been explained by Farina et al. (2011a) as a soundtope when produced by a local community of acoustic animals.

Foreground sound, for its dynamic characteristics, triggers reactions in animals, and humans also seem more independent by the structure of land use and land cover. Foreground sounds become an information structure that has a relevant role in shaping local acoustic communities.

1.8 Hi-Fi and Lo-Fi Soundscapes

If the distinction between sound and noise is quite simple, it is not so easy to define a high-fidelity soundscape and a low-fidelity soundscape.

In the first case we have a soundscape in which every component is distinctly heard without significant effects of masking, but in the second case the overlap is so important that we have difficulties in distinguishing the individual sounds.

These two characteristics are not necessarily associated with noise or acoustic pollution. Along a mountain stream, the sound of the flowing water reduces the

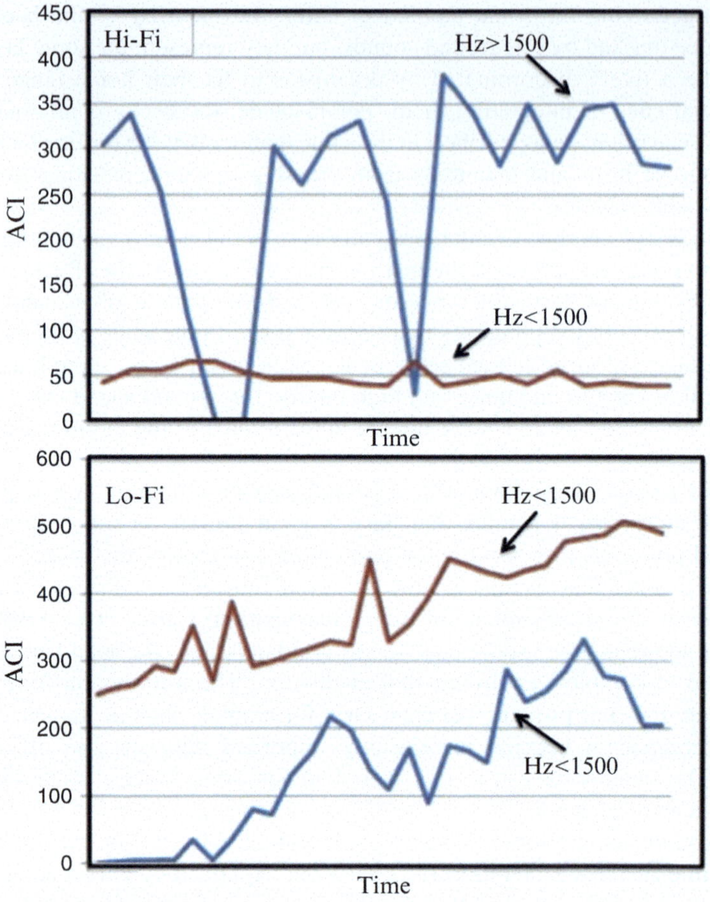

Fig. 1.6 Example of high-fidelity (Hi-Fi) sonotope in a remote Mediterranean maqui (Deiva Marina, Liguria, Italy) and of low-fidelity (Lo-Fi) sonotopes recorded in the Devesa Park of Girona (Catalogna, Spain). The Hi-Fi and Lo-Fi sonotopes have been distinguished according to the amount of information calculated with the acoustic complexity index (ACI) (see Chap. 9), for 2-h audio files, pertaining to frequencies below 1,500 Hz (mostly noisy background) and frequencies higher than 1,500 Hz (mostly biophonies) (From Farina 2011, unpublished data)

capacity to distinguish the song of a dipper (*Cinclus cinclus*), or when the mistral (wind) is blowing on the Provence Maqui, it is not possible to hear songbirds. At the same time, during the dawn chorus of a bird community it is hard to distinguish the individual species.

Of course, when the environment is complicated by anthrophonies, the quality of the emerging sound is strongly reduced. In Fig. 1.6 are depicted two examples of Hi-Fi and Lo-Fi soundscapes, the first from the Mediterranean Maqui in Cinque Terre (Italy) and the second from a urban park in Girona (Spain). The first soundscape is characterized by distinct sounds from individual species where the background noise does not mask the biophonies.

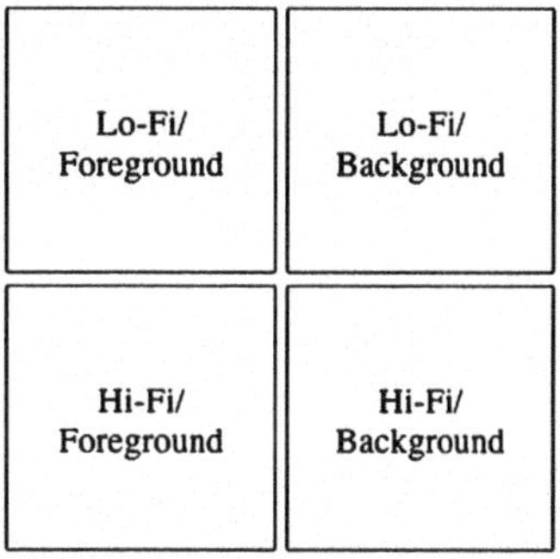

Fig. 1.7 Combination of foreground and background sonic environment and Hi-Fi and Lo-Fi categories created further subcategories of a sonic context

In the second case, the noise produced by car traffic enters the urban park located at the periphery of Girona (Catalogna, Spain) and interferes with the morning biophonies produced by birds. Bird song is largely masked by such background noise.

Foreground and background sonic environment and Hi-Fi and Lo-Fi are two different perspectives with which we distinguish some important sonic patterns. The combination of these two perspectives as illustrated in Fig. 1.7 produces potential subcategories of sonic conditions.

1.9 Landscape: Emerging Patterns and Dynamics

1.9.1 Some Definitions and Relevant Patterns

To better understand the soundscape dynamics that emerge from landscape structures, it is necessary to describe what is a landscape and what are the emerging patterns that allow using this epistemological domain to explain the complexity of our world.

Several definitions of landscape have been produced during the years since the German geographer Carl Troll realized that the orthogonal vision of the Earth from an airplane provided important information on the functioning of the landscapes.

Since that time in Europe and later in the United States, landscape ecology flourished as an important ecological discipline from both theoretical and applied aspects.

At least three main currents of thought were quickly established. The first restricted the concepts to the description of a geographic landscape. A second line of thought takes into consideration the ecological valence of the landscape interpreted as a spatial arrangement of ecosystems. Finally, a third line considered the landscape as a cognitive object that is first created in the mind of the observer. In a retrospective analysis, all these components are parts of a unified conceptualization of the landscape. In conclusion, each perspective is indispensable to develop or to integrate the two others.

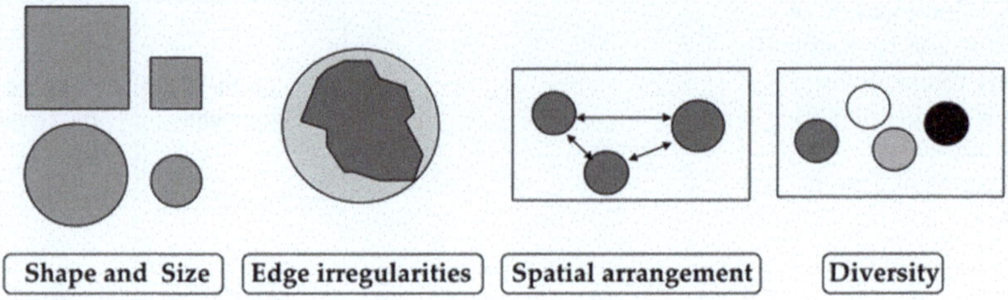

Fig. 1.8 Main attributes of patches: shape and size, edge irregularity, spatial arrangement, diversity

By geographic landscape, we mean the description of a specific region in terms of land form, stream and river configurations, distribution of lakes, dry depressions, slope characteristics, and associated patterns. The study of the landscape from a geographic perspective has utilized the hierarchical theory by which elementary units, the patches, are aggregated in further higher-order units that in turn are components of a superior system. Often across such hierarchies is possible to observe a fractal structure. According to the geographic perspective are distinguished, at least at a specific scale, patches from a more homogeneous matrix.

Patches represent the spatial unit of a land mosaic and their main characteristics depend on shape and size, edge irregularities, spatial distribution in the mosaic, and diversity (Fig. 1.8), and consider fragmentation as a resultant process of disturbance.

The relationship between the geographic landscape and the ecological landscape are evident when we attribute functions to the observed spatial patterns. In these terms, size, dimension, internal structure, and distance of each patch from others are attributes that in turn affect distribution and diversity of the organisms (plants, animals, virus, bacteria, algae, etc.).

It is well known that the dimensions of patches are essential to maintain a stable population or a more diverse community and that the lengths of the borders attract ecotonal species and discourage habitat-interior (core) species. The fragmentation of homogeneous stands strongly affects the survivorship of interior species and facilitate the intrusion of alien species and predators. Fragmentation in the different phases, as outlined in Fig. 1.9, may have several effects on distribution and movements of species discouraging non-ecotonal species. However, margins in nontropical ecosystems are very attractive systems for many species and in this apparent paradox ecology can contribute to explaining the different constraints.

The ecotones represent important structures to understand the functioning of a landscape; in fact, the margin between one system and another is a tension zone through which energy, nutrients, organisms, and information pass. Ecotones exist at all ecological scales from a few millimeters to thousands of kilometers across the biomes. The ecological frontiers created by the juxtaposition of different systems originate gradients with which species enter into contact and can suffer or benefit from the ecological constraints.

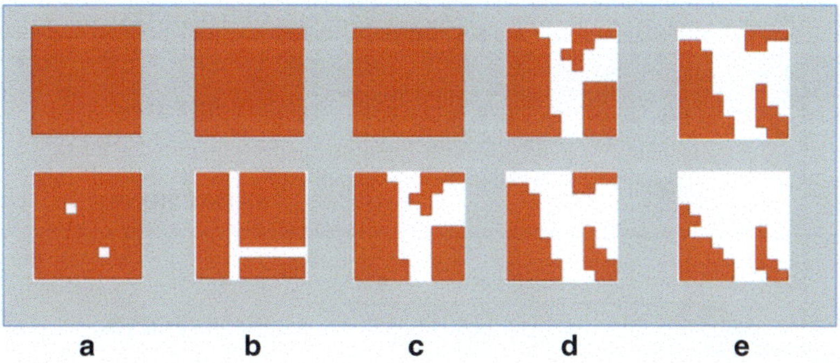

Fig. 1.9 Fragmentation results in a complex process that occurs when a disturbance changes the value of specific cover and creates information at the edge of the disturbance process. The different levels of fragmentation are indicated in terms of severity: perforation (the first seeds of a disturbance) (**a**); dissection (a first separation along a significant extension, such as a road in a tropical forest) (**b**); fragmentation sensu stricto (a significant reduction of cover and insularization) (**c**); shrinkage (further reduction of the size of patches) (**d**); and rarefaction (reduction in number and size of fragments) (**e**) (Reproduced with permission from Bogaert 2000)

1.10 Sonic Patterns in Soundscape Ecology

In landscape ecology, patterns and processes are phenomenological categories that allow better understanding of the complexity of this ecological domain. Heterogeneity and ecotones are two examples of spatial patterns; fragmentation and edge effect are examples of processes (Farina 2006). This epistemic approach can be used as well in soundscape ecology to distinguish sonic patterns from sonic processes.

1.10.1 Definition of Sonotope

The landscape is patchy at several scales, and this patchiness reflects the heterogeneous distribution of resources (see "General Theory of Resources"; Farina 2012). Animals aggregate according to such distribution and they interact acoustically on the basis of their distribution, habitat preferences, density, and intra- and interspecific competition.

It is reasonable to hypothesize that according this pattern the sonic environment created by the blending of geophonies, biophonies, and anthrophonies can also be equally shaped.

As in landscape ecology, where the ecotope is the coincident overlap of geotopes and biotopes, in the same way the sonotope is a distinct sonic unit or sonic patch produced by the overlap of geophonies, biophonies, and anthrophonies. Every source of sounds has a different spatial and temporal extension, and consequently

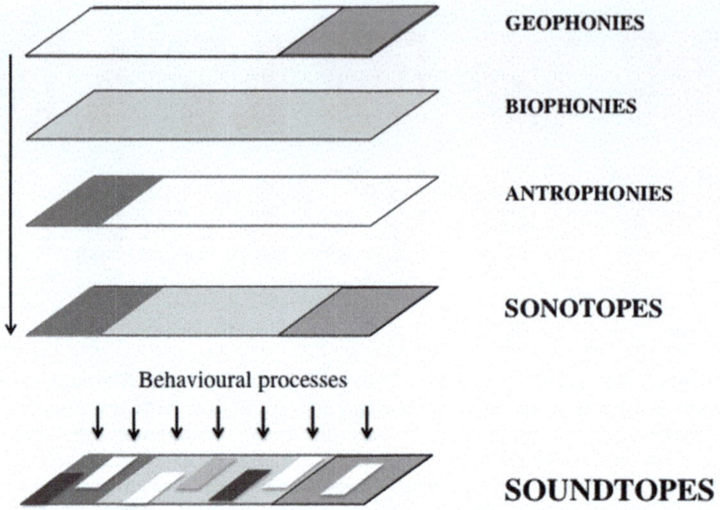

Fig. 1.10 The spatial overlap of geophonies, biophonies, and anthrophonies creates sonic patterns or sonotopes. Every sonotope is characterized by a peculiar blend of sounds. The behavioral processes connected with the acoustic performance of vocal animals create in turn distinct soundtopes inside each sonotope

the sonotopes are the result of interscaled overlapping of geophonies, biophonies, and anthrophonies.

We can expect that geophonies have the largest extension, which decreases for biophones and finally for anthrophonies. As in the geographic landscape, sonotopes can be characterized according to several sets of scales and according to the approach and the purposes.

The sonotopes refer to the distribution in time and space of the different "sonic patches" that reflects quality, disposition, occurrence time, and spatial overlap of sound sources (geophonies, biophonies, and antrophonies) (Fig. 1.10).

The sonotopes may be considered patterns created by biotic and abiotic sonic agents. Sonotopes become the areas in which some sonic characters are recognized in space and time as distinct. The distribution of sound around a waterfall is an example of a sonic pattern of abiotic origin.

Generally, sonic patterns are persistent in time and space, as the sound of a waterfall, or cyclic, as the sounds of tidal waves or daily breezes, animal choruses, and urban noises, but their persistence depends on the characteristic of the sonic source.

Sonotopes become the planning unit of a soundscape, and each sonotope can create feelings to meet the needs of the visitors or elicit their annoyance. Some sonotopes can play the role of sonic refuges. In these refuges, the visitor identifies a great variety of sounds that really represent an acoustic diversity.

Sonotopes can be identified spatially using a net of sound recorders and the amount of acoustic energy and its quality can be mapped.

1.10.2 Definition of Soundtope

The soundtope is the acoustic result of a volunteer organized (coordinated) assemblage of vocalizing animals that has as prevalent proxies the behavior of the members. Soundtope can be considered a coordinated vocalization used as a strategy to reduce the risk for intrusion of foreign individuals in a local community.

Soundtope is a potential subdivision of a sonotope and is created only by the contribution of biophonies (Fig. 1.10). In fact, is the result of active intra- and interspecific communication uttered especially at times of dawn and dusk.

A soundtope allows an active and passive exchange of information among collaborative members of a community. Intensity, duration, and interaction of the vocalizations depend on complex mechanism of semetic interactions (sensu Hoffmeyer 2008). Soundtopes are particularly evident at dawn and at dusk (for birds) and can be considered the behavioristic interpretation of a chorus. Soundtopes are fundamental sonic patterns that are created by members of a community with specific goals and which seem in some measure independent from environmental constraints. In birds, soundtope structure emerges especially at dawn and at dusk, known by the general name of choruses.

A chorus is a dimensionless phenomenon composed of different soundtopes that are characterized by temporary, coordinated vocalizations. When a soundtope is active, a large amount of information is transmitted between the members of this aggregation. Soundtopes have a core area and a buffer area; at the border we can distinguish the "soundtope sonotone" in which transient eavesdropping individuals set the strength of the soundtope.

According to a biosemiotic perspective, the soundtope represents any type of sonic eco-field (sonic field) for the location of some specific resources, such as "abundance of resources" or "abundance of nesting sites."

Soundtopes have more information than the sonotopes because the spatial arrangement of vocal animals is not simply determined by environmental constraints (vegetation, exposition, etc.) but also by behavioral processes (competition, courtship, etc.). For their epistemology it is necessary to refer to the eco-field hypothesis (Farina and Belgrano 2004, 2006).

The hypothesis of the existence of soundtopes is confirmed by the diffusion of acoustic coordinated cues adopted by the majority of animals in different environments and conditions (Malavasi and Farina 2013).

1.10.3 Definition of Sonotone

The ecotones are patterns created by the encounter of different patches, or different types of vegetation. The ecotone conceptualization is a central paradigm in landscape ecology, as confirmed by an impressive literature on this subject (Farina 2006, p. 195).

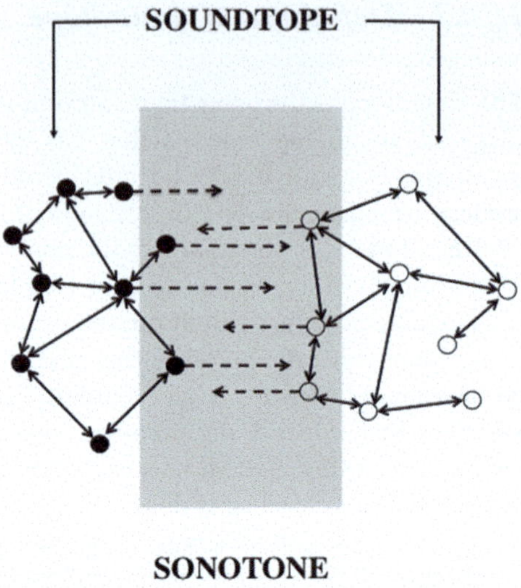

Fig. 1.11 At the border of soundtopes, the sonic environment results from a combination of sounds coming from different soundtopes, which creates sonotones

The term ecotone is the result of two words, eco (oikos) and tone (tension). At the ecotones, many processes find a strong constraint called the edge effect. The edge effect in most cases enhances biodiversity but also other processes such as predation. Ecotones are also considered ecological traps or sink areas (sensu Pulliam 1988) in which species attracted by their structure and food abundance experience a higher predatory pressure or habitat disturbance.

It is reasonable to consider that, as for vegetation and animal distribution, ecotones exist also for other phenomena such as the sonic environment. The sonic environment is considered in a heterogeneous medium composed of distinct sonotopes, and such sonotopes at their margins enter into contact with other sonotopes. Thus, we can expect a mixing of different acoustic characteristics that, blending in a stochastic way, create a very complex sonic area in which it is difficult for acoustic signals to be decodified appropriately to avoid masking and overlap (Fig. 1.11).

In bird in particular, singing at the ecotones means receiving acoustic information also from different types of environment, and we can expect unwanted masking effects and interspecific interferences. The acoustic novelties encountered at the sonotones have to be considered as a tangle in the acoustic codification/decodification process.

The sonic edge effect is expected to influence many acoustic functions such as ranging, mating choice, territory patrolling, and acoustic partitioning. Definitive effects are expected according to the acoustic niche hypothesis for species that enter into sonic contact with acoustic novelties coming from adjacent sonic environments.

1.11 The Relationship Between Landscape Structure and Soundscape Patterns

As presented in Fig. 1.12, at least three models are proposed to illustrate the relationship between landscape and soundscape.

In the first model, the sonic patterns are coincident with the landscape forms; in other words, for every configuration of the landscape there is a coincident configuration of the soundscape. The margins of the landscape are coincident with those of the soundscape, which means that the spatial scale of the two elements is the same. This model is very improbable, but in some cases possible, as on small islands that are surrounded by water where the soundscape ends at the island border (Fig. 1.12a).

In the second hypothesis, the soundscape has larger borders than the landscape, starting from the fact that sound cannot be confined inside individual patches in the landscape heterogeneity. This second model is more probable and realistic than the first and takes into consideration the diffusive property of the sounds (Fig. 1.12b).

The third model (Fig. 1.12c) assumes that the borders of a landscape are affecting the distribution of the soundscape but that further subdivision of the soundscape is possible inside a patch of the landscape. This aspect assumes that the soundscape is created by species that intercept the spatial heterogeneity and diversity of the landscape but that the soundscape is also the result of patterns created inside and independently by the physical environment. Thus, the soundscape can be produced apparently at a finer scale compared with the underlying landscape.

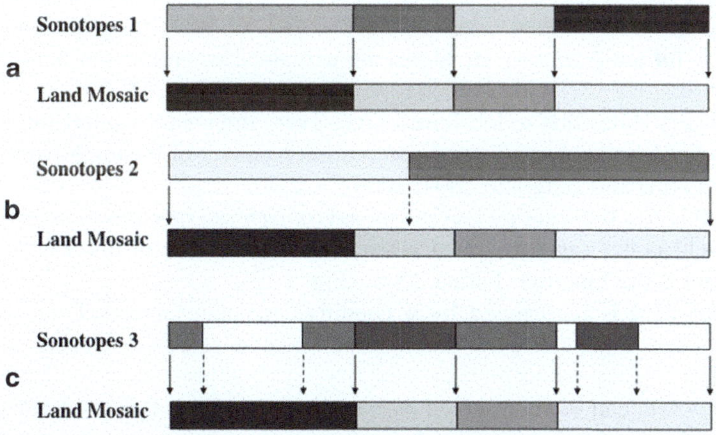

Fig. 1.12 Three main hypotheses about the spatial overlap of land mosaic and soundscape: sonotopes are coincident with the borders of the land mosaic (**a**); sonotopes are larger than the land mosaic (**b**); sonotopes are more fine grained and are not coincident with the resulting land mosaic, which is more coarse grained (**c**)

1.12 The Sonic Components of the Environment and the General Theory of Resources

The word resource is used in both ecology and economics, and literally it means "re-surge," from the Latin "spring (surgere) again (re)." This term can be applied to every natural element (abiotic and biotic) as well as to any manmade elements, material or immaterial, which, after having been used/consumed, regenerates itself by way of its internal, independent mechanisms. Some sounds (especially music) or some sonotopes such as the bird chorus in an urban park, or the silence of a remote location in the wild, can be considered acoustic resources.

According to his "general theory of resources" (Farina 2012), the individual sounds, sonotopes, and soundtopes may be considered resources for a receiver when these objects are requested to satisfy a need. For instance, in territorial defense a song from a competitor represents a ranging resource to establish the less costly behavior in territory maintenance. A soundtope of a roosting flock of starlings (*Sturnus vulgaris*) represents information on the location of abundant food for eavesdropping individuals. A sonotope created by sea waves against a coral reef represents the location where the pelagic larvae of reef fish move (Tolimieri et al. 2000).

1.13 Definition of a Sonic Eco-Field

Sonic processes refer to the dynamics and the effects that sounds produce on living organisms such as reverberation, masking, annoyance, stress, mating selection, and phonotaxy. The sonic processes are subjectively perceived and as a consequence they always require a focal species or aggregation (population, community) to be identified. Some important sonic processes activated for mating, for territory defense, or for antipredatory strategies are explained when the eco-field theory is adopted.

This theory states that to locate resources every organism requires the presence of an eco-field, defined as a spatial configuration carrier of meaning for a specific function (Farina and Belgrano 2004, 2006).

According to a biosemiotic Percian model (*object, interpretant, representamen*), the eco-field is the *representamen* where the *object* is the resource whereas the *interpretant* is the function (Farina 2008) (Fig. 1.13).

The eco-field is recognized by a cognitive template activated by a specific function when a physiological need for a specific resource exceeds a certain threshold.

The eco-field can be represented by a spatial arrangement of fixed objects such as trees, mobile animals such as herds, or sonic energy. In this last case we prefer to name this specific eco-field a sonic eco-field. The spatial distribution of broadcast sonic signals is used to locate specific resources.

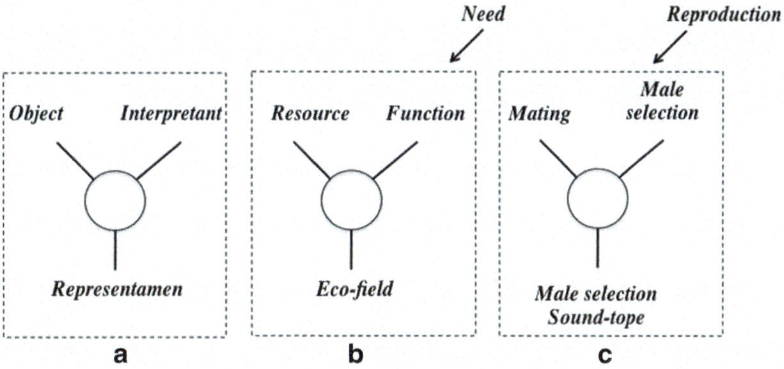

Fig. 1.13 The triadic structure of the segnic process according the Percian semiotic interpretation is utilized to confirm the eco-field hypothesis. (**a**) The three components of the segnic process (object, *interpretant*, *representamen*) concur with the segnic process. (**b**) The eco-field hypothesis has been explained using the Percian triadic sign process wherein is the need to initiate the segnic process. Resources are located as a result of the identification of a spatial configuration carrier of meaning (the eco-field) using a function as *interpretant*. (**c**) An example of mating (object–resource), male selection soundtope (*representamen*–eco-field), and male selection (*interpretant*–function)

The sonic eco-field may be represented by a spatial arrangement of contemporary coordinated vocalizations or soundtopes and in this case can be utilized to evaluate density of individuals, richness of the composing community, or intra- and interspecific competition.

The theory of the eco-field can be used to explain the function of the soundtope that can be defined as a set of sonic eco-fields utilized for tracking resources such as mating or food location.

For instance (Fig. 1.13c), the "resource" is represented by the "mating," the (behavioral) "function" is represented by "male selection," and the "eco-field" is represented by the "male-selection soundtope."

The activity-centric standpoint of Jennings and Cain (2013) is very close to the idea of the sonic eco-field, but the arguments considered by these authors do not consider the theory of resources and the eco-field hypothesis.

1.14 Landscape Fragmentation and Soundscape Processes

Population decrease is generally caused by the fragmentation and rarefaction of habitats under a growing worldwide human impact. Fragmentation is one of the most common causes of habitat degradation and species extinction. Small and isolated populations are under a stronger pressure of extinction than large populations, mostly dependent on inbreeding depression, demographic stochasticity, genetic drift, and Allee effect. In social animals in particular, reduction of population size curtails cultural transmission and large group cooperation.

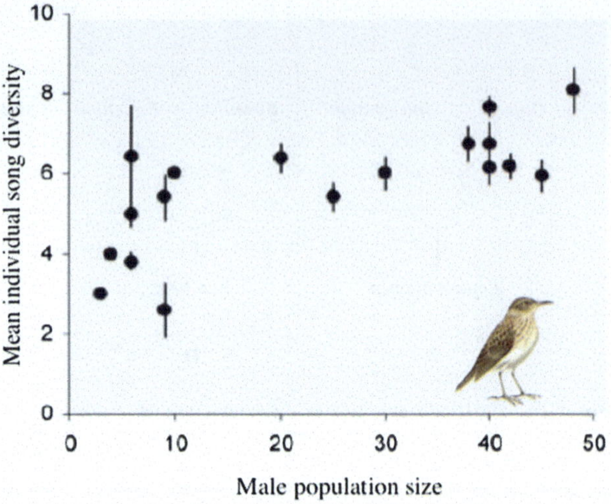

Fig. 1.14 Relationship between mean individual song diversity and male population size of Dupont's lark (*Chersophilus duponti*) (Reproduced with permission from Laiolo et al. 2008)

There is growing evidence that fragmentation affects not only the distribution and abundance of species but also their canorous repertoire.

In oscine birds, the cultural transmission of the acoustic repertoire seems a very efficient tool to reinforce individual and population fitness. The complexity of the acoustic cues in these animals is manifested across several life traits such as song rate and spectro-temporal performance.

Dialects represent variation on the geographically spaced theme and are important to balance local adaptation in the face of barriers (Laiolo 2008). Finally, birds emit with their song honest signals of well-being for local variability.

For instance, Dupont's lark (*Chersophilus duponti*), a threatened lark in Southern Europe, is a lark with a patchy distribution in Peninsula Iberica in residual steppeland. This species persists in open habitats although such are deeply fragmented by urbanization, road infrastructures, and intensive agriculture. In this species, studied by Laiolo et al. (2008), a significant relationship between the structure of a landscape and the complexity of its song repertoire has been found. The overall song diversity results from the variation among populations (beta-cultural diversity). Song dissimilarity and distance between subpopulations evidently agreed, and the same result was found between individual song diversity and patch area. A distance of only 5 km was found sufficient to characterize a specific cultural tradition in song repertoire. The smaller patches had the poorer song repertoire.

These authors have found that the acoustic repertoire of populations of small size was related to the dimension of the male population size (Fig. 1.14). For the first time, how culture in nonhuman organisms is strictly connected with the structure of a landscape has been demonstrated. The bioacoustics method utilized is extremely efficient to detect subtle changes in song repertoire and allows monitoring a population using a rigorous approach. The dimension of the male population is an efficient proxy to be correlated with the population viability, and such an approach can find useful application in the fields of ecological and biological conservation. A large male pool contributes to the diversification of the acoustic repertoire and, vice versa, reduction of the number of males curtails the potentiality of the song repertoire. This effect has been observed also in small populations of humans in reducing the diversification of the language.

A recent investigation carried out by Briefer et al. (2010) has demonstrated that fragmentation of habitats of the skylark (*Alauda arvensis*), a species in strong decline in Europe, causes an increase in sharing syllables in populations living in more fragmented conditions than in populations living in continuous habitat. On the other hand, changes in song complexity were not found when a population living in fragmented habitat was compared with a stable population living in a continuous habitat. Probably the most cultural component of the acoustic repertoire of skylarks is affected by spatial repartition of subpopulations across a fragmented landscape. In conclusion, habitat fragmentation caused by human barriers increases the amount of song sharing but not the complexity of the song.

As stated by Laiolo and Tella (2005), "bird vocalizations might become an early warning system detecting the effects of fragmentation over relatively short times and before other indicators (such as genetic markers)."

Summary

The sonic environment is important for most organisms that perceive, in a species-specific way, the acoustic information coming from different components of the ecosystem.

The soundscape is an important theme that can be approached in several distinct ways, ranging from a simple physical perspective to the cognitive perspective of the human mind.

Acoustic phenomena have so different and complex degrees of specifications and time/space scaling that is not possible to concentrate in a few sentences the overall conceptualizations that accompany this epistemic domain. In particular, when we try to connect the soundscape to the landscape, we find the maximum information possible.

A soundscape may represent a context or a complex message or asset of random individually based acoustic signs.

The effect of the sonic environment on organisms represents a promising field of ecological research, and in particular we are concerned with soundscape ecology. This new ecological discipline was presented to a landscape ecology congress in 2009 with a symposium titled "Soundscape Ecology: Merging Bioacoustics and Landscapes," where different components such as acoustic ecology, landscape ecology, bioacoustics, urban and environmental acoustics, behavioral ecology, and biosemiotics were recognized as primary roles.

Soundscape ecology finds important applications in assessment of the environmental quality of parks and protected areas, in urban planning and design, in ethology and anthropology, and finally in long-term monitoring of the effects of climatic changes.

The soundscape is the result of the overlap of three distinct sonic sources: geophonies, biophonies, and anthrophonies.

The geophonies are the result of sonic energy produced by nonbiological natural agents such as wind, volcanoes, sea waves, running water, rain, thunderstorms, lightning, avalanches, earthquakes, and flooding.

The biophonies are the results of animal vocalizations (song, calls, alarms, voices).

The anthrophonies are the result of all the sounds produced by technical devices (engines, blades, wheel revolutions, industries, etc.).

According to distributive attributes, we distinguish a background and a foreground sonic environment. The background refers to every sound mixed all together that has information that cannot be more distinct in the composing parts. The background sonic environment is strictly connected with the characteristics of geographic landscapes. The foreground sonic environment refers to sound close to the perception point (listener) and becomes an information structure that has a relevant role in shaping local acoustic communities.

Hi-Fi and Lo-Fi are further possibilities to distinguish categories of the sonic environment. Hi-Fi (high fidelity) means a sonic context in which is possible to distinguish many acoustic details such as the song of birds or the chirp of crickets. The Lo-Fi (low fidelity) is a sonic ambient in which the noise component prevails and is not possible to distinguish individual sonic signals and where a sonic message must be reinforced in amplitude to be received. Foreground/background and Hi-Fi/Lo-Fi are concepts of the sonic environment that have a common epistemology, and different combinations of these four categories determines sonic environments with distinct characters.

The landscape represent the basic structure from which sounds originate, and to better understand sonic dynamics it is necessary to have well in mind the basic concepts of the landscape (ecology). At least three different perspectives can be used to describe the landscape and its associated soundscape: a geographic perspective, an ecological perspective, and a cognitive perspective.

At least three models can explain the relationships between soundscape and landscape: the first model considers a perfect overlap between the geographic components of the landscape and the sonic components, a second model considers the sonic environment at a coarser scale compared with the landscape structure, and the third model recognizes a finer resolution to the soundscape.

From this last model is derived the definition of sonotope as the result of spatial overlap of geophonic, biophonic, and anthrophonic patterns. Each sonotope for effect of behavioral constraints can be further subdivided into soundtopes. In the zone of contact between different soundtopes, tension zones of acoustic uncertainty are called sonotones.

The sonic component of the environment can be consider a resource, and according to the general theory of resources the sonic component represents a specific resource necessary to organisms to survive. For instance, a male frog calling is for a female a mating resource indispensable for concluding the reproductive process. Soundtopes represent sonic eco-fields necessary to track specific resources such as mating or foraging.

Alteration of landscape structures, as in fragmentation, has important effects on the sonic environment, and in particular, diversity and quality of individual song repertoire have been proven to be impacted by the reduction of population size of vocal species.

References

Bogaert J (2000) Quantifying habitat fragmentation as a spatial process in a patch-corridor-matrix landscape model. PhD dissertation, Faculteit Wetenschappen, Departement Biologie, University of Antwerpen

Bouzid OM, Tian GY, Neasham J, Sharif B (2013) Investigation of sampling frequency requirements for acoustic source localization using wireless sensor networks. Appl Acoust 74:269–274

Briefer E, Oiejuk TS, Rybak F, Aubin T (2010) Are bird song complexity and song sharing shaped by habitat structure? An information theory and statistical approach. J Theor Biol 262:151–164

Brown AL, Kang J, Gjestland T (2011) Towards standardization in soundscape preference assessment. Appl Acoust 72:387–392

Bucur V (2006) Urban forest acoustics. Springer, Berlin

Carles JL, Lopez Barrio I, de Lucio JV (1999) Sound influence on landscape values. Landsc Urban Plan 43:191–200

Davies WJ (2013) Special issue: applied soundscapes. Appl Acoust 74:223

Farina A (2006) Principles and methods in landscape ecology. Toward a science of landscape. Springer, Dordrecht

Farina A (2008) The landscape as a semiotic interface between organisms and resources. Biosemiotics 1:75–83

Farina A (2012) A biosemiotic perspective of the resource criterion: toward a general theory of resources. Biosemiotics 5:17–32

Farina A, Belgrano A (2004) The eco-field: a new paradigm for landscape ecology. Ecol Res 19:107–110

Farina A, Belgrano A (2006) The eco-field hypothesis: toward a cognitive landscape. Landsc Ecol 21:5–17

Farina A, Lattanzi E, Malavasi R, Pieretti N, Piccioli L (2011a) Avian soundscapes and cognitive landscapes: theory, application and ecological perspectives. Landsc Ecol 26:1257–1267

Farina A, Pieretti N, Piccioli L (2011b) The soundscape methodology for long-term bird monitoring: a Mediterranean Europe case-study. Ecol Inform 6:354–363

Farina A, Lattanzi E, Piccioli L, Pieretti N (2012) SoundscapeMeter, Bioacoustic Software: Verion 1.0 Beta. www.disbef.uniurb.it\biomia\soundscapemeter

Gasc A, Sueur J, Jiguet F, Devictor V, Grandcolas P, Burrow C, Depraetere M, Pavoine S (2013) Assessing biodiversity with sound: do acoustic diversity indices reflect phylogenetic and functional diversities of bird communities? Ecol Indic 25:279–287

Hedfors P, Berg PG (2003) The sounds of two landscape settings: auditory concepts for physical planning and design. Landsc Res 28(3):245–263

Hoffmeyer J (2008) The semiotic niche. J Mediterr Ecol 9:5–30

Jennings P, Cain R (2013) A framework for improving urban soundscapes. Appl Acoust 74:293–299

Krause B (1993) The niche hypothesis. Soundsc Newsl 6:6–10

Krause B (2012) The great animal orchestra. Little, Brown, New York

Laiolo P (2008) Characterizing the spatial structure of songbird cultures. Ecol Appl 18 (7):1774–1780
Laiolo P, Tella JL (2005) Habitat fragmentation affects culture transmission: patterns of song matching in Dupont's lark. J Appl Ecol 42:1183–1193
Laiolo P, Vögeli M, Serrano D, Tella JL (2008) Song diversity predicts the viability of fragmented bird populations. PLoS One 3(3):e1822
Malavasi R, Farina A (2013) Neighbours' talk: interspecific choruses among songbirds. Bioacoustics 22(1):33–48
Morton E (1975) Ecological sources of selection on avian sounds. Am Nat 109(965):17–34
Naveh Z, Lieberman AS (1984) Landscape ecology: theory and application. Springer, New York
Payne SR, Davies WJ, Adams MD (2009) Research into the practical policy applications of soundscapes concepts and techniques in urban areas. DEFRA report NANR200, June 2009. http://randd.defra.gov.uk
Pieretti N, Farina A, Morri D (2011) A new methodology to infer the singing activity of an avian community: the Acoustic Complexity Index (ACI). Ecol Indic 11(3):868–873
Pijanowski B, Farina A (2011) Introduction to the special issue on soundscape ecology. Landsc Ecol 26:1209–1211
Pijanowski BC, Villanueva-Rivera LJ, Dumyahn SL, Farina A, Krause B, Napoletano BM, Gage SH, Pieretti N (2011a) Soundscape ecology: the science of sound in the landscape. Bioscience 61(3):203–216
Pijanowski BC, Farina A, Dumyahn SL, Krause BL, Gage SH (2011b) What is soundscape ecology? Landsc Ecol 26(9):1213–1232
Pulliam D (1988) Sources, sinks, and population regulation. Am Nat 132(5):652–661
Scarre C, Lawson G (eds) (2006) Archaeoacoustics. MacDonald Institute for Archaeological Research, Cambridge
Schafer RM (1977) The tuning of the world. Knopf, New York
Sueur J, Pavoine S, Amerlynck O, Duvail S (2008) Rapid acoustic survey for biodiversity appraisal. PLoS One 3(12):e4065
Tolimieri N, Jeffs A, Montgomery JC (2000) Ambient sound as a cue for navigation by the pelagic larvae of reef fishes. Mar Ecol Prog Ser 207:219–224
Truax B, Barrett GW (2011) Soundscape in a context of acoustic and landscape ecology. Landsc Ecol 26:1201–1207
Wright J, Cotton PA (1994) Song versus food: trade-off decisions in a breeding colony of European starlings *Sturnus vulgaris*. Bird Study 41:95–101

Chapter 2
Sonic Characteristics of the Landscape

2.1 Introduction

Knowledge of the sonic characters of the landscape is important to evaluate the modalities of propagation of polluting noise, to manage noise abatement actions, to investigate and better understand animal communication for conservation purposes, and to assure sustainable development and a standard of well-being for humans.

Physical and biological patterns across a landscape are the principal drivers of the sonic environment and related dynamics. Topography, weather, and vegetation are, at a local scale, the most important factors of sonic characterization. We have to add to these characters other geographic components (e.g., flatness, steepness) and anthropic components (e.g., houses, infrastructures).

In the urban landscape, obstructions such as walls, houses, bushes, and highway barriers have refractive properties on sounds. However, the effect on sound propagation is minimal for buildings and similar structures because the obstruction to air flow is sufficient to raise the viscous and thermal boundary layer to the vicinity of the superior part of the buildings; for instance, paved roads reduce ground absorption more than occurs in natural flat lands.

2.2 Weather Conditions and Sound Propagation

The effect of weather on sound propagation is not only an important physical issue but has many connections with vegetation structure and animal communication modalities.

Meteorological effects may be severe for signal transmission, but this largely depends on the type of landscape and on climatic variability (Albert 2004).

Investigation of this aspect is quite old and has fascinated generations of physicists (Ingard 1953). Sound is attenuated more by wind gustiness than by humidity, fog and rain, temperature, and wind refraction. Ground attenuation is

A. Farina, *Soundscape Ecology: Principles, Patterns, Methods and Applications*, DOI 10.1007/978-94-007-7374-5_2, © Springer Science+Business Media Dordrecht 2014

important as gusting, especially when transmitters and receivers are close to the ground and function in the frequency considered. The maximum attenuation close to the ground pertains to the frequency between 200 and 500 Hz.

The dominant source of pressure fluctuation at a microphone is represented by the intrinsic turbulence in the flow (Morgan and Raspet 1992). The noise produced by wind is a problem for collecting biophonies but on the other hand is important information on environmental dynamics. Rustling leaves can be an important cause of background noise over a wide range of frequencies.

2.3 Climatic Patterns and Sounds

Very few papers are concerned with climatic changes and their consequences on the sonic ambience and in particular on the acoustic performance of animals. It is expected that change in climate conditions has effects on sound transmission and consequently on vocal communication. Evidence on the relationship between sonic patterns and climate pertains to indirect effects of climatic shifts and calling phenology.

For instance, the calling phenology of frogs has been documented to be changed during the past century, as reported by Gibb and Breisch (2000) near Ithaca, New York. Four species of frogs have been observed to be calling 10–13 days earlier than in the period 1900–1912. Two species have unchanged phenology, but none of the species is calling later in the season. This observation indicates that the global warming has produced biological responses in these populations (Table 2.1).

Different effects can be predicted: the rise in temperature increases the absorption of lower frequencies but enhances the higher frequencies, although this effect co-varies with atmospheric pressure and relative humidity.

The changes of climate modify the phenologies of plants, and an early leafing can modify the sonic ambience during the vocal performance of many species of birds. A comparison of height at which birds select song posts, carried out in the period 1986–1989 and 2010 by Moller (2010), demonstrated that birds have increased the height at which they sing. In this period, spring and summer

Table 2.1 Climate warming and calling phenology of six frogs near Ithaca, New York, 1900–1999

Average first calling date				
Species	1900–1912	1990–1999	Mean change (days)	*p*
Spring peeper	April 4	March 20	−13.6	0.038
Wood frog	April 9	March 29	−13.0	0.007
Bullfrog	June 5	May 22	−11.5	0.098
Gray treefrog	May 4	April 14	−10.5	0.030
American toad	April 18	April 11	−1.5	0.303
Green frog	May 10	May 16	5.5	0.808

Reproduced with permission from Gibb and Breisch (2000)

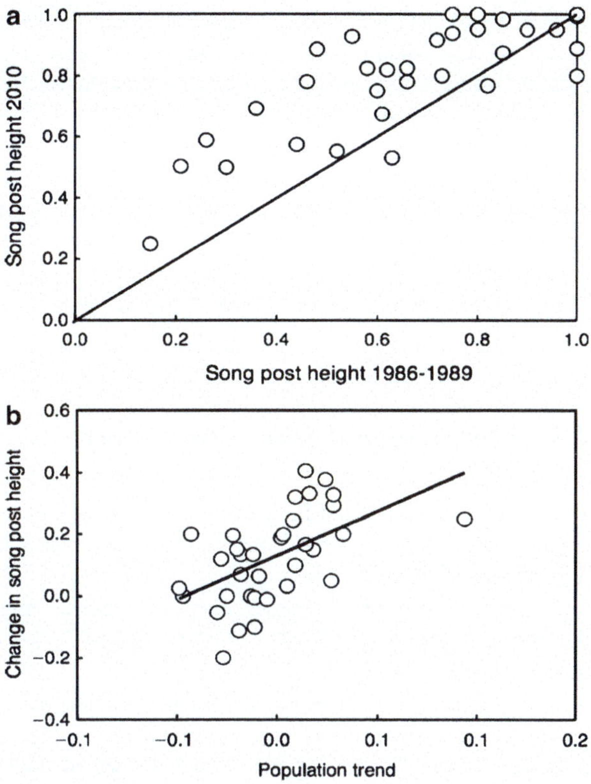

Fig. 2.1 (**a**) Song post position of different bird species in 2010 in relationship to position in 1986–1989. The *line* is $y = x$. (**b**) Regression line and increase in song post position between 1986–1989 and 2010 in relationship to population trend of different species of birds (Reproduced with permission from Moller 2010)

temperature and precipitations have increased by 20 and 30 %, respectively. The average song post was 18 % higher (1.2 m on average), and this change has been observed more clearly in species singing in trees when compared with species living in shrub vegetation. Sexually dichromatic species were found to vary more in height than monochromatic species, and finally species more susceptible to sparrowhawk (*Accipiter nisus*) predation were moving less in height than species that were less susceptible. Also, populations with high density have been found to move higher in the song posts. In conclusion, species with higher intraspecific competition, such as dichromatic species and species with populations with higher density, are more sensitive to the height of the song post: the first type because they perform a more intense sexual display and the second from the need to reduce population competition (Fig. 2.1). These results suggest that the singing sites of birds change rapidly, and this variation affects sonic ambience, and consequently the design of songs, mating success, and prey–predator interactions.

In several insects temperature has a great influence on the biological activity. In the cicada *Diceroprocta olympus* (Homoptera: Cicadidae), the hypothesis of thermoregulation seems proved by the regression of the T_b (body temperature), ambient temperature (T_a), and perch temperature (T_p), but in contrast with other acoustic insects there are no significant relationship between the calling song and T_b (Sanborn and Maté 2000). Ambient temperature regulates the sound power of cicada calling songs. Investigations conducted by Sueur and Sanborn (2003) on three species of European cicadae of genus *Tibicina* have found a direct correlation between ambient temperature and sound power in *Tibicina corsica fairmarei*, *T. garricola*, and *T. tomentosa* (Fig. 2.2). Similar results have been obtained by Fonseca and Revez (2002) working on three Portuguese cicadas (*Tettigetta argentata*, *Tettigetta josei*, and *Tympanistalna gastrica*), demonstrating that the neuromuscular structure used in the timbal cycle and the production of the echeme succession (an uninterrupted burst of sound) are temperature dependent.

2.4 Elevation and Song Complexity

Elevation is an important factor that can modify the sonic ambient of species. Less oxygen and more frequent and variable wind and breezes can affect the song repertoires of species.

Harder conditions at higher elevations favor monogamy, a similar level of parental care between females and males, and lesser variation in mating success between sexes.

A comparative study on extant finches carried out by Snell-Rood and Badayaev (2008) demonstrated that species breeding at the lower elevations sang more elaborate songs, producing a louder song when compared to their sister species breeding at higher elevation (Fig. 2.3). The species breeding at lower altitudes have longer songs with a greater number of notes.

These results demonstrate how the song repertoire of birds is affected by environmental conditions and that this influences sexual selection. The structure of a landscape is not the only proxy in determining the sonic ambience: other factors connected with the availability of resources seem to have an important role.

Botero et al. (2009) have discussed an important issue related to climatic unpredictability that seems to act as a driving force to the elaboration of signals of quality in animals. In particular they have studied the song display in mockingbirds using a pool of 1,738 recordings of mimids from which they elaborated sounds from 29 species, using the following information:

1. Number of syllable types
2. Mean syllable type consistency
3. Tandem repetition index
4. Mean song duration (in seconds)
5. Mean syllable duration (in seconds)
6. Song rate (songs·min^{-1})
7. Syllable switching rate
8. Syllable versatility
9. Mean number of syllable types in a song

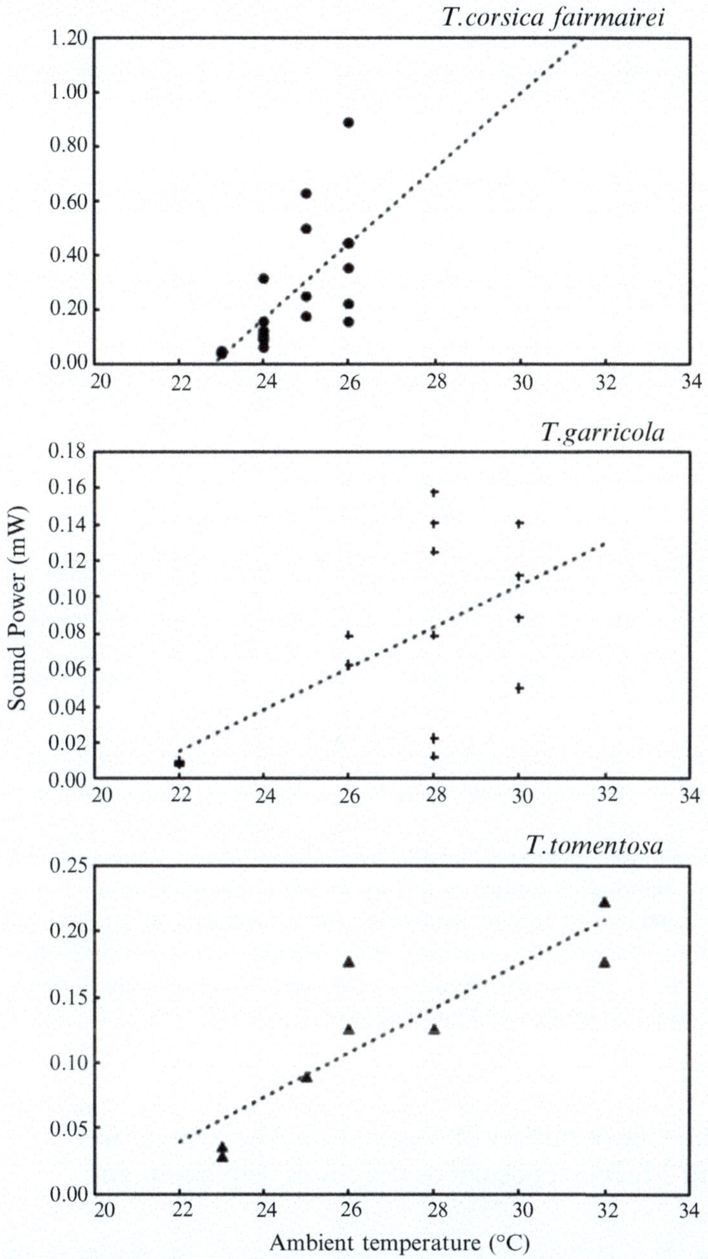

Fig. 2.2 Relationship between ambient temperature and sound power of the calling song of *Tibicina corsica farmairei*, *T. garricola*, and *T. tomentosa* (Hemiptera: Cicadidae: *Tibicina*) (Reproduced with permission from Sueur and Sanborn 2003)

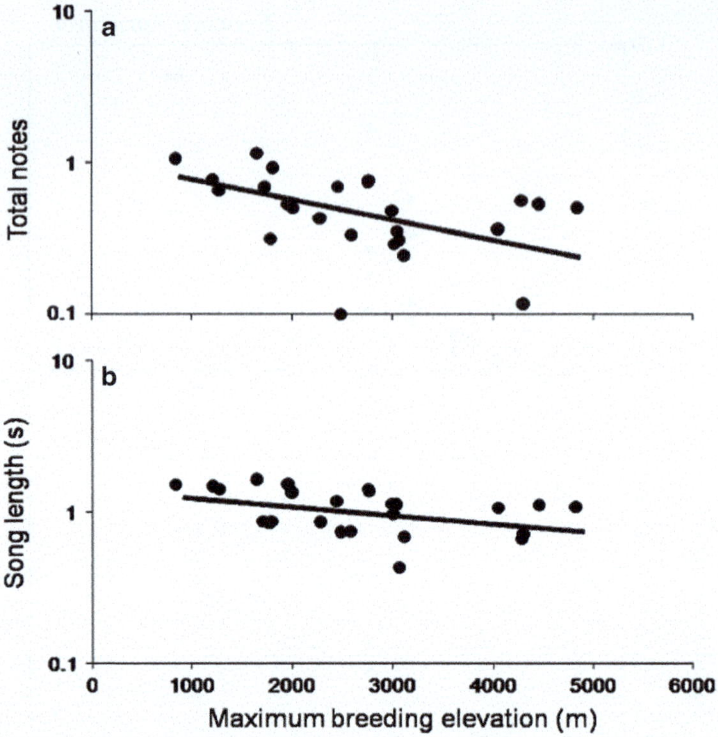

Fig. 2.3 Relationship between breeding elevation and song length and total notes of extant cardueline finches (Reproduced with permission from Snell-Rood and Badayaev 2008)

A more variable and unpredictable environment seems to enhance learning and innovation, elaborating signals of quality in this group of birds.

This result seems in contrast with the evidence collected by Briefer et al. (2010), who have found a reduced sound repertoire among the species of finches living at higher altitude. Apparently these two groups of organisms follow different strategies under environmental constraints.

2.5 The Contribution of Vegetation Microstructure and Macrostructure to the Sonic Environment

Vegetation is a complex system composed of soil, air, water, stems, trunks, branches, and foliage of herbs, shrubs, and trees. Reverberation, absorption, and scattering are some effects on acoustic signals caused by vegetation.

In forests, sound propagation is strictly connected with the ground effect, scattering from tree trunks and branches, and absorption by leaves. Scattering, the return of echoes from an object that intercepts a signal, is linked to the presence of trunks and branches and increases with the increase of frequency. Absorption by leaves is important only at higher frequencies.

The ground effect produces an attenuation at lower frequencies (<500–800 Hz). This effect depends on interference and impedance.

Vegetation functions as a sound screen, especially for high-frequency sounds. The attenuation observed in dense vegetation is largely caused by the scattering effect, which is driven by leaf width and leaf area per unit volume of canopy.

Sound energy that invests the leaves is in minimal part trapped in the leaf tissues and converted to heat: because of the large number of leaves present on a tree, the final result may be quite important and the excess attenuation substantial.

Measurements conducted by Aylor (1972) on a stand of perennial reed (*Phragmites communis*) with a density of 59 ± 10 plants/m^2 and an height of 2.5 m found that attenuation in the frequency range 500 Hz to 10 kHz increases at a rate of 4–4.5 dB per doubling of frequency. In particular, tests carried out on leaves by Martens and Michelsen (1981) using a laser-Doppler vibrometer system have proved that leaves behave as linear mechanical systems when invested by sound and noise at sound pressure of up 100 dB re 20 μPa in a range of frequencies between 0.5 and 5.5 kHz. Of course, this result largely depends on the orientation of the leaves in relationship to the sonic energy source.

At high frequencies when the dimension of leaves is close to the dimension of the sonic waves, high scattering reduces sound propagation.

2.6 Reverberation and Attenuation in Broadleaf and Pine Forests

Sound attenuation through trees is a common and well-known phenomenon. In particular, vegetation filters out some frequencies, and this has opened the way for the acoustic adaptation hypothesis in vocal animals.

When sound spreads close the soil, it is possible to observe a peak of excess of attenuation spectrum below 500 Hz and a corresponding dip at midfrequency (~1 kHz). For higher frequencies the excess of attenuation depends on the scattering of trunks and branches and by the scattering and absorption of leaves, as indicated in the results from the investigation by Price et al. (1988) in broadleaf forests. Huisman and Attenborough (1991) have demonstrated the interactions between ground effect and trunk scattering in a pine forest. Burns (1979) has measured the level of absorption of sound by pine trees, resulting a moderate frequency attenuation of individual boughs from thermoviscous absorption in the boundary layer in the surrounding air. This study has confirmed Embleton's anomalous gain results (see Embleton 1963).

Reduction of the deleterious effect of reverberative decay and frequency attenuation is considered an important factor in the evolution of birdsong. Also are welcome the studies of the acoustic properties of the different types of vegetation across different successional stages. In research in two Australian forests, Padgham (2004) has demonstrated with experiments the importance of distance as a truly influential factor in sound degradation. In a range of frequencies between 1 and 3 kHz the reverberation is less sensitive to distance. The attenuation of the signal is greater in structurally denser forest. When the sound source is positioned in an elevated site, this condition reduces the reverberation. The higher position of the receiver reduces the effect of attenuation, and this could be a quite complex condition for accurate ranging. These experiments have investigated the way in which sound propagates across forests with different structure, species composition, and history, and ultimately the consequences on the behavior of the acoustic species and on their long-term evolutionary adaptation.

In an important experiment Aylor (1971) measured the transmission of random noise through dense corn, a dense hemlock plantation, an open pine stand, a dense hardwood brush stand, and in an overcultivated soil. In particular the relationship between attenuation and frequency can be modeled and the model applied to all other typologies of vegetation. Each cover has a different threshold to excess attenuation: for example, corn 6 dB/100 ft for each doubling of frequency between 500 and 4,000 Hz. The stems of hemlock, pine, and brush have an excess of attenuation of about 5 dB/100 ft at 4,000 Hz. Bare ground has an attenuation effect at frequencies between 200 and 1,000 Hz, depending on the amount of air in the soil (permeability). Tilling soil reduces the frequency of peak attenuation from 700 to 350 Hz. Scattering and ground attenuation represent the main actors of sound attenuation by vegetation, but this depends on the distance to the sound source.

2.7 Vibrations and Leaves

Leaves are used by some animals, such as spiders, as a medium to enhance vibratory signals employed in animal communication. Magal et al. (2000) have investigated the role of leaf structure in vibration propagation of apple leaves. Using the impact of small spheres to produce waves on the leaves, three components of leaf structure were identified: The midvein, minor vein, and the interspaced homogeneous region compose the leaf lamina. The midvein absorbed vibratory energy at a level of 80 % at the leaf base and 40 % at the apex. Minor veins had a decrease of 70 % at the leaf base and 31 % at the apex. The loss of energy at the homogeneous area was 40 %. The midvein acted as a low-pass filter. The energy loss was dependent not on vein type but on vein diameter: diameter less than 0.2 mm was considered as equivalent to a homogeneous region.

2.8 Sound Propagation at Forest Edges

The impact of the forest edge on acoustic propagation is not well known, it is reasonable to expect a strong discontinuity of sound propagation from the abrupt changes in microclimate, ground cover, and biomass present between interior forest, edges, and open areas (rural or natural). Some recent measurements have been carried out by, for example, Swearingen et al. (2007).

The hypothesis of a habitat-dependent noise condition has been evaluated by Slabbekoorn (2004) in two tropical forest types (a primary and secondary rainforest and a gallery forest) (forest ecotone). In the rainforest the spectral profile is characterized by an increasing peak of frequency from category 1 to 7 (kHz), but in the forest ecotone the spectrum was very monotonous with two peaks at 6 and 8 kHz, and the differences were significant (Fig. 2.4). The rainforest is characterized by distinctive frequency bands, but in the forest ecotone the peaks, probably resulting from cicada activity, are present only at the higher frequencies. The differences between the two types of forest probably are the result of wind-generated noise and animal-borne sounds. The wind noise results are more important in the forest ecotone.

The distribution of noise in specific frequencies probably affects the acoustic selection of species, because animal signals have a great plasticity and can be modified in a very short time according to environmental conditions. In conclusion, it is reasonable to hypothesize that ambient noise may drive signal evolution. In fact, ambient noise can mask the singing activity of males, concurring to reduce the margin of affordability of female recognition of honest signals.

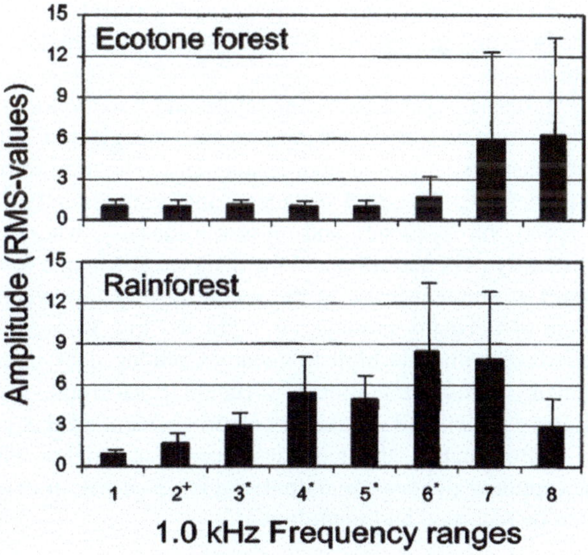

Fig. 2.4 Distribution of relative amplitude level per 1.0-kHz interval from 0 to 8.0 kHz along eight sites of rainforest and ecotone forest of Cameroon (Central Africa) (Reproduced with permission from Slabbekoorn 2004)

In conclusion, ambient noise must be considered not "just noise" but a reactive sonic context in which animals adapt their acoustic signals to improve communication, reduce energy cost at the breeding site, develop range for competitors and predators, and access food information by social exchange.

2.9 The Jungle Soundscape

The term jungle refers to adventure novels, but this environment, distributed from tropical latitude to equatorial latitude, is the coffer of biodiversity.

The soundscape in jungles is an event that creates mystery and surprise in the listener for the amazing variety of sounds produced simultaneously by thousands of species in a few square meters. This soundscape results in the richest sonic environment that we can find on the Earth. Because of the dense vegetation that is expanded in several layers until an height greater than 70 m, hearing is a more efficient tool to communicate and to acquire information than sight.

Eyring (1946) investigated how sound can pass through dense vegetation, measuring transmission loss, terrain loss coefficient, and humidity loss coefficient.

The ambient sound of the jungle is described by Eyring as follows.

> A jungle far removed from human activities at times may be deathly silent. At other times it is filled with animal sounds: humming, buzzing, chirping, noisy and musical "mate" calls; with the rustle of palms under wind action; with the sound of dropping and rushing water, the result of heavy dew and rain; and even with the sound of thunder. Seldom, if ever, do all possible jungle sounds join in a grand finale. When rain and wind are at their height, animal life is usually quiet; when the weather is fine, birds sing at daybreak; but insects, which have maintained a chorus all night, slowly bring their night calling to a close.

Birds and amphibians present a cyclical increase of their songs and calls during the breeding season, but for these last there is major variability linked to weather conditions.

In dense vegetation, temperature and wind velocity gradients are so small that their effect on sound transmission can be neglected. Humidity produces an increase of transmission loss at high frequencies. The terrain loss, defined as the loss in dB between two specified distances, increases linearly with distance. This increase depends on frequency and on terrain density, where terrain density measures the difficulty of penetration and the distance at which a foreign object can be seen. The level of ambient noise in the jungle is very low when measured during the quiet time of acoustic animals. At night the low frequencies produced by the breeze decrease while the high frequencies produced by insect choruses increase. In the jungle, sound orientation is difficult to ascertain, with an average error of 20°. Reverberation and scattering concur to maintain this judgment at a high rate.

Unfortunately, the acoustic diversity of jungles, which is the expression of the exceptional richness of animal species, is at risk of extinction before such "voices" can be recorded and identified.

2.10 The Sound of Fire

The sound of fire probably is an unknown sound from nature. Few people have experience of such a sound when it represents a danger or a risk.

However, organisms that live in fire-prone habitats have developed a special sensibility that allows them to benefit from the consequences of burns or to escape the risk of being killed by flames using olfactory or visual sensations. For instance, buprestid beetles are attracted by fire that is localized by using infrared sensors and antenna olfactory receptors. Ungulates move in burnt areas to find fresh vegetable masses sprouting from ash-fertilized soil. Birds are attracted by fires to find fleeing insects in the proximity. Some animals fly away, but others, such as ticks, after sensing smoke, drop to the ground and search for a safe shelter.

No animals until now have been found to use sound for fleeing or for finding benefits from burn events. However, Grafe et al. (2002) have observed the red frog (*Hyperolius nitidulus*) to flee from a fire quite far from smoke or heat perception. This West African species spends the summer time in savannah in dry exposed positions. The idea that this frog was fleeing using the sound produced by a fire was confirmed by playback experiments on juveniles. Frogs when exposed to the crackling sound of a fire raised their head, scanning the surroundings, and then jumped toward adjacent fire-resistant vegetation. This species reacts to the airborne sound of a fire and uses acoustic cues not only for reproductive purposes but also for other vital functions. In conclusion, in some anurans acoustic perception has evolved not only for sexual selection but also for orienteering and escaping threats.

Summary
The sonic characters of the landscape are important to better understand animal communication in the perspective of natural conservation and possible sustainable actions to assure a standard of well-being for humanity.

Weather conditions and in general climatic context have a great influence on the sonic environment, affecting the sound activity of vocal species. Wind depresses biophonies but on the other hand is a source of important information on atmospheric turbulence.

Most vocal species are directly affected by changes in climatic and weather conditions, confirming the importance of sonic ambience as an indicator of changes in ecosystems.

The structure of vegetation and its density are important drivers of sonic propagation. Vegetation interferes with sonic energy by reverberation, absorption, and scattering.

In a forested landscape sound propagation is strongly affected by ground effect, by scattering from tree trunks and branches, and by absorption by leaves.

Wind-generated noise and animal-borne sounds are differently distributed when core and edge areas are compared. In particular, forest edges represent not only an ecological discontinuity and a tension zone but is also a cause of sonic discontinuity.

The jungle environment represents a more spectacular example of the sonic environment at the highest acoustic diversity. The dense vegetation, the complexity of the vertical layers of vegetation, the high humidity, and the lack of wind create a unique system to which vocal species have been adapted for a long time.

It seems surprising that the sound produced by a wildfire can represent a signal for some species. It is the case documented also experimentally in the red frog (*Hyperolius nitidulus*), which reacts to the sound of an approaching fire by searching for a fire-free refuge.

Sounds from stressed nature are not only limited to fire but also can be caused by drought. Thus, plants suffering from drought stress produce special sounds by the collapse of single cells from dehydratation, and these sounds seem to be intercepted by some species of bark beetle to orient their infestation in a stressed tree.

References

Albert DG (2004) Past research on sound propagation through forests. US Army Corps of Engineers, ERDC/CRREL TR-04-18

Aylor D (1971) Noise reduction by vegetation and ground. J Acoust Soc Am 51(1):197–205

Aylor D (1972) Sound transmission through vegetation in relation to leaf area density, leaf width, and breadth of canopy. J Acoust Soc Am 51(1):411–414

Botero C, Boogert NJ, Vehrencamp SL, Lovette IJ (2009) Climatic patterns predict the elaboration of song displays in monckingbirds. Curr Biol 19:1–5

Briefer E, Oiejuk TS, Rybak F, Aubin T (2010) Are bird song complexity and song sharing shaped by habitat structure? An information theory and statistical approach. J Theor Biol 262:151–164

Burns SH (1979) The absorption of sound by pine tress. J Acoust Soc Am 65(3):658–661

Embleton TFW (1963) Sound propagation in homogeneous deciduous and evergreen woods. J Acoust Soc Am 35:1119–1125

Eyring C (1946) Jungle acoustics. J Acoust Soc Am 18(2):257–270

Fonseca PJ, Revez MA (2002) Temperature dependence of cicada songs (Homoptera, Cicadoidea). J Comp Physiol A 187:971–976

Gibb JP, Breisch AR (2000) Climate warming and calling phenology of frogs near Ithaca, New York, 1900–1999. Conserv Biol 15(4):1175–1178

Grafe TU, Dobler S, Linsemair KE (2002) Frogs flee from the sound of fire. Proc R Soc Lond B 269:999–1003

Huisman WHT, Attenborough K (1991) Reverberation and attenuation in a pine forest. J Acoust Soc Am 90(5):2664–2677

Ingard U (1953) A review of the influence of meteorological conditions on sound propagation. J Acoust Soc Am 25(3):405–411

Magal C, Schöller M, Tautz J, Casas J (2000) The role of leaf structure in vibration propagation. J Acoust Soc Am 108(5):2412–2418

Martens MJM, Michelsen A (1981) Absorption of acoustic energy by plants leaves. J Acoust Soc Am 69(1):303–306

Moller AP (2010) When climate change affects where bird sing. Behav Ecol 22(1):212–217

Morgan S, Raspet R (1992) Investigation of the mechanisms of low-frequency wind noise generation outdoors. J Acoust Soc Am 92(2):1180–1183

Padgham M (2004) Reverberation and frequency attenuation in forests – implications for acoustic communication in animals. J Acoust Soc Am 115(1):402–410

Price MA, Attenborough K, Heap NW (1988) Sound attenuation through trees: measurements and models. J Acoust Soc Am 84(5):1836–1844

Sanborn AF, Maté S (2000) Thermoregulation and the effect of body temperature on call temporal parameters in the cicada *Diceroprocta Olympus* (Homoptera: Cicadidae). Comp Biochem Physiol Part A 125:141–148

Slabbekoorn H (2004) Habitat-dependent ambient noise: consistent spectral profiles in two African forest types. J Acoust Soc Am 116(6):3727–3733

Snell-Rood E, Badayaev AV (2008) Ecological gradient of sexual selection: elevation and song elaboration in finches. Oecologia 157:545–551

Sueur J, Sanborn AF (2003) Ambient temperature and sound power of cicada calling songs (Hemiptera: Cicadidae: Tibicina). Physiol Entomol 28:340–343

Swearingen MS, White MJ, Guertin PJ, Mifflin JA, Onder TE, Albert DG, Decato SN, Tunick A (2007) Acoustic propagation through a forest edge. US Army Corp of Engineers Engineer Research and Development Center, ERDC SR-07-03, Vicksburg, MS, US

Chapter 3
Bioacoustics Theories

The structure and complexity of biophonies are the result of evolutionary pressure. Physical structure of the organisms, adaptive mechanisms in sound production and transmission, reduction of acoustic competition, and behavioral processes are possible integrative strategies to enhance sonic context. Four major hypotheses are considered relevant in the perceiving and communicating processes common to all animal species: the morphological adaptation hypothesis (MAH), acoustic adaptation hypothesis (AAH), acoustic niche hypothesis (ANH), and species recognition hypothesis (SRH).

The morphological adaptation hypothesis (MAH) refers to the role of body size as a biological constraint of the vocalization organs and their acoustic performance, confirming an inverse relationship between acoustic frequencies and body size.

The acoustic adaptation hypothesis (AAH) states that the environment is an important cause of modification and alteration of acoustic signals. Dominant frequencies and other long-distance calls are the result of an interaction between the animals and the environment to maximize the efficiency of the emitted sounds. Frequency and structure of acoustic repertoire are the plastic traits that can be modified according to the environmental constraint.

The acoustic niche hypothesis (ANH) states that every species has a unique acoustic space in which to structure the sonic species-specific signature to reduce interspecific competition and to optimize intraspecific communication mechanisms.

The species recognition hypothesis (SRH) supposes that species living in sympatry try to reduce the risk of utilizing similar sonic traits that could confound species in reproduction and create the risk of hybridization. This set of hypotheses has epistemic relationships to form a meta-bioacoustic theory.

A. Farina, *Soundscape Ecology: Principles, Patterns, Methods and Applications*,
DOI 10.1007/978-94-007-7374-5_3, © Springer Science+Business Media Dordrecht 2014

3.1 Introduction

Biophonies are integral and substantial aspects of the soundscape, and their variety and dynamics concur with overall environmental biodiversity. In this chapter we illustrate and discuss some hypotheses originated by empirical indoor/outdoor research in different fields of the physiology, behavioral, and cognitive bioacoustics disciplines devoted to explaining nature and the role of acoustic performances of vocal animals. At least four hypotheses contribute to the bioacoustics theoretical framework: the morphological adaptation hypothesis (MAH), the acoustic adaptation hypothesis (AAH), the acoustic niche hypothesis (ANH), and the species recognition hypothesis (SRH). Testing and discussing these hypotheses contributes to better understanding of how natural selection operates and what strategies are used by animals when performing acoustic signals (indeed, the majority of cases of these hypotheses have been tested on birds). Briefly, the morphological adaptation hypothesis (MAH) refers to the embodiment constraints, the acoustic adaptation hypothesis (AAH) considers the effects of environmental stressors, the acoustic niche hypothesis (ANH) is formulated under a competitive context, and the species recognition hypothesis (SRH) refers to a behavioral context (Fig. 3.1). All these phenomenological contexts represent concurrent proxies that shape acoustic performances and which often operate in tandem, as argued by Seddon (2005).

3.2 The Morphological Adaptation Hypothesis (MAH)

Different forces in action in natural selection and vocalizations (song, call, alarms, voice) are phenotypical characters with high plasticity used to better define species-specific competencies.

Most vocal animals have complex organs of vocalization; for instance, in mammals the sound is produced by the vocal folds and modulated by the tongue. In songbirds the sound is produced by the syrinx and then modulated by the vocal tract and beak (after Fletcher 2007).

Larger animals such as elephants communicate at a greater distance than smaller animals such as wildcats, and the same occurs for birds such as cockatoos when compared with the distance traveled by the songs of smaller birds. However, the acoustic energy produced in proportion by larger species is only one component of the vocal signal. It is reasonable to speculate that for long-distance communication animals utilize also the best range of frequencies to reduce sound degradation, but it remains unknown what are the evolutive proxies that produce specific differentiation.

The morphological adaptation hypothesis (MAH) refers to the role of body size as a biological constraint of the vocalization organs and their acoustic performance.

According to this hypothesis, organisms with large body mass utter vocalizations with lower frequency than smaller species. For instance, in many species of

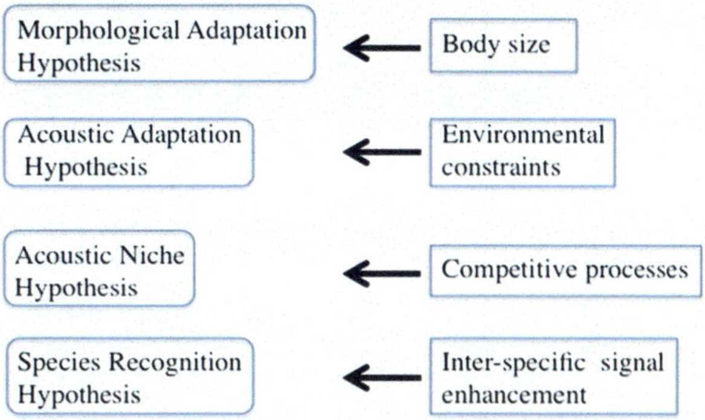

Fig. 3.1 The four bioacoustics hypotheses and their major constraints

birds body size is correlated with frequency. This correlation is explained in birds by the capacity to vibrate the syringeal membrane, which can produce high-frequency sound only in a small apparatus.

The relationship between body mass and frequencies utilized in vocal performances has been proved by Wallschlager (1980) in 90 European passerine birds and by Fletcher across different species of birds and mammals (Fletcher 2004) with a relationship between frequencies and body mass expressed by a power law $f \propto M^{-0.33}$ (f = frequency, M = body mass) (Fig. 3.2).

Similar results have been found by Boeckle et al. (2009) from 76 species of frogs where 25 % of the variation in dominant frequency has been explained by body size.

In birds, besides body size the dimension of the bill is a further important element to modulate sounds, especially high-frequency sounds. In fact, the change in bill size and its morphology are not only correlated with foraging habits and the structure of the habitat, but the suprasyringeal tract that functions as a resonating space is directly related to bill dimension and shape. Small bills can produce more modulations than large bills, and this morphological trait is negatively related with the trill rate. Smaller bills produce trills at a higher rate and definitively larger bills are less versatile and less able to produce sound than smaller bills. In particular, the bill is gaped more during the production of high-frequency notes, and the speed with which the bill is opened or closed depends on the dimension of the bill. Podos (1997), investigating 34 species of birds belonging the family Emberizidae, has found that the increase of trill rate corresponds to a reduction of frequency bandwidth utilized.

3.3 The Acoustic Adaptation Hypothesis (AAH)

A community of acoustic animals is arranged according to the acoustic signals that individuals emit to fulfill different functions. Long-range and short-range communications occur intensively during specific active times. Density, receiver psychology, signal used, and environmental conditions are the major factors that

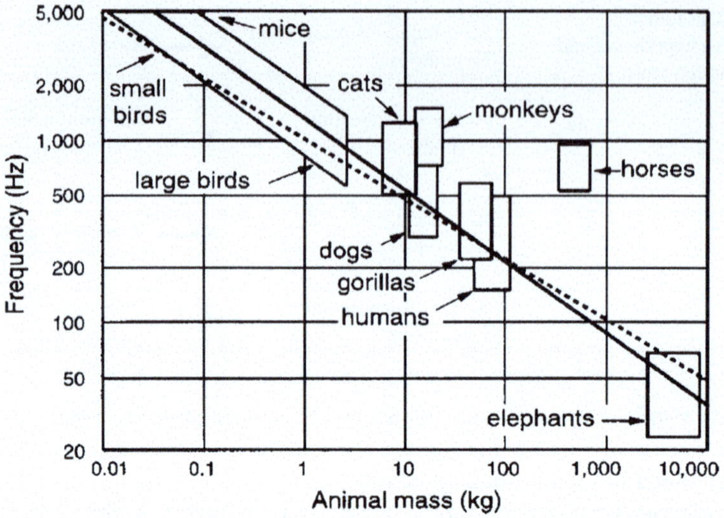

Fig. 3.2 Correlation between frequency and animal mass. At the two extremes are the smallest birds (e.g., the firecrest, *Regulus ignicapillus*) and the largest mammals [e.g., the elephants (*Loxodonta* spp.)]. The *dashed line* represents the regression between frequency at body mass $f = M^{-0.33}$ (simple linear scaling); the *solid line* is the predicted relationship $f = M^{-0.4}$ (Reproduced with permission from Fletcher 2004)

influence the acoustic network. The environmental condition of the habitat constrains the communication efficiency and efficacy between a sender and a receiver. Environmental conditions affect the quality of the signal and also the capacity by a receiver to range, that is, to estimate the distance to the emitters.

The environment is an important cause of modification and alteration of the acoustic signals.

The song when used to communicate over long distances is prone to modifications by environmental constraints such as topography, vegetation cover, and natural or anthropogenic noise masking.

The absorption of some frequencies that modify the structure of a signal and its meaning for a receiver is a common effect.

The acoustic adaptation hypothesis (AAH) is based on the assumption that the sounds used in long-distance communication are correlated with habitat acoustics. This hypothesis, first formulated by Morton (1975) on birds, states that dominant frequencies and other long-distance calls are the result of an interaction between the animals and the environment to maximize the transmission efficiency of the emitted sounds. In this way different habitats interfere differently, according to the sound frequency bands, and callers try to occupy the frequency bands that have lower degradation. This consideration holds not only for sounds in air but also sounds in the waters of the open ocean.

Furthermore, temporal patterns of a sound design can also be affected by environmental constraints. For instance, low-frequency signals travel through the vegetation

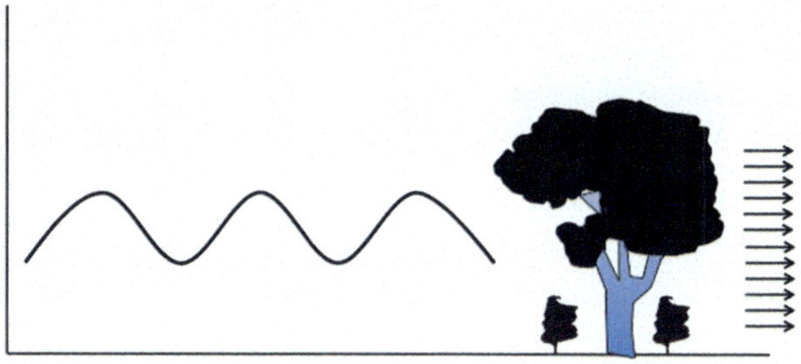

Fig. 3.3 Schematic representation of the effects of vegetation sounds according to frequencies: low-frequency sounds are more able to penetrate dense vegetation

better than high-frequency signals, and songs that contain short notes repeated at longer intervals travel more efficiently in dense vegetation (Fig. 3.3, 3.4 and 3.5).

At the same time, acoustic signals propagate at different rates from ground to canopy level, and according to this property the height at which animals vocalize and sing can be modified to adapt to the best transmission rate.

Song notes, calls, and contact notes are modulated according to the habitat, and adaptation is expected in every species.

The sound reaction to habitat conditions is not constant in different situations. For instance, in open habitats there is a weaker relationship between attenuation and frequencies, but in closed habitats (dense vegetation coverage) this effect is enhanced.

The acoustic adaptation hypothesis is based on the assumption that sound when crossing a homogeneous frictionless medium diverges spherically from a source that has a small dimension compared with the sound's wavelength. The far sound decreases at a rate of 6 dB at each doubling of distance for the inverse square law. The amount of energy that is emitted by a punctual source is distributed radially on a surface that is increased by the square of the radius where the surface corresponds to the $s = 4\pi r^2$. However, the environment is more complex than an ideal space and the inverse square law presents anomalies. In fact in the real world sound energy can be reduced by several factors such as absorption by air, ground, and vegetation or reflected and diffracted, and we expect that such disturbances vary according to the length of the sound propagation path.

The limit in field experiments of the AAH by playback procedures results in part from the unknown level of impedance of trees, rocks, grasses, water, and so on. The excess of attenuation (EA) that measures the amount of energy captured by the medium outside the limit of the inverse square law is linked to frequency by a positive correlation. Higher frequencies are reflected more by leaves and branches in forests than low frequencies, but local conditions can create different conditions for sound transmission.

A method based on playback and on recording sessions able to capture the frequencies that in a specific habitat experience the least excess attenuation has been first applied by Morton (1975).

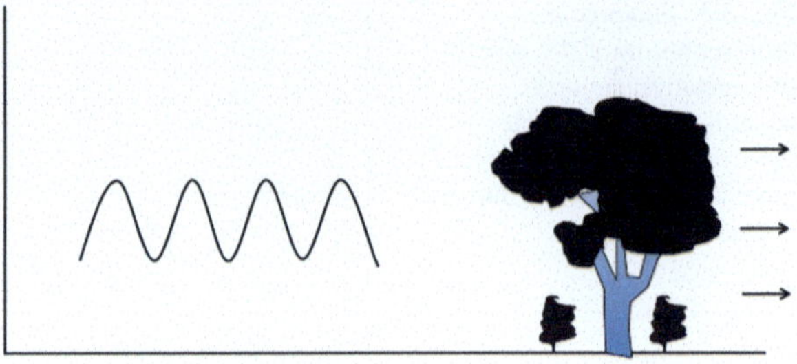

Fig. 3.4 Schematic representation of the effects of vegetation on sounds according to frequencies: high frequencies encounter more friction in dense vegetation

In Fig. 3.4 is indicated the excess of attenuation level for random noise and for pure tone in forest, edge, and grass conditions. This author after playback experiments has confirmed that forest habitats differ from edge, and open areas in a narrow range of frequency (1,585–2,500 Hz) have a lower attenuation than lower or higher frequencies. The attenuation increases sharply above 2,500 Hz. Birds that inhabit the lower forest level have predominantly pure tones averaging 2,200 Hz.

The edge habitats have a broad range of frequencies, and the attenuation rate does not differ significantly, suggesting that the selection pressure is not in action in these habitats. In grassland there is a positive correlation between increasing frequency and increasing attenuation.

The results presented by Morton (1975) have been confirmed by Marten and Marler (1977) in temperate habitats. Animals vocalizing at an height greater than 1 m can transmit at a longer distance when low frequencies are used. When the sound is emitted very close to the soil, where the attenuation of lower frequencies is higher, this allows us to state that the distance at which the sound may be transmitted depends more on the height at which a bird is singing than by the frequencies used.

Simulations conducted by Brown and Handford (1996) have confirmed the acoustic adaptation hypothesis. They constructed two different types of "source signals," rapid amplitude modulation (AM) (trills) for open habitats and low AM (whistles) in close habitats.

The degradation of such signals was performed artificially by various echo treatments or by various amplitude decrease treatments. The results are consistent with the AAH because in closest habitats whistled signals transmit with better quality than trilled signals. In open habitats trilled signals are transmitted with a lower variability of quality than whistled signals and for their redundancy are transmitted with more efficiency.

Further experiments conducted by these two authors (Brown and Handford 2000) have demonstrated that whistles degrade less than trills in open and forest habitats but in open habitats trills are received with higher quality than whistles (Fig. 3.5).

The AAH has been proven by Patten et al. (2004) at the subspecies level in the song sparrow (*Melospizia melodia heermanni* and *M. m. fallax*) on the Pacific slopes of California. These two species occupy riparian habitats, but the first

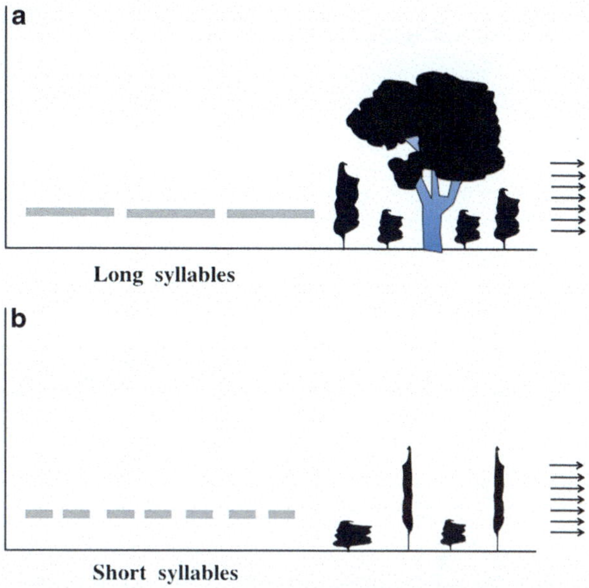

Fig. 3.5 Long syllables, such as whistles (**a**), can cross dense layers of vegetation with minimal interference, whereas short syllables (**b**), such as trills, are more likely to be transmitted in open environments

subspecies lives in a denser understory with greater vertical heterogeneity. According to the AAH, the song of *M. m. heermanni* has lower pitched and more widely spaced elements than that of *M. m. fallax*.

As an adaptation to the conditions of dense rainforests, birds use low-pitched whistle-like songs.

Narrow-frequency bandwidth notes (NFB) are pure tones that do not have great changes in frequency; these tones are emitted by birds that live in dense forests and differ from frequency-modulated notes. These notes benefit from reverberation, allowing the transmission of a longer and louder signal as demonstrated by playback experiments conducted by Slabbekoorn et al. (2002) on the green hylia (*Hylia prasina*), a bush warbler passerine (family Cettiidae) living in dense forests of tropical Africa. Its song is quite simple, composed of two successive NFB.

Longer notes produced a stronger behavioral response, demonstrating that reverberation can improve signal transmission when the signal design is adapted as in NFBs. The positive effect of reverberation has been tested on the white-necked thrush (*Turdus albicollis*) and on the sympatric species cocoa thrush (*Turdus fumigatus orinocensis*) by Nemeth et al. (2006). The first species has song elements that are double those of the cocoa thrush, and an experiment of long-distance transmission showed that the white-necked thrush, which has a longer sound, has a less-attenuated signal (Fig. 3.6).

The authors suggest that the positive effect of reverberation favors longer songs.

A more complex scenario, depicted by Ziegler et al. (2011), showed that the call attributes of the frog *Hypsiboas pulchellus* are strongly affected by habitat structure in terms of temporal parameters of the call and that temperature affects the size of individuals and consequently the dominant call frequency. These authors

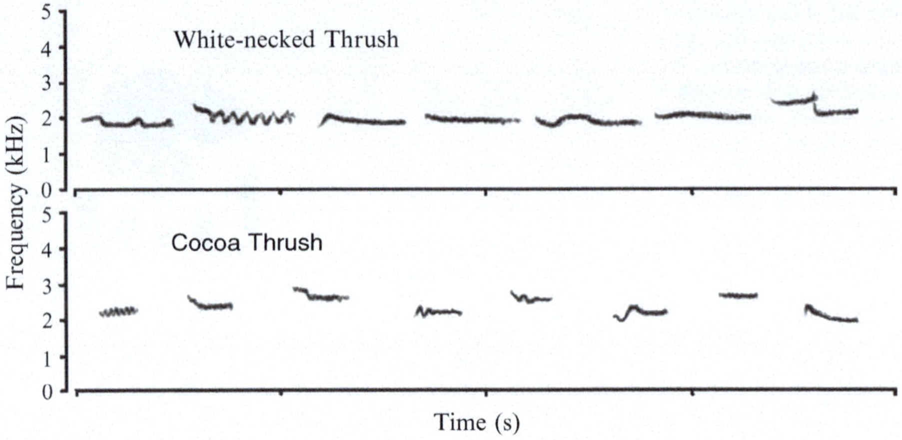

Fig. 3.6 Spectrogram of the white-necked thrush and cocoa thrush that illustrates well the difference in syllable structure (Reproduced with permission from Nemeth et al. 2006)

suggest the role of phenotypic plasticity is incorporated in the connection between environmental structure and individual call attributes (Fig. 3.7). In nature, animals living in a natural noisy background such as a stream adopt this strategy, as does the dipper *Cinclus mexicanus* (Kingery 1996) or *Phylloscopus magnirostris* (Dubois and Martens 1984), which have sound repertoires quite distinct from other bird species living in quieter habitats.

Boeckle et al. (2009) have found that the skipper frog (*Staurois latopalmatus*) that lives along the waterfalls of Bornean streams has a sound repertoire well adapted to this exclusive environment. In fact, this species seems to avoid quiet stream environments. This frog has two types of calls, a short single-note and a long multi-note call. Both calls have a sound pressure that is less than the noise produced by a waterfall. The dominant frequency is around 5 kHz, and from the correlation between the body size (snout–vent length) of 74 Ranidae and their dominant frequency it appeared that this species should have a lower frequency, and the results suggest an acoustic adaptation to a low-frequency-dominated noise. These authors suggest that microhabitat characteristics can act as a strong selective pressure on the patterns of the acoustic signals.

There are increasing evidences that body and bill size, associated with habitat structure, influence the communication mechanism, at least in birds. Recently Derryberry (2009) focused on the variation of trills in the white-crowned sparrow (*Zonotrichia leucophrys*). Trills that are repetitions of the same syllable seem more affected by the mechanisms of sound production and by habitat-dependent selection. Reverberation, defined as the amount of energy residual from a note and reverberated by the vegetation or other objects that fill the silent interval between notes, affects the transmission of frequency-modulated trills. Vegetation density, transmission distance, frequency, element repetition rate, and element position within the trills are factors influencing the reverberation effect. In dense vegetation

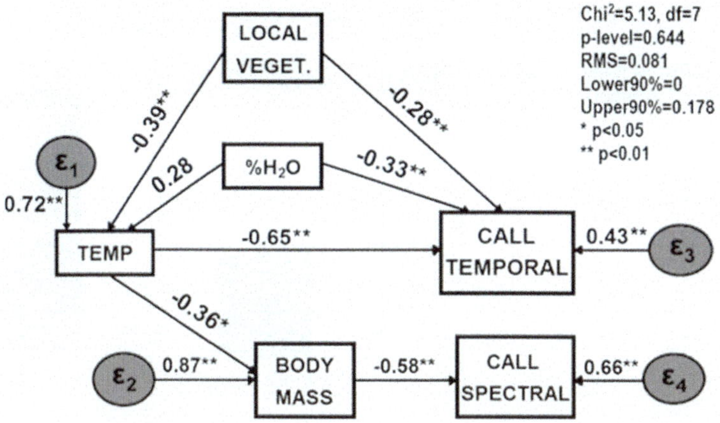

Fig. 3.7 Path analysis regarding a causal model utilized to explain the relationships between abiotic and vocal variables. *Epsilon* represents the amount of variances unexplained by the model patch coefficients, and their significance between variables is indicated (Reproduced with permission from Ziegler et al. 2011)

this species produces trills at a slower rate and a lower minimum frequency. This analysis has examined the white-crowed sparrow sounds recorded in 1970 and in 2005. In the studied area the vegetation showed an increase of cover and density. The hypothesis that this species has modified their acoustic repertoire according to the changes occurring in their habitats has been proved to be consistent. This is the first case of evidence of acoustic plasticity of birds under the constraint of environmental changes. We can expect a modification of the soundscape of entire regions to result from vegetation changes favoring some sonic components. With increase of the vegetation cover, the reverberation increases, and consequently the songs will be modified to be produced more slowly and at lower frequencies.

We consider this fact to be relevant because the rapid changes of landscape structures everywhere affected by human disturbances could be easily monitored applying the sonics tools. Vegetation cover can affect definitively temporal and spectral characteristics of the sonic ambience.

Experiments conducted by Naguib (2003) have demonstrated that fast trills transmit less well than slow trills, with effects on long-range communication. In Fig. 3.8 and in Fig. 3.9 is shown how reverberation increases with increasing vegetation cover and in relationship with the frequency of the trill element. Low-frequency notes slowly repeated are less affected by reverberation and can assure long-range communication better than high-frequency notes quickly repeated, as has been observed in birds living in open and densely vegetated habitats.

The animal phenotype is strongly affected by the environment in which a species resides. For instance, acoustic variations in bird vocalization have been proved to be the result of environmental constraints, which suggests an evolutionary process that has great consequences on overall biodiversity (Slabbekoorn and den Boer-Visser 2006). The recent dispersion of urban settlements has created novel habitats for

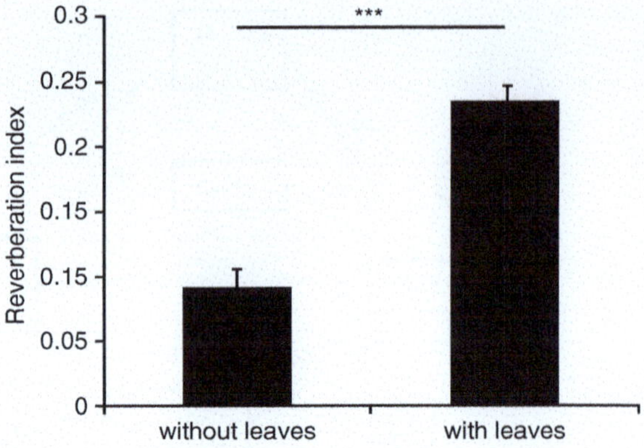

Fig. 3.8 Average reverberation index for deciduous forest in winter before foliage has emerged (*left*) and in summer with full vegetation (*right*). (Reproduced with permission from Naguib 2003)

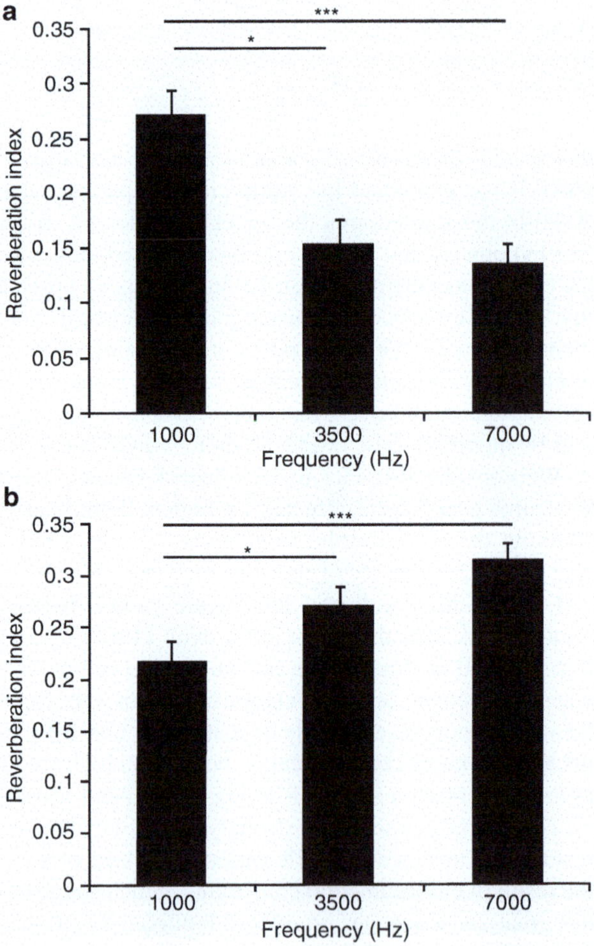

Fig. 3.9 Level of reverberation of trills with element repetition rates of 10 ms obtained at different frequencies and playback in deciduous forest before foliage emerged (November) (**a**) and deciduous with foliage (July) (**b**) (Reproduced with permission from Naguib 2003)

many species (from plants to animals) that may find resources more abundant than in natural conditions. The management of urban space to also preserve biodiversity is not an extravagant idea but an important issue for the long-term vision of biodiversity on this planet. In many species of birds, males sing to attract females and to defend their territory. If such communication is masked by background noise, as is very common in urban areas, the males must fight directly with potential intruders, vesting energy without the biosemiotic support of a sign process, and females may be not attracted or attracted with greater difficulty.

In the great tit (*Parus major*), one of the most common birds adapted to live in urban areas, differences in song performance have been observed in different cities and habitats. Slabbekoon and den Boer-Visser (2006) have investigated the song repertoire of this species in ten large cities across Europe, comparing their songs with those of adjacent forested areas. Interesting modifications were found in strophes, which are a repeated set of phrases where each phrase is composed by two, three, or four notes that finally produce a song type. In the urban environment, this species produced only 80.7 % of song types compared with 90.8 % in forested areas and sang more rare song types.

A comparison between urban versus forest areas show that the song repertoire is different and that in urban areas the average minimum frequency is significantly higher than in forested areas.

It could be a dimension effect on the frequency shift if this species should have a change in body size with a lower body size in urban area compared to forest areas, but this is not been found up to now, so effects resulting from different body size cannot explain the differences in frequencies.

Change in song performance has been observed not only in the frequency domain but also in the temporal domain. Songs of "city birds" are shorter in duration with shorter inter-song intervals.

The first note of the phrase is shorter in city singers, but not the other notes, and this is in agreement with the fact that urban areas are not only noisier than forest habitat but are open areas when compared with forest. To assure better long-distance communication the signals should be of lower frequency, shorter, and produced at a fast rate. These authors have argued that the modification of the first note creates an advantage because the first note is critical for detection and recognition. The phenotypic plasticity found in the great tit could accelerate genotypic divergence in the future, with potential urban speciation.

The changes in great tit song performance have been observed in many countries and in many noisy environments.

The habitat conditions outside tropical areas are quite variable, similar to deciduous habitats where, according to the season, there are different acoustic conditions for the presence or absence of foliage cover and for the status of leaf maturation which alter the acoustic conditions. These differences produce effects on the acoustic impedance and on reflection and diffraction of the acoustic signal on the leaf surfaces. In temperate biomes the seasonal growth of plants such as ferns (e.g., *Pteridium aquilinum*) or other annual tall forbs modifies the sonic environment with an increase of sound degradation of the high frequencies. Turbulence produced by seasonal winds and breeze and different temperatures modify the sonic environment as well.

Blumenrath and Dabelsteen (2004) have studied the influence of foliation in the acoustic performance of the great tit (*Parus major*), a species that lives in many types of habitats and with the capacity of individual recognition of acoustic signals from conspecifics. The males are able to discriminate the distance at which another male is singing on the basis of song degradation. These authors have investigated song degradation before and after foliation using recorded song and a playback technology. The experiments consisted of recording the effect of foliage/no foliage on isolated songs. In absence of vegetation the same song degradation observed when there was a dense cover of leaves was obtained by doubling the distance from the loudspeaker. In conclusion, the presence of foliage limits long-distance communication in this species, with effects on the territorial organization and maintenance.

Reviews on the acoustic adaptation hypothesis have been proposed by Ey and Fischer (2009) and a meta-analysis was presented by Boncoraglio and Saino (2007), but despite these attempts a clear discriminatory prospect does not emerge between species, and rules and exceptions persist, diminishing the explicative capacity of this hypothesis. It seems that this theory does not explain the communication constraint everywhere. Although the AAH has been tested in 121 species of Australian birds from 41 families by Blumstein and Turner (2005), only a modest support for the hypothesis was found.

To test the acoustic adaptation hypothesis, frequency range and trill rate of chipping sparrow (*Spizella passerina*) were investigated in two different habitats by Brandley and Burns (2007) and Burns (2007) without finding significant results. As the authors stated, however, the small sample size, the closeness of sites, and the minor habitat differences could have altered the results.

Similar conclusions have been drawn on the long-distance alarm call of four different species of marmots of North America (the yellow-bellied marmot, *Marmota flaviventris*; the Olympic marmot, *M. olympus*; the hoary marmot, *M. caligata*; and the woodchuck, *M. monax*), which failed to be proved according to the Daniel and Blumstein results (1998). The authors suggest that other factors in this specific case are responsible for the evolution of species-specific marmot alarm calls.

3.4 The Acoustic Niche Hypothesis (ANH)

Ecological theory states that every species has a unique niche in terms of habitat and resources to reduce competition with other species (Hutchinson 1959). Ecological niche partitioning is primarily fixed in morphology and then genetically in behavior, but recent experiments have demonstrated that a learning process could produces separation in some components of the ecological niche. Slagsvold and Wiebe (2007), manipulating the young of the great tit and blue tit by cross-fostering in the wild, transferring eggs of the blue tit into the nest of the great tit and vice

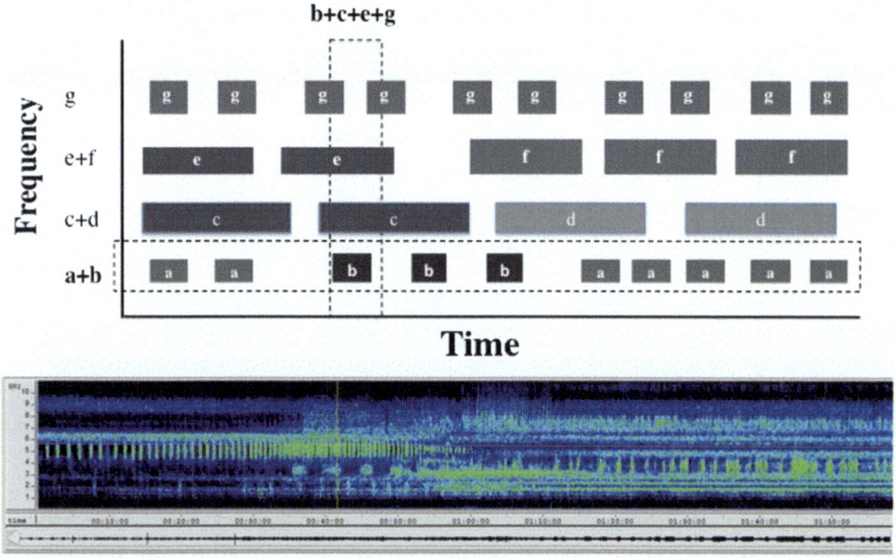

Fig. 3.10 Schematic representation of acoustic repartition across frequencies and time of the individual repertoire of different vocal species (*a–g*) and a spectrogram (Reproduced with permission from David Monacchi) from an African tropical forest (*lower*) showing the dense repartition of sounds that belong to the frequencies

versa, found that the feeding niche was changed and that the new habits were conserved during life in experimental individuals. This work opened a grand vista to the mechanism that allows species to coexist.

Habitat selection and behavior can modify call properties in animals. In environments with a high diversity of vocal animals, spectral and temporal overlap should be reduced to avoid masking and confused signals (Krause 1993) (Fig. 3.10).

The acoustic niche hypothesis, as named by Bernie Krause in 1980 (Krause 2012), posits that the sound spectrum is a limited resource and that species try to reduce and minimize acoustic competition. The acoustic niche can be defined as partitioning into a frequency space the sounds of animals to reduce the interspecific masking effect. This frequency partitioning has been documented in bats, for example.

The acoustic niche hypothesis is related to the evolution of habitats and the time lag of coexisting species, as recently pointed out by Malavasi and Farina (2013).

The acoustic niche hypothesis has been proved in birds, frogs (at least some), and in some species of cicadas. In this last group of insects Sueur (2002) has observed reduction of acoustic heterospecific overlap.

However, this expectation was not demonstrated in a recent investigation by Planquè and Slabbekoorn (2008) in a Peruvian bird assemblage. These authors did not find a negative correlation between spectral and temporal overlap. Exploring the acoustic behavior of 20 vocally dominant species from 14 families, they did not find a significantly lesser correlation between spectral and temporal overlap. The frequencies more often used by more species did not show significantly less overlap as was expected if all species were singing randomly. However, short-term auditory feedback mechanisms can adjust bird song to reduce interspecific acoustic interference.

The acoustic character can represent an important trait that drives the survivorship of many animals. The spatial and temporal character of a site can modify the responses of organisms to the conspecific or heterospecific acoustic cues.

The mechanisms of partitioning may be either spatial or temporal. Spatial mechanisms require species to move in a way so as to not be masked by other concurrent species. Temporal strategies present more complex mechanisms: for instance, a temporal segregation (calling at a specific time of the day, e.g., dawn, dusk, midday), or a jamming avoidance (inhibition of calling when another species is calling at the same frequency band).

Evolutionary mechanisms can produce shifts of sounds into highest or lowest frequencies that are unoccupied by the majority of species, as the case of the cave-eared torrent frog (*Amolops tormotus*) from China, which produces song below 12 kHz (Feng et al. 2006).

Sinsch et al. (2012) have studied an anuran community inhabiting an agricultural wetland near Butare, Ruanda. They found that niche segregation was produced mainly by advertisement call features and that spatial and temporal niche dimensions contributed secondarily. Significant acoustic resource partitioning has been found for an average niche overlap (0.30) that was intermediary between a random overlap (0.51) and the minimum niche overlap (0.11). From this study appears, interestingly, the wide overlapping of *Ptychaneda* spp., a recent invading group of frogs.

Playback experiments simulating the invasion of the American bullfrog (*Lithobates catesbeianus*) on a native species, the white-banded tree frog (*Hypsiboas albomarginatus*), have been conducted by Both and Grant (2012) who demonstrated the immediate shifting of frequency of native species during the playback sessions (Fig. 3.11). This shift was maintained also after the stimulus associated with a decrease of signal duration. Because of the broad band of the bullfrog calls, it is expected that the arrival of this species in new areas could produce severe effects on many other frogs that could be masked. The importance of frog calls for mating selection is well demonstrated, and an event such as call masking could produce deleterious effects on the entire frog community.

In some species of anurans, the acoustic niche hypothesis is only partially verified. For instance, Vasconcelos and Rossa-Feres (2008) have found that some anurans presented different advertisement calls that, combined with the physical calling sites, create an acoustic partitioning. Similarly Henry and Wells (2010) have found that in the *Crysoperla carnea* group (Insects, Neuroptera) sympatric species will partition acoustic niche space to reduce mating mistakes. Using an F_1 hybrid between two very closely related sympatric species, *Chrysoperla plorabunda* and *C. adamsi*, they showed that acoustic discrimination between parents and hybrid is weaker and the hybrid phenotype overlaps with that of the parents, thus confirming that the intermediate song phenotype cannot persist.

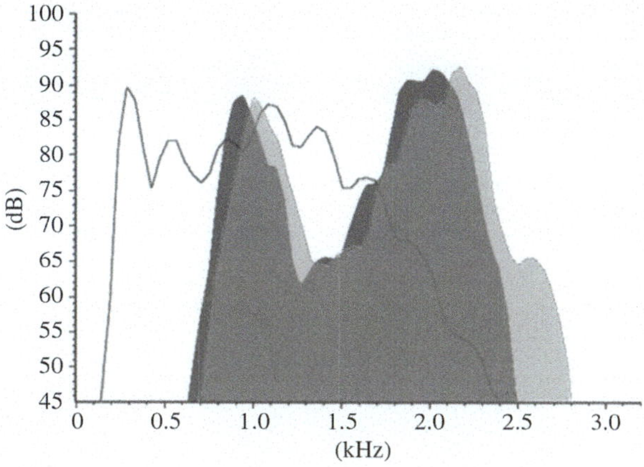

Fig. 3.11 Example of acoustic shift between two interacting species, represented by power spectra of the white-banded treefrog before (*dark grey*) and during (*pale grey*) the stimulus of a broadcast American bullfrog advertisement call (*white*) (Reproduced with permission from Both and Grant 2012)

The necessity to have distinct and well-recognized sounds is essential in the mating system of many songbirds. In many species the presence of masking by biological origins can produce an increase of energy to counteract the sonic level. In cases of nonbiological masking, such as produced by anthropogenic noise, some species can move in silence, waiting for a quiet moment. In other cases animals can shift their vocal output to time periods in which the acoustic masking is lower. Brumm (2006) has proved that male nightingales (*Luscinia megarhynchos*) when exposed to the playback of heterospecific song are able to start to sing in the silent intervals of the interspecific songs (Fig. 3.12). This flexibility enables birds to insert their songs in a silent interval, to save energy, and to improve the efficiency of signal transmission, as observed during frog choruses. There are two possible strategies to reduce masking effects: increase vocal sound pressure or avoid concurrent songs. In the nightingale Brumm (2006) has observed that a male starts singing 880 ms after the last male has finished singing. The same was observed in the coqui treefrog (*Eleutherodactylus coqui*) by Zelick and Narins (1985): in this species males started calling within a 750-ms interval when exposed to a playback noise. This interval is necessary for a male to be sure that the other male has completed his acoustic performance. In other cases the voluntary overlap of two songs by two competitive species is an aggressive signal, as argued by Naguib and Mennill (2010).

Anuran choruses are very complex soundtopes created by contemporary calls of several males at the same time. In a neotropical frog, *Hyla microcephala*, Schwartz (1993) has conducted experiments of playback by using an interactive computer-based system able to simulate the call interruptions of the males. A male when interrupted by the call of another male increases the spacing between his notes. Using synthetic calls with a silent gap, this author has found a vocal reaction of males to sound intensity. The acoustic overlap between males was less than 10 %. When males were calling in groups, the amount of sound interference between the males was highly variable, ranging from 34 to 92 %.

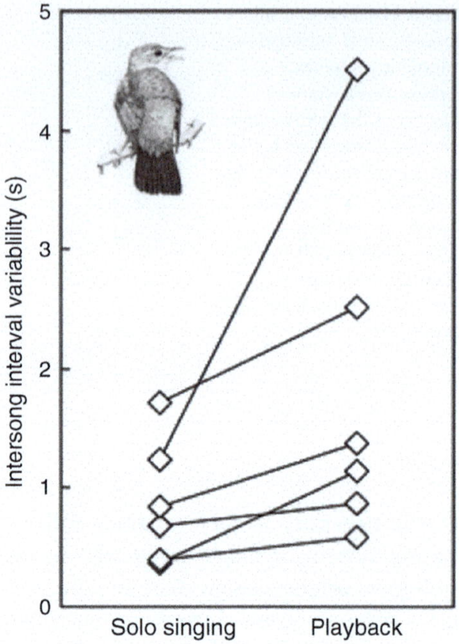

Fig. 3.12 Intersong variability expressed after a playback experiment on caged males of nightingale (*Luscinia megarhynchos*). Solo singing refers to songs spontaneously produced without playback (Reproduced with permission from Brumm 2006)

3.5 The Species Recognition Hypothesis (SRH)

Between the different hypotheses that are in action in the natural selection arena, the species recognition hypothesis (SRH) supposes that species living in sympatry try to reduce the risk of utilizing similar sonic traits that could confound species in reproduction, thus increasing the risk of hybridizations.

The species recognition hypothesis has been confirmed in neotropical suboscine passerines in which the song of sympatric pairs of closely related species has a more divergent acoustic pattern than that of allopatric pairs (Seddon 2005).

Definitively, three hypotheses have been tested by Seddon (2005): the morphological adaptation hypothesis, the acoustic adaptation hypothesis, and the species recognition hypothesis, by studying a group of antbirds (*Thamnophilidae*). These birds are ideal subjects because their songs are not culturally influenced by learning, have a very simple loudsong, and the differences in song performance can be associated with variation of environmental conditions.

According the species recognition hypothesis sympatric species should have a more distinct song repertoire than allopatric species. We define sympatric species as species that are living in the same habitat and overlap at least 25 %. Allopatric species are species that live in habitats that are geographically separate.

In this investigation, Seddon has found that bill morphology was positively correlated with the dimension of birds, according to the morphological adaptation hypothesis, but that also the other two hypotheses were confirmed. This finding suggests that loudsong in this family of birds is driven by a set of environmental and behavioral characteristics.

According to the acoustic adaptation hypothesis antbirds that sing on the ground produce a higher-pitched song than the species singing in the forest midstory, confirming that song properties are fixed also by habitat structure. Near ground, with dense vegetation reduces sounds intensity and reverberation is reduced in a song of 2–5 kHz.

In conclusion, Seddon (2005) affirms that "ecological adaptation may indeed drive signal evolution."

Summary
The structure and complexity of biophonies are the result of evolutionary pressure. Four major hypotheses are considered relevant for this process. Physical structure of the organisms, adaptive mechanisms in sound production and transmission, reduction of acoustic competition, and behavioral processes are possible integrative strategies to enhance sonic context.

The morphological adaptation hypothesis (MAH) refers to the role of body size as a biological constraint of the vocalization organs and their acoustic performances, confirming an inverse relationship between acoustic frequencies and body size.

The acoustic adaptation hypothesis (AAH) states that the environment is an important cause of modification and alteration of acoustic signals. Dominant frequencies and other long-distance calls are the result of interaction between the animals and their environment to maximize the efficiency of the emitted sounds. Frequency and structure of acoustic repertoire are plastic traits that can be modified according to environmental constraints.

The acoustic niche hypothesis (ANH) states that every species has a unique acoustic space in which to structure the sonic species-specific signature to reduce interspecific competition and to optimize intraspecific communication mechanisms.

The species recognition hypothesis (SRH) supposes that species living in sympatry try to reduce the risk of utilizing similar sonic traits that could confound species reproduction and create the risk of hybridization.

This set of hypotheses has epistemic relationships to form a meta-bioacoustic theory.

References

Blumenrath SH, Dabelsteen T (2004) Degradation of great tit (*Parus major*) song before and after foliation: implications for vocal communication in a deciduous forest. Behaviour 141:935–958

Blumstein DT, Turner AC (2005) Can the acoustic adaptation hypothesis predict the structure of Australian birdsong? Acta Ethol 15:35–44

Boeckle M, Preninger D, Hödl W (2009) Communication in noisy environments. I: Acoustic signals of *Staurois latopalmatus* Boulenger 1887. Herpetologica 65(2):154–165

Boncoraglio G, Saino N (2007) Habitat structure and the evolution of bird song: a meta-analysis of the evidence for the acoustic adaptation hypothesis. Funct Ecol 21:134–142

Both C, Grant T (2012) Biological invasions and the acoustic niche: the effect of bullfrog calls on the acoustic signals of white-banded tree frog. Biol Lett 8:714–716

Brandley N, Burns M (2007) Effects of habitat and location on chipping sparrow song characteristics. http://hdl.handle.net/2027.42/57457

Brown TJ, Handford P (1996) Acoustical signal amplitude patterns: a computer simulation investigation of the acoustic adaptation hypothesis. The Condor 98:608–623

Brown TJ, Handford P (2000) Sound design for vocalizations: quality in the woods, consistency in the fields. The Condor 102:81–92

Brumm H (2006) Signaling through acoustic windows: nightingales avoid interspecific competition by short-term adjustment of song timing. J Comp Physiol A 192:1279–1285

Burns M (2007) Song divergence of chipping sparrow in mixed forest and open habitats: testing the Acoustic Adaptation Hypothesis. http://deepblue.lib.umich.edu/bitstream/2027.42/57458/1/Burns_Melissa_2007.pdf

Daniel JC, Blumstein DT (1998) A test of the acoustic adaptation hypothesis in four species of marmots. Anim Behav 56:1517–1528

Derryberry EP (2009) Ecology shapes birdsong evolution: variation in morphology and habitat explains variation in white-crowned sparrow song. Am Nat 174:24–33

Dubois A, Martens J (1984) A case of possible vocal convergence between frogs and a bird in Himalayan torrents. J Ornithol 125:455–463

Ey E, Fischer J (2009) The "acoustic adaptation hypothesis:" a review of the evidence from birds, anurans and mammals. Bioacoustics 19:21–48

Feng AS, Narins PM, Xu C-H, Lin W-Y, Yu Z-L, Qiu Q, Xu Z-M (2006) Ultrasonic communication in frogs. Nature 440:333–336

Fletcher NH (2004) A simple frequency-scaling rule for animal communication. J Acoust Soc Am 115:2334–2338

Fletcher NH (2007) Animal bioacoustics. In: Rossing TD (ed) Springer handbook of acoustics. Springer, New York, pp 785–802

Henry CS, Wells MM (2010) Acoustic niche partitioning in two cryptic sibling species of *Chrysoperla* green lacewing that must duet before mating. Anim Behav 80:991–1003

Hutchinson GE (1959) Homage to Santa Rosalia, or why are there so many kinds of animals? Am Nat 93:145–159

Kingery H (1996) American dipper: *Cinclus mexicanus*. Academy of Natural Sciences, Philadelphia

Krause B (1993) The niche hypothesis. Soundsc Newsl 6:6–10

Krause B (2012) The great animal orchestra. Little, Brown, New York

Malavasi R, Farina A (2013) Neighbours' talk: interspecific choruses among songbirds. Bioacoustics 22(1):33–48

Marten K, Marler P (1977) Sound transmission and its significance for animal vocalization. Behav Ecol Sociobiol 2:271–290

Morton E (1975) Ecological sources of selection on avian sounds. Am Nat 109:17–34

Naguib M (2003) Reverberation of rapid and slow trills: implications for signal adaptations to long-range communication. J Acoust Soc Am 113:1749–1756

Naguib M, Mennill DJ (2010) The signal value of birdsong: empirical evidence suggests song overlapping is a signal. Anim Behav 80:e11–e15

Nemeth E, Dabelsteen T, Pedersen SB, Winkler H (2006) Rainforests as concert halls for birds: are reverberations improving sound transmission of long song elements? J Acoust Soc Am 119 (1):620–626

Patten MA, Rotemberry JT, Zuk M (2004) Habitat selection, acoustic adaptation, and the evolution of reproductive isolation. Evolution 58(10):2144–2155

Planquè R, Slabbekoorn H (2008) Spectral overlap in songs and temporal avoidance in a Peruvian bird assemblage. Ethology 114:262–271

Podos J (1997) A performance constraint on the evolution of trilled vocalizations in songbird family (Passerifomes: Emberizidae). Evolution 51:537–551

Schwartz JJ (1993) Male calling behavior, female discrimination and acoustic interference in the neotropical treefrog *Hyla microcephala* under realistic acoustic conditions. Behav Ecol Sociobiol 32:401–414

Seddon N (2005) Ecological adaptation and species recognition drives vocal evolution in neotropical suboscine birds. Evolution 59(1):200–215

Sinsch U, Lumkemann K, Rosar K (2012) Acoustic niche partitioning in an anuran community inhabiting an Afromontane wetland (Butare, Rwanda). Afr Zool 47(1):60–73

Slabbekoorn H, den Boer-Visser A (2006) Cities change the songs of birds. Curr Biol 16:2326–2331

Slabbekoorn H, Ellers J, Smith TB (2002) Birdsong and sound transmission: the benefits of reverberations. Condor 104:564–573

Slagsvold T, Wiebe KL (2007) Learning the ecological niche. Proc R Soc B 274:19–23

Sueur J (2002) Cicada acoustic communication: potential sound partitioning in a multispecies community from Mexico (Hemiptera: Cicadomorpha: Cicadidae). Biol J Linn Soc 75:379–394

Vasconcelos TS, Rossa-Feres DC (2008) Habitat heterogeneity and use of physical and acoustic space in anuran communities in Southeastern Brazil. Phyllomedusa 7(2):127–142

Wallschlager D (1980) Correlation of song frequency and body weight in passerine birds. Experientia (Basel) 36:412

Zelick R, Narins PM (1985) Characterization of the advertisement call oscillator in the frog *Elutherodactylus coqui*. J Comp Physiol A 156:223–229

Ziegler L, Arim M, Narins PM (2011) Linking amphibian call structure to the environment: the interplay between phenotypic flexibility and individual attributes. Behav Ecol 22(3):520–526

Chapter 4
Communication Theories

A communication theory represents a general framework used to describe the plethora of observations, rules, and exceptions described in the variegated context of animal communication.

A central principle in the communication theory states that biological signals are honest, which means for an acoustic signal such as a song that the complexity of the repertoire and all the secondary characters associated are expressions of natural selection.

In communication there is a sender and a receiver connected by an informative signal. A third subject is the eavesdropper or the involuntary receiver that utilizes the information which circulates between the two. Eavesdropping represents an important evolutionary phenomenon. Communication is costly in term of energy budget allocated by vocal individuals, and a well-designed communicative framework should be the result of selective pressure.

Communication is an important component of animal life, contributing to reproduction and to survival life traits. Acoustic signals are used in several groups of species, humans included, to find resources, to avoid predators, and to find or adapt locations for reproduction.

To improve the efficacy of a signal it is important to reduce the loss of energy on one hand and on the other to optimize the information contained in the signal. The efficiency of a signal largely depends on the internal design and on the capacity of an emitter to separate the signal from background noise.

Analysis of the complexity of the acoustic communication codes and specifically the acoustic codes represents an interesting perspective. Every acoustic message is the result of a sequence of elementary parts whose syntax represents a message. Acoustic codes operate at intraspecific level but also at interspecific level, improving the efficiency of the communicative behavior.

A. Farina, *Soundscape Ecology: Principles, Patterns, Methods and Applications*,
DOI 10.1007/978-94-007-7374-5_4, © Springer Science+Business Media Dordrecht 2014

4.1 Introduction

Communication is an important component of animal life, and in particular communication mediates reproduction and survival, as stressed in the introductory remarks of the book *Animal Communication Networks* by McGregor (2005). Communication is a process between a signaler and a receiver in which information in the form of structured energy transients between the two sides. Communication represents a semiotic strategy used by a sender to contact a receiver via a semetic relationship (sensu Hoffmeyer 2008). However, this dyadic vision is overpassed by the reality of groups of individuals that simultaneously exchange information. In this way, a communication network is created and maintained in a specific space and for some time. The network model of communication seems to have more reasons to exist because the reality often is not confined to a couple of contenders but includes also a multitude of individuals sparsely located in the foreground where they are waiting to enter into the action. Such individuals can use an eavesdropping-like strategy to grasp some of the transient information between a transmitter and a receiver.

To communicate means to connect a source with a receiver, and this depends on the energy with which a signal is emitted and by the environmental conditions through which such signals pass (Forrest 1994). Attenuation and degradation are common effects that modify the signal that passes through a physical medium (air, water, soil, biological tissues) and a biological codification and decodification. It has been demonstrated that the environmental modification of an acoustic signal largely depends on the frequency at which the signal is emitted but also on the fluctuation of amplitude and reverberations. For instance, spherical spread, absorption, refraction, reflection, and diffraction are phenomena associated with sound transmission through the atmosphere that are relevant in the communication process (Wiley and Richards 1978).

We distinguish between broadcast information and received information. The information at the source is greater than the information at the receiver. The broadcast information depends on the signal that is produced; the received information depends on the capacity of the receiver to decode the signal and to interpret such signal for its purposes.

Noise is a property of the communication channel, and such noise contributes to decoding the signal by the receiver. Errors can occur in the transmission of the signal but also at the reception. At the receiver, errors may be of two types: an error in discriminating between a perturbation and a real signal and an error in not distinguishing variation in signals. The two errors are connected. For instance, random degradation of a signal during the transmission increases the second error. The capacity to discriminate relevant variations in signals decreases with the distance between source and receiver.

4.2 The Communication Theory

We can admit that a general theory should exist to provide a framework in which to place the different epistemic objects that are adopted in the acoustic realm and in communication.

Communication has been an extraordinary event in the history of evolution (Margoliash and Hale 2008). Communication consists in the transfer of signals from an emitter that convey information to a receiver.

This theory can be explained adopting a biosemiotics approach, although more recently code biology, which represents an innovative perspective to investigate and interpret the communicative process, offers a new theoretical support.

The communicative process probably has been submitted to an evolutionary process by which the signals have been improved in terms of efficiency of communication and reliability of manipulation by the receiver. On the other end, the receiver mechanisms have evolved with increasing reliability and efficiency of information reception. The reception of the signals must be sufficiently clear and distinct to produce in the receiver the requested behavioral or physiological decision.

During the transfer, part of the signal is modified by the effect of environmental constraints (absorption, distortion, interference) and minimization of reception by emitter exploiters (Xe) (predators, signal-homing parasites). To receive the signal in the most efficient way, the receiver must filter the noise present with in the information channel, must reduce the effects of subsidiary emitter (Es) and receiver exploiters (Xr) such as Batesian mimics or males with inferior genes (Endler 1993).

Signal generation represents the first step in the signaling process. The evolution of a signal depends on physical, chemical, and biophysical constraints. To produce biological signals is always energetically costly and time consuming, and, for instance, there is empirical evidence of such energetic cost in frogs and birds. To reduce such cost, a specific singing design can be produced with the effort of muscles involved in the singing process, but timing, intensity, and form of signals can come under the selective pressure of other environmental factors. Generally a trade-off between complexity, reliability, and efficiency is achieved.

Several factors affect signals during communication:

Signal generation and emission (biophysical limits energetic limits, biochemical limits, energy storage, timing and location, environmental effects on emission, information content versus clarity, information density of rate, genetic variation)

Transmission through the medium (background noise, interfering signals, attenuation, blocking, absorption, reflection and refraction, distance, spectral properties, temperature, self-interference, information density, timing, and location)

Signal reception and processing (biophysical and biochemical limits to reception, sensory adaptive state and attentiveness, need to be attentive, need for alerting signals, short reception time, other signals, noise, jamming, signal reception rate, information processing rate, pattern recognition needs)

Decisions based on the perceived signal (other signals, choice time wasted, reasons for choice, need for choice, predator risk, parasite risk, physiological state, quality of signal, quality of different signal components, signal channel use)

As we have already stated, communication requires energy, which represents fatigue, a physiological stress. In birds this represents a serious constraint that in part can be avoided with a varied repertoire that engages different muscles at a time.

Signaling can be restricted to a specific time of season by an environmental constraint such as water availability in deserts for frogs. Signaling activity in this case will be confined to the period of water availability.

The signal transmission through the medium represents a further evolutive step in the communication theory. In fact, signals should reduce the effect of background noise and interfering signals from other species. Under a selective constraint signals should avoid attenuation, blocking, absorption, reflection, and refraction. We have already discussed this aspect in Chap. 3 in the section devoted to the acoustic adaptive hypothesis.

The intensity of the signal faces attenuation with distance, and a signal that reverberates can arrive with a rate of degradation that reduces its efficiency if poorly designed. These effects are more evident if the signal has a high content of information or when such information is transmitted at a high rate. Thus, the complexity of the signals that we can observe is the result of a trade-off between the maximum possible of information that can be transmitted and all the environmental and biological constraints that limit this efficiency. Signaling behavior is under such evolutionary constraint with preferences for time and space to minimize the distortion; this is the case of animal choruses that probably are concentrated in the most favorable period of the day when the air is calm and noises are at a minimum.

It is reasonable to admit that explosive breeders and microhabitat specialists use more predictable signal design than species that are continuous breeders and are microhabitat generalists.

In signal generation, many constraints affect complexity and quality of the signal, and the same occurs in signal reception: there are biophysical, biochemical, and energetic limits in receptor design.

For instance, if the background noise is predictable enough this noise can be filtered out or the place or time could be selected to reduce the effect of the noise.

The use of multiple systems of signals, for instance, visual combined with acoustic, could solve problems of noise interferences or discrimination. The reduction of the effects of noise in a signal can be solved with the use of a filter or with a signal design easy to distinguish.

In evolutionary terms, wrong receivers have to be reduced, as subjects with the capacity of eavesdropping can use the information of the emitter or the receiver for personal benefit. Such as a case is documented in cicadas that are parasitized by sarcophagid flies. These flies are attracted by the cicada male courtship song. In this way flies exploit cicadas, and fly larva destroy the sound generation organ of the cicada, which reduces the risk that another fly can parasitize the same individual, and fly behavior definitely prevents further exploitation of the cicada's signal.

The success of a signal largely depends on the physiological status of the receiver engaged in more strategic functions. In this case attentiveness to the signal could be lower and the signal not considered. The receptiveness of the receiver may be influenced by other conditions or concurrent functions. Thus, in time of scarcity of foraging resources the priority is to find food and not to reply to a mating or social signal. Species under predator or parasitoid pressure can evolve a compromise. Aggregation can reinforce signals, as in frogs, but this could attract more parasitoids than would isolated males.

A signal quality has two components: clarity and value (efficacy and strategy can be used as synonyms). A signal component can be low in quality because is difficult to receive and to process, or because it gives false or misleading information about the emitter (strategy).

The quality of a signal assures that the receiver can decode the signals appropriately, but a low-quality signal creates difficulties in discrimination. When a type of signal such as the acoustic one is difficult to achieve, another type of signal, for instance, a visual cue, can be utilized at integration to maintain an open communicative channel between subjects.

Signal, sensory, and cognitive systems are the major components of the communicative system, and a change in one of these may have fitness consequences for the other two. In particular, this aspect is very important in sexual selection.

For instance, males that emit more efficient signals to attract females will have an evident chance to perpetuate their lineages. This attitude is known as "sensory exploitation," that is, the capacity of a male to manipulate the sensory reception of the females.

In conclusion, the evolution of the sensory system will favor the signals that are easy to receive, to distinguish, and easiest for basing decisions, and the sensory system, signals, and signaling behavior thus evolve contemporarily.

Species living in the same environmental condition can evolve similar signals as expression of convergent evolution of some plastic traits. We can call this process a "sensory drive," which means that different species under the same environmental conditions evolve similar signals that can be predicted starting from predictable conditions.

Territorial birds use sound with a multitude of functions, but territorial defense seems one of the most common strategies in this group. In Fig. 4.1 are represented different hypothetical combinations of spatial dislocation of two concurrent males as a theoretical example of the complexity of communication mechanisms (Naguib 2005).

4.3 Eavesdropping and Interspecific Communication

Eavesdropping is a way in which an individual voluntarily or involuntarily intercepts a communication broadcast for another individual. Peake (2005) reports "In the context of animal communication, the term has been used in a number of different ways that can be summed up by a general definition: the use of information in signals by individuals other than the primary target."

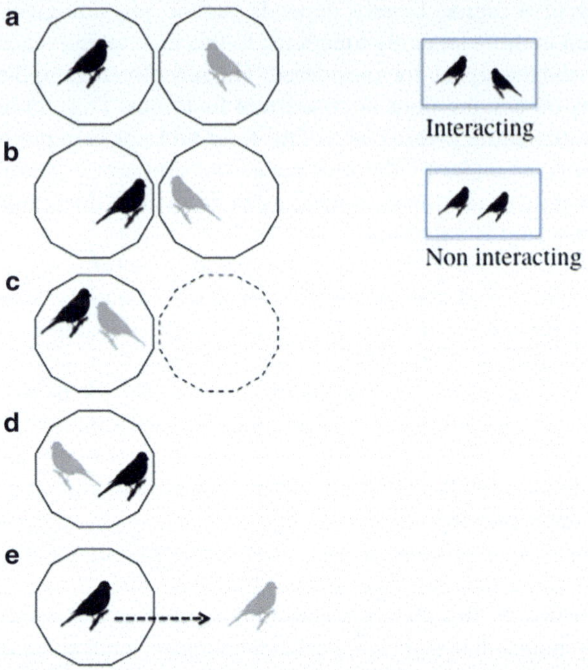

Fig. 4.1 Schematic representation of acoustic interaction between male songbirds. (**a**) Two long-distance interacting males (*left black*, *grey right*) that have established territories. (**b**) Short-distance interactions at the borders between two males with established territories. (**c**) Short-distance interactions between a resident male and an intruder coming from a nearby territory. (**d**) Short-distance noninteracting males (a male of an established territory and a nonterritorial stranger). (**e**) Long-distance noninteracting males (a territorial male and a far stranger that has not yet established a territory) (Modified with permission from Naguib 2005)

Environmental heterogeneity is a relevant component for animals, and to acquire information seems the best strategy to reduce the associated uncertainty and to make the right choices (Emmering and Schmidt 2011). In particular, birds use a multitude of information sources for the different vital functions in their selected habitat. The biosemiotic mechanism of the eco-field (see Chap. 1) is the basis for tracking resources. For instance, in a closely forested habitat the safety eco-field (where the focal resource to be located is represented by the safety from predators) in a dynamic system where animals are in continuous movement is principally represented by the spatial distribution of acoustic cues that can indicate the presence or absence of a predator. Thus, a species is able to recognize the signals that are emitted by a predator during its activity or the alarm calls of potential prey that first reveal the threat. Experiments conducted by Emmering and Schmidt (2011) on the ovenbird (*Seiurus aurocapilla*) and the veery (*Catharus fuscescens*), two ground-nesting passerines that are predated by the eastern chipmunk (*Tamias striatus*), have demonstrated how species have the capacity of eavesdropping on the signals of other species, to interpret the signals attributing a specific meaning. In playback experiments three types of acoustic stimuli were used (chipmunk vocalization, frog control, no playback). Ovenbirds and veery were found to reduce their activity up to 30 m from the playback station and placed their nesting territory far from the playback station. The reaction to the simulated predator presence was twofold stronger in the ovenbird than in the veery. Interspecific eavesdropping represents an important step in the integration of species inside a community and represents a necessary step for the creation of ecological communities.

The effect of predation risk has been verified in the Siberian jay (*Perisoreus infaustus*) by Eggers et al. (2006), simulating by playback the acoustic result of a corvid presence close to a breeding territory of this species. The reaction of the Siberian jay was to increase the protective covering of the nests and to reduce the clutch size.

A further example of eavesdropping was presented by Übernickel et al. (2012) in three neotropical bats during search phase calls and feeding buzzes using a playback methodology. The greater fishing bat (*Noctilio leporinus*) and the lesser bulldog bat (*Noctilio albiventris*) reacted very well to playback of search phase calls and feeding buzzes of conspecific and congeneric bats. The greater fishing bat was attracted by the acoustic cues of *Noctilio albiventris*, but no reaction was apparently found in the sac-winged *Saccopteryx bilineata*. Eavesdropping behavior has been proved in the two species of *Noctilio*, and this largely depends on species-specific traits (e.g., foraging style, social behavior) more than on call design per se. The advantage to eavesdropping largely depends on the information on abundance of prey that can be shared between species.

Eavesdropping is an important activity in prey–predator and parasitoid relationships. Eavesdropping has been observed in different species of bats in the Southern Hemisphere and has also been observed in the Palearctic after the investigation of Jones et al. (2011).

In particular the signaling mechanism between tettigoniid bush crickets and bats is well known where these insects have also an ultrasonic repertoire that can be tracked by bats in a range of 20–80 kHz. In Europe, the lesser mouse-eared bat (*Myotis blythii oxygnatus*) search their prey in dense grass habitats. This species is morphologically similar to *Myotis myotis* but is ecologically distinct. *Myotis myotis* forages in forest and requires open space, its preferred prey being carabid beetles, whereas *M. blythii oxygnatus* forages on tettigoniids.

The two species can be identified by the analysis of their fecal samples.

There is evidence that the simulated echolocation calls of *M. myotis* produce in *Tettigonia viridissima* an escape response and that repetitive ultrasonic 30-kHz pulses can cause cessation of song in this insect.

The strategies to localize the two main types of distinct prey were investigated for these two species:

Tettigonia viridissima for *M. blythii oxygnatus* and the rustling sound of carabid beetles for *M. myotis*.

The two species of bats were utilized in experimental broadcasting of the sounds of *Tettigonia* and of the carabids. During the playback experiment using *Tettigonia* sounds, both bats were attracted by the call of *Tettigonia*, which was a real surprise because *M. myotis* was also attracted by the sound of *Tettigonia*. However, spectral species differences appeared during the experiment: *M. blythii oygnathus* had a stronger predatory response to the calling song of the two species of Tettigonia, and *M. myotis* showed a higher predatory response to the beetle rustling and to one tettigoniid species. This experiment has demonstrated eavesdropping on prey communication signals and also sensory partitioning between the two sympatric bat species.

4.4 Echolocation: A Special Case of Communication

Echolocation is a winner strategy in bats. The efficiency of this sonic mechanism depends by the distance at which a bat can intercept a prey by isolating background and vegetation effects. For this it is important to investigate the distance at which bats can detect a prey and separate the echoes from rocks, water surfaces, trees, and meadows. A main topic that remains to be explained is if bat echolocation distance of prey depends on the distance of the prey from the background. Two different theories have tried to explain this matter: first, bats produce impulses that are long enough that there is time for returning echoes before the emission of a successive pulse; and in the second hypothesis, bats alternate between different call types and thus ambiguous identification is strongly reduced. Frequency of calls and atmospheric condition are the limiting factors over which audible echoes are reflected from a target to an echolocating bat. The frequency is a very important component of echolocation: low-frequency signals are used for long-range echolocation, and high frequencies are used for short-range echolocation. Bats that use low frequencies and a high source level (SL) forage in forested areas, but bats that use a high-frequency range and lower SL search their prey in open areas. Bats living in tropical forests where humidity is very high that are foraging inside dense vegetation have a high-frequency signal and very low SL (Stilz and Schnitzler 2012).

4.5 Heterospecific and Conspecific Attraction

It has been observed that some species are attracted by a portion of habitat already occupied by other species. This phenomenon was been first coined by Mönkkönen et al. (1990) as "heterospecific attraction." For instance, migratory birds have been observed to be attracted by resident species that probably have more information about available resources. Breeding habitat selection is a fundamental step in the life of a species, and if a species has a migratory habit good knowledge of the more useful habitat is quite reduced. For this reason, resident species can operate as involuntary informers for such species.

Population density, habitat structure, floristic, acoustic cues, predatory pressure, and conspecific and heterospecific contacts are some of the environmental variables that concur in breeding habitat selection. Some species have been proved to prefer to settle close to other species in a way that is reasonable to think there is some type of benefit such as antipredator behavior of consociated species, use of information about the quality of the shared habitat, abundance of social stimuli that accelerate and improve the breeding cycle, and increase in efficiency to a communal defense from predators. In particular, many species of birds have the capacity to utilize the acoustic information of other birds.

Heterospecific attraction may be possible if the resources are honestly signaled. Resident species have more time and energy to explore the local resource and to cope better than migratory birds with local environmental peculiarities.

Residents have already avoided the risky portion of a suitable habitat or have been eliminated by a predator, so the presence of residents in a specific area has high probability to be a signal of habitat safety.

If competition is present between residents and nonresidents, the visitors will try to avoid the habitat patches already occupied by residents. In contrast, if there is no competition between species, this will favor the selection by visitors of the patch already occupied and tested from the aspect of the resources by the residents.

Observational and experimental evidence is in accord to recognize that the heterospecific attraction is not an artifact but a diffuse phenomenon, especially in areas in which resident and migratory species have common habitat. For instance, there is experimental evidence that titmice density attracts other passerines after a change in local density produced artificially by providing food to titmice in winter and nest boxes during the breeding season (Mönkkönen and Forsman 2002).

Further experimental evidence by Doutrelant et al. (2000) has proved that interspecific competition may have an important role in the macrogeographic variation of birdsong. In the blue tit (*Cyanistes caeruleus*) there are variations in song repertoire according to geography, so there are populations that have a trill and other populations without the trill. This characteristic is correlated with the breeding density of the great tit (*Parus major*) and not with other possible environmental factors such as vegetation structure or blue tit density. These authors tested the hypothesis that the trill of the blue tit was representing a character shift to reduce competitive interactions with the dominant great tit. Experiments conducted in Corsica, mainland France, and Denmark consisted of presenting a blue tit trill to the great tit and also untrilled song and great tit song. The great tit responded somewhat well to male great tit song and to blue tit untrilled song; its reply was weaker to trilled blue tit song. This observation confirms how the modification of a character can reduce interspecific competition via an acoustic communication change.

Conspecific attraction plays an important role in habitat selection and on reproductive performance (habitat copying). The hypothesis that birds aggregate according to social cues more than environmental cues is consistent with the results obtained by Redmond et al. (2009). The eastern kingbird (*Tyrannus tyrannus*) has been observed by these authors to occupy only a part of the suitable habitat during the breeding season at Malheur National Wildlife Refuge in southeastern Oregon. Kingbirds have a tendency to settle where there is a high conspecific density (Fig. 4.2). It is true also that individuals settle in the areas where in the past year more fledglings have been produced (habitat copying).

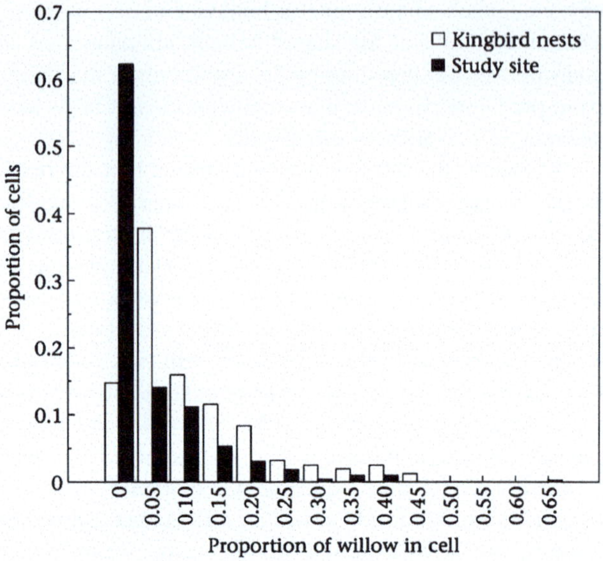

Fig. 4.2 Proportion of willow within 1-ha cells centered on kingbird nests (*open bars*) and 1-ha cells used to characterize the overall availability of willow habitat within the study area (*closed bars*) (Reproduced by permission from Redmond et al. 2009)

4.6 Male–Female Song Recognition

The acoustic recognition of a partner is fundamental for successful reproduction. In birds this capacity has been observed in zebra finches (*Taeniopygia guttata*), dunnocks (*Prunella moularis*), pied flycatchers (*Ficedula hypoleuca*), and eastern whipbirds (*Psophodes olivaceus*). Recently another species, the great tit (*Parus major*), has been added to this list (Lind et al. 1996). Using a playback protocol, females of the great tit were "called out" during the clutch period (Fig. 4.3). The male invites the female to leave the nest and often offers food to the female. This behavior is common during the incubating time. For the experiment, songs of the partner and songs of other males were utilized. Females reacted by abandoning their incubation only after the song of the partner. The recognition of the partner may be important for a female because the partner often provides food that the female has difficulty in finding for herself during the incubation period. Male calling could represent also an important signal of the absence of predators in the surroundings.

4.7 Sound Degradation and Ranging Hypothesis

Acoustic degradation is one of the most conspicuous processes that occur between a transmitter and a receiver. Acoustic degradation, which occurs more in the close forested environment than in open areas, has a strong influence on communication efficiency between vocal animals (Wiley 1998). Sound attenuation by distance and vegetation scattering degrades sounds and reduces the efficiency of acoustic communication.

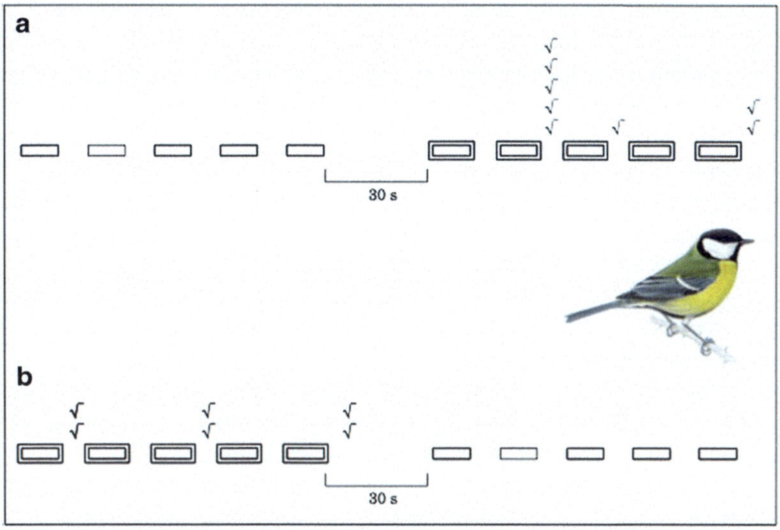

Fig. 4.3 Experimental design to test the female latency of great tit (*Parus major*) from the playback of mate's song (*doubled-lined box*) and of stranger mate (*single rectangle*) and female at a distance away from the nesting box (√). (**a**) Playback of the stranger male precedes playback of mate. (**b**) Playback of the mate precedes playback of stranger. The latency of response of female great tits to playback of male song during incubation is shown (Reproduced with permission from Lind et al. 1996)

Territorial birds and vocal mammals can use auditory cues to evaluate the distance at which another conspecific is calling or singing. This attitude is defined as "ranging." Ranging capacity is reduced by many physical and behavioral elements.

When a bird listens to another individual of the same species, singing or produces an immediate reaction of alert if the signal is close, or if the signal is distant it can be ignored. The signal changes characteristics for effects of amplitude variation and environmental degradation according to distance.

One disputed issue regards the capacity to react to the signal according to a learning or instinctive process.

This concept opens a wide vista on this aspect of communications. Some species of birds seem to have an innate capacity to distinguish the distance of conspecific signals and to behave accordingly. For instance, the young of the Carolina wren (*Thryothorus ludovicianus*) prefer to cope with undegraded exemplars (Morton et al. 1986). The learning mechanism is associated with memory; an experience on a specific stimulus is preserved as a memory.

A probable recent experience is considered as familiarity. If a subject is tested with a familiar stimulus without apparent responses, there are good reasons to consider such stimulus is not included in the learning mechanism.

To formulate correctly the "ranging hypothesis" it is important to investigate if the ranging process is learned or innate. In fact, this author has proposed that a bird before ranging must learn the structure of each song pattern. In other word, a bird first must be able to produce the same pattern.

Local populations try to preserve song types as true dialects. This effect degrades with distance or because of an important barrier. Investigations carried out by Morton (1987) in the Carolina wren (*Thryothorus ludovicianus*) have demonstrated that individuals at the same site share 66 % of their songs (range, 49.5–89.5 %) with a negative linear correlation between shared songs and distance, but males resident on islands share only 20 % of their song repertoires. The author reports that 3 km water was equivalent to 145 km in the mainland.

The ranging hypothesis follows the discovery that some cells of the forebrain in passerines respond at the highest rate to a subject's own song (Margoliash 1983; Margoliash and Konishi 1985).

Birds could act either in temporal or in frequency domain to evaluate with precision the level of degradation of a song and consequently range accurately. However, familiarity does not seem improve the ranging process.

Difficulty in interpreting correctly the ranging mechanisms is represented by the multiplicity of explanations on lower responsiveness to degraded songs.

For instance, sensitivity to variation in amplitude is not surprising; a subject reacts better to a high-amplitude signal than to one of low amplitude. Other types of degradation such as frequency dependence attenuation or reverberation can lower signal detectability.

Another important aspect is represented by the expectation from the receiver for a signal. Familiar signals are detected more than unfamiliar ones. As a consequence of this fact a signal degraded and unfamiliar could be detected less often, escaping direct ranging.

A problem that arises during playback experiments is that a repeated song from a loudspeaker can be judged by simple triangulation by an attracted subject; this can be easily solved by broadcasting the signal only for a very short period of time. Despite some evidence, the ranging hypothesis remains far from common acceptance, especially in term of accuracy by which individuals are ranging. For instance, it is not clear if ranging accuracy is the same for the three forms of degradation: amplitude, frequency-dependent attenuation, and reverberation. It is not clear if ranging accuracy changes according to species and type of environment. Finally, the capacity of ranging and competitive or cooperative interactions between territorial neighbors remain open issues.

Birds immersed in a variable sonic environment receive signals from foreground to background. Birds have the capacity to discriminate the quality of the signals and to decide to react when a sound is emitted by a threatening intruder or to ignore the signal from distant conspecifics. It is reasonable to expect a strong reaction from the receiver when the signal is not degraded (Fig. 4.4). The assessment of the auditory distance (ranging) is a mechanism utilized by species that have to defend territory and habitat resources. Perceiving the level of degradation requires the receiver to have the capacity to decodify the signal in real time. Naguib (1996) has tested the ranging behavior in the Carolina wren (*Thryothorus ludovicianus*) by broadcasting artificially degraded sounds (Fig. 4.5). In this species the capacity has been proved

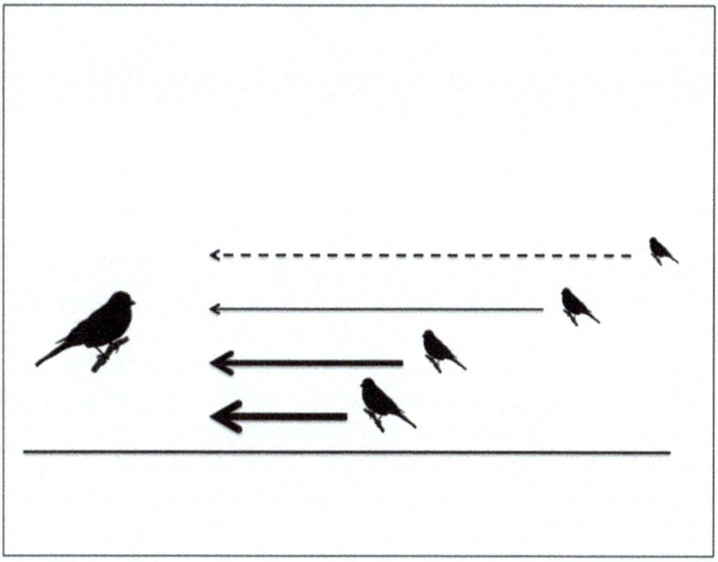

Fig. 4.4 Representation of the effect of amplitude variation and environmental degradation (represented by size of *arrows*) on receiver with increase of distance of the sender

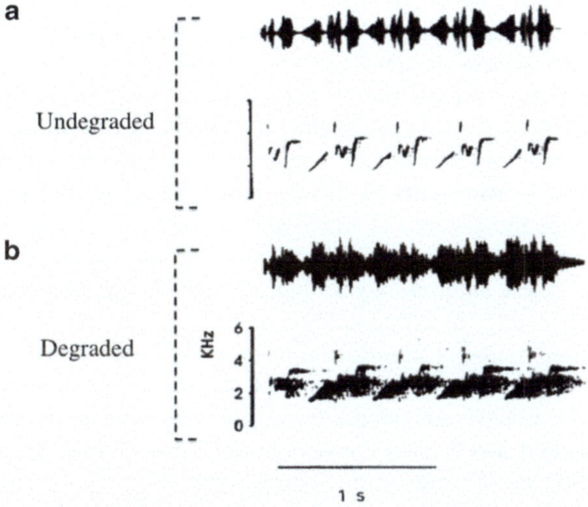

Fig. 4.5 Undegraded (**a**) and artificially degraded (**b**) song of the Carolina wren (*Thryothorus ludovicianus*). The degradation is similar to a song perceived at a distance of 50 m (Reproduced with permission from Naguib 1996)

to distinguish the overall amplitude, reverberation, and the relative intensity attenuation of high frequencies (Naguib 1995). In this way this species has at least two mechanisms available to assess ranging in a variable environment and to defend territories efficiently in habitats with different sonic properties.

The same effect has been found in the winter wren (*Troglodytes troglodytes*) when experimentally exposed to playback of song degraded by long-range propagation and undegraded song. The wren is sensitive to degradation, with a less intense territorial response to the stimulus, but with a change in song post that enables the bird to better improve assessment of long-range propagation and the receiver's ability to hear the opponent's song (Mathevon and Aubin 1997).

Several playback experiments have been carried out in recent years, and in particular the ranging behavior of the chaffinch (*Fringilla coelebs*) has been studied by Naguib et al. (2000). Playback songs degraded for emulating a distance of 0, 20, 40, 80, and 120 m were reproduced and the distance of territorial males from the loudspeaker was noted. The results confirmed previous studies that birds use acoustic cues to evaluate the distance of singing conspecifics. The experiments demonstrated that chaffinch males are able to evaluate the distance of the singing conspecific and to regulate their behavior accordingly.

Location of conspecifics, especially when individuals are hidden in dense vegetation or when the light is not sufficient to use visual cues, can be determined using acoustic cues. There is evidence that birds are able to evaluate the distance of a conspecific competitor (male) according to the quality of the sound signal. The acoustic adaptation hypothesis (AAH) emphasizes that birds are adapted to minimize signal degradation by adapting their acoustic cues to the character of the environment. Acoustic adaptation to the habitat offers at least the advantage to signal to competitors from a great distance, and the emission of a sound that copies the structure of habitat results in an acoustic message that could be interpreted as being closer. The AAH has been proved in many conditions and by experiments such as those designed to compare the magnitude of degradation of song propagated in native versus foreign habitats. For example, Fotheringham et al. (1997) have used playback experiments on four species of wood warblers (Parulinae). Males were found to respond to near songs in the same way as a conspecific intruder, but with less intensity to distant songs. These results suggest that males are able to range the distance at which another male is singing, and their reaction is based on the level of degradation with which an individual perceives the conspecific song.

These authors suggest that selection should favor individual males that structure songs to reduce degradation and thus create an advantage because the rivals are discouraged or intimidated by an undegraded signal.

At least two mechanisms of ranging are contemporarily in action: the signaler tries to have a nondegraded signal to be transmitted and to intimidate the rivals. A second mechanism consists in the capacity by a receiver to estimate the precise distance of a conspecific competitor.

The ability to perceive song degradation may be useful either for signaler or for receiver by reducing the uncertainty in behavior that could affect individual fitness.

There are different ideas about the use of cooperative behavior. In fact it is more reasonable to admit that selection favors the acoustic cues that manipulate or misinform the receiver.

Environmental and social conditions shape communication in songbirds and humans as well creating a continued challenge between signalers and receivers. In natural systems the unpredictability of events often creates a permanent evolutionary tension in communicative life traits.

In some conditions such as the tropical forest the dense vegetation is a characteristic that probably has persisted for thousands of years, and this environmental characteristic probably has shaped communication in several acoustic species. In such an environment Mathevon et al. (2008) have studied the effect of vegetation on the singing pattern of the white-browed warbler (*Basileuterus leucoblepharus*) (Aves: Passeriformes: Fringillidae, Parulini). This species, which lives in the undergrowth of Atlantic forests where there are dense vegetational barriers, has an acoustic repertoire composed of two major signals: a high-pitched contact call and a territorial song. The song is used to attract females and to maintain territorial interactions. Despite the simplicity of the acoustic repertoire, this species has been proved to ensure with its vocal signals species-specific recognition, individual identity, and the location of the sender. Species-specific information is performed for a wide audience, but the individual identity is a signal more susceptible to propagation and is reserved to neighbors. The receiver can locate the signaler according to the degradation of the acoustic signal. The signal emitted by this species is degraded in the higher frequencies just at 100 m distant from the broadcaster. The low-frequency modulation allows species to maintain information for species identification and seems tolerant to sound simplification resulting from vegetational constraint. Individual recognition is largely the result of the distribution of gaps in frequencies between notes. A playback experiment conducted by these authors has demonstrated that these subtle changes are sufficient to identify familiar from unfamiliar neighbors. By considering that this part of the song repertoire disappears with distance, this type of recognition is reserved to individuals that are close others and which have created a private sonic environment. In other playback experiments using two loudspeakers, one to attract a male and a second used to emulate another distant male, Mathevon et al. (2008) have observed that the investigated male is able to range the presence of a second male according to the level of an ad hoc degraded song. Decrease in frequency of the succession notes may enable males to range the distance with good precision. Thus, considering the regular structure of the tropical undergrowth forest, the amount of remaining notes is directly correlated with the distance from which a signal is emitted. An azimuth assessment is necessary to localize the direction from which the signal is emitted. Experimental variation of amplitude modulation has demonstrated that birds made many errors when tested with a signal without amplitude modulation. Thus, ranging and direction are based on degradation cues and on amplitude modulation characteristics.

In a coniferous forest the small leaves are more efficient in scattering higher frequency sounds than those of broadleaf species. In the pied flycatcher (*Ficedula hypoleuca*), the female spends a shorter time inside the nesting cavity when compared with other nesting cavity birds. This fact was explained by experimentation by Lampe et al. (2007) showing that from the inside the nest box the degradation of the male song is high. In this species the male is polyterritorial, although monogamous males increase the breeding efficiency of the female. Song degradation with distance has been proved to be greater in broadleaved and coniferous forests.

In the whitethroat (*Sylvia communis*), Balsby et al. (2003) have found that the five types of vocalization were constrained differently according to habitat considered and behavior. Sound degradation was found to degrade with distance along a hedgerow but to be negligible in open meadows. Song flights seem not different in terms of efficiency when compared with the other types of vocalizations. These authors have argued that a song flight may have a different purpose, such as being a visual location by potential receivers or surveillance by the territory owner. Thus, a different level of degradation for each sonic type is expected.

Birds that defend territories generally are more reactive toward songs that are played in the middle of their territory than those at the periphery or outside the territory. This effect has been observed in different species such as the great tit (*Parus major*), gray catbird (*Dumetella carolinensis*), or white-throated sparrow (*Zonotrichia albicollis*). This fact is interpreted as a capacity of ranging from birds that recognize, using sound degradation, the distance at which another individual is singing. This ability enables an individual to react accordingly when the intruder is close or inside the territorial limit or to ignore such individuals considered to be outside the territory and not a threat.

Amplitude and degradation are the two cues probably implicated in the ranging process, but amplitude is under control of the singer whereas degradation depends on distance and seems a more effective cue for correct ranging. In particular, it seems that ranging between two singers that use the same song type (McGregor et al. 1983) is more accurate and that this occurs also in the presence of undegraded song rather than degraded song. This observation confirms the hypothesis of Richards (1981) as tested on the Carolina wren (*Thryothorus ludovicianus*) that degradation cues are used for ranging and that ranging is more accurate when the song sung is of the same type.

Richards (1981) found that this species uses distance information of sound to react differently: attacking when the singer is close and countersinging when the singer is far. Degraded sound means that a bird is outside the territory and this produces a singing reply. When the sound arrives undegraded this means that the conspecific singer is close and the reaction is a direct attack. This capacity allows wrens to conserve time and energy to be used in territorial defense.

The position of a singer in dense vegetation and the position of a receiver are important because sound degrades according to the amount of vegetation that interferes with it. To bypass this problem some birds select a high perch for singing, such as the blackcap (*Sylvia atricapilla*), a common warbler in Europe dwelling in densely vegetated habitats. Mathevon et al. (2005), using playback techniques, investigated the effect of the selection of high perches on the receiver and conducted an experiment in which the height of speaker and microphones were changed. Four acoustics parameters that are sensible to sound degradation were measured: signal-to-noise ratio, excess attenuation, distortion within the sound produced as a blur ratio, and tail-to-signal ratio. These measures indicated that song degradation was decreasing with the increase of height at which speaker and receiver microphone were placed. Decrease was quite great when the microphone (receiver) was placed high in the vegetation in respect to the speaker. This experiment has demonstrated how the blackcap benefits by staying in a higher perch when it functions as the receiver rather than as the sender.

The conspecific recognition of the acoustic signal (song or call) is fundamental to assure a channel of communication is always active. All the information present in an acoustic signal (song or call) could appear necessary for communication. In reality, the background noise, the masking effect, and interference with other subjects can reduce the efficiency of this acoustic message. Mathevon and Aubin (2001) have tested in the blackcap (*Sylvia atricapilla*) the hypothesis that in some songs the meaning of the message is not lost when the song structure is degraded. In this species, the male responds even to a highly manipulated model of conspecific song. The reply is suspended when frequency modulation is suppressed or the number of syllables is reduced drastically. In terms of code biology, a receiver male when a manipulated song is administrated is able to intercept the fundamental code that persists. In other words, the capacity showed by the blackcap male to decode the conspecific song even when manipulated could represent an adaptation to transmitting acoustic signals in a forest environment where these signals are degraded by vegetation.

Ranging seems present in monkeys such as the pygmy marmoset (*Cebuella pygmea*) that, according to the distance separating sender and receiver, utilizes three different vocalizations (trill, J calls, and long calls), adapting also to the characteristics of the habitat where they are calling (Fig. 4.6). The first two calls have a higher frequency and degrade rapidly, so they are used in short-distance communication; the long-distance call that is a less pulsatile vocalization is utilized to communicate over longer distances (Fig. 4.7) (De La Torre and Snowdon 2002).

4.7.1 Amplitude Fluctuation and Reverberation and Ranging Efficiency

Amplitude fluctuations and reverberations are two important variables affecting the identification and localization of acoustic signals uttered by birds. Amplitude fluctuation is the result of the scattering from air turbulence that varies with weather conditions (wind, temperature) and hour of day. This factor produces difficulties in birds in accurate ranging; the distance at which a species detects a sound and consequently the intelligibility of the signal are falsified.

Reverberation represents a further distortion of the signal in a forest habitat. This factor is easily observed in a spectrogram. Reverberation produces a more confused design in the spectrogram, and we expect identical interference in the process of identification of the signal by the receiver is likely. Amplitude fluctuation and reverberation are operating in the forest landscape, reducing the efficiency of the communication network, but on the other hand this represents an acoustic buffer to

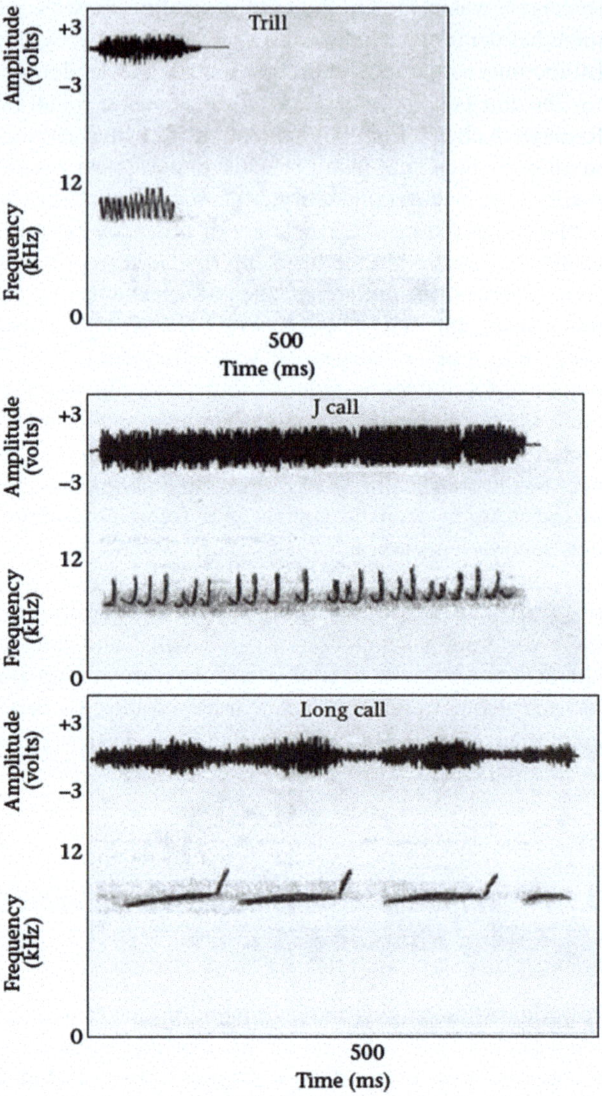

Fig. 4.6 The three types of calls of marmoset (*Cebuella pygmaea*) (trill, J call, long call) (Reproduced with permission from De La Torre and Snowdon 2002)

reduce predatory pressure. In fact, a trade-off between conspecific and intraspecific communication and acoustic localization of individuals at a long distance should exist. Considering that a predator generally operates at a broader scale than its prey, this scaled efficiency in acoustic communication could represent an advantage.

Amplitude fluctuation caused by atmospheric turbulence masks the amplitude modulation of a signal, whereas reverberation caused by scattering from topography and vegetation acts on frequency modulation.

Irregular amplitude fluctuation masks low frequencies of amplitude modulation in signals. Reverberation is more sensitive to frequency range. It is less evident at

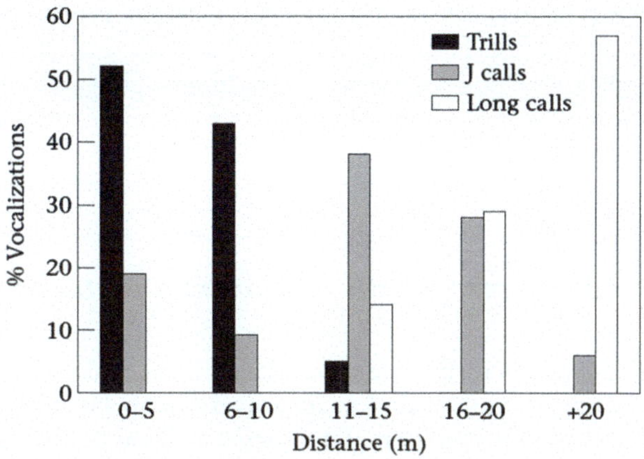

Fig. 4.7 According to the distance from a receiver the marmoset (*Cebuella pygmaea*) changes the type of call utilized. Trill is used at short distance, J call at an intermediate distance, and long-distance call at distance >11 m (Reproduced with permission from De La Torre and Snowdon 2002)

frequencies between 2 and 8 kHz. At low frequency, reverberation assumes the form of echoes, but at higher frequency it produces a decay of acoustic energy, masking high rates of amplitude modulation and repetitive frequency modulation. Most acoustic animals use the intermediate frequencies (2–8 kHz), as experimentally evaluated by Richards and Wiley (1980).

4.7.2 Blurring, Attenuation, and Signal-to-Noise Ratio and Ranging Efficiency

Detectability, locability, and readability of bird song are degraded with the distance propagated through the environment. According to physical theory and principles the degradation can be represented by three aspects: blurring, attenuation, and signal-to-noise ratio.

Blurring of an amplitude and frequency pattern is the result of filtering acting on the frequencies by reverberation and atmospheric turbulence.

Attenuation is produced by absorption, multiple scattering, and geometric attenuation.

The signal-to-noise ratio largely depends by the quality of the signal but also on the level of background noise.

These three concurrent aspects of sound degradation have been confronted with the common blackbird (*Turdus merula*), a species that has such a complex song repertoire that is possible to find in this species acoustic elements common to other species. This species seems to be able to reduce the sound degradation by selecting the best-adapted perch site and direction in which to sing. Experiments using the broadcast song of this species were conducted by Dabelsteen et al. (1993). Blurring, attenuation, and signal-to-noise ratio measures were shown to be correlated and decreased with distance. The components of the blackbird song with higher frequencies were degraded more rapidly than the low-frequency song repertoire.

When the receiver (a tape recorder) was changed in position from lower to highest positions, the resultant sound degradation was also reduced. This observation explains the position chosen on the highest perches by a singing male, which favors the propagation all around of the song that has a definitive meaning of territorial demarcation.

4.7.3 Some Caveats on Playback Methodology

The playback methodology to assess the potential ranging in animals today represents a standard method and is intensively used, but this methodology can introduce biases by simplification of the sonic condition during the experimentation.

For instance, Penna et al. (2012) warn that the use of broadcast signals for a transmission experiment may introduce a simplification of the conditions in which animals communicate in nature, as recently tested on the Iberian midwife toad (*Alytes cisternasii*).

A recent study conducted by Thomas et al. (2003) on the response of robins (*Erithacus rubecula*) during nighttime playback of conspecific songs shows a quite interesting pattern: two populations of robins (Irish and Welsh) responded differently to nocturnal playback experiments. The Irish population rarely reacted to playback, but the Irish birds were reactive to playback stimulus. The unexpected difference between these two populations is not found during the day. The differences in nocturnal playback response must be explained excluding the intensity of illumination or air temperature. An explanation can be found in the different predatory pressures of these two populations in Ireland and Wales. In Wales, the long-eared owl (*Asio otus*) is common and diffuse but in Ireland it is rare, and we know that this predator localizes, as does every nocturnal predator, their prey by ear. This study demonstrates how prudence must be used to plan playback experiments because of unexpected differences in interpopulation behavior.

4.8 Distress Calls

Among several behavioral traits used by animals to signal the presence of predators, functioning as public information, distress calls are used by many birds to warn other species about the menace and simultaneously to confound predators. In fact, distress calls are extremely difficult to locate aurally. The distress calls have a basic structure: a carrier frequency with numerous harmonics, slowly modulated in frequency.

Distress calls have a complex structure and suffer in their propagation when uttered in dense vegetation. To test the hypothesis that distress calls of birds living in vegetated habitat are better adapted than calls of species living in an open field,

Mathevon et al. (1997) compared the distress call of the common blackbird (*Turdus merula*) and the starling (*Sturnus vulgaris*). The distress calls of these two species have the same characteristic structure: a carrier frequency with harmonics slowly modulated in frequency, but in the starling the spectrum is broader than that of the blackbird. Playback experiments conducted in dense vegetation at a distance of 40 m have indicated a significant difference between the two species. As predicted by the hypothesis, the distress call of starling was strongly affected by vegetation with a modification of the frequencies above 4 kHz which are strongly attenuated. The distress calls of blackbirds conserve their structure, confirming the hypothesis of a better adaptation to a long-range transmission in dense vegetation. Despite the less efficient transmission of the starling distress call, the frequency modulation and message decoding are preserved, although at lower reliability (Fig. 4.8).

The difference in propagation across vegetation could be discussed also from an etho-ecological aspect. The two species have a different social structure. Starlings live in flocks and probably use also other antipredatory signals, whereas blackbirds live in a very scattered population and require more acoustic signals.

4.9 Soundscape Orientation and Spatial Source Segregation

Most animals use conspecific signals to find a partner or to avoid a competitor, although heterospecific semetic relationships are also extensively used. For instance, bats use frog choruses to locate rich areas of aquatic insects, and some birds recognize the alarm calls of other birds when a predator is nearby. Sound can be used to orient a species in a landscape (Slabbekoorn and Bouton 2008). In this way, birds, pelagic fish larvae, and newts extensively use sounds to find sites for breeding or for foraging (Fay 2009).

In people who have visual restriction, acoustic cues are used to move around indoors and in the outdoors. For blind people, soundmarks such as the noise of a road, the sound of a bell, etc., are fundamental to self-orienting. Acoustic cues are used by animals to move in the dark, in dense vegetation, or in turbid waters. Frogs use conspecific choruses to find a common pond. Nocturnal migratory birds can find landmarks in the call of other species of birds or frog choruses.

Bee (2007) has proved experimentally that the female grey treefrog (*Hyla chrysoscelis*), which has a phonotaxis for the male chorus, is able to segregate sound sources in space.

In newts, a phonotactic produced by the chorus of frogs has been proved experimentally, although a definitive confirmation is still required.

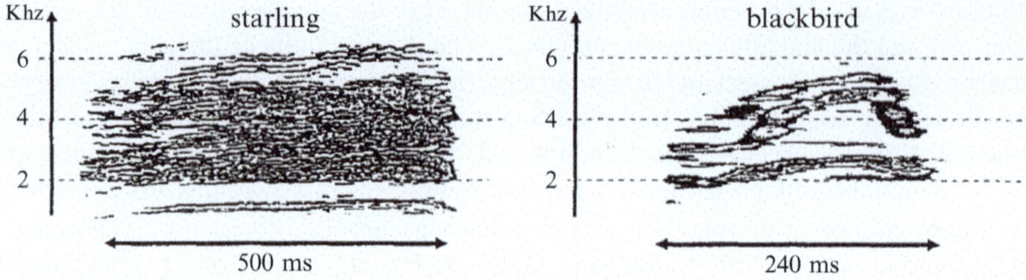

Fig. 4.8 Spectrogram of distress calls of starling and blackbird (Reproduced with permission from Mathevon et al. 1997)

4.10 Ecological Codes

The ecological complexity of natural systems is based on interactions among landscapes, ecosystems, communities, populations, and individuals. Most of these interactions operate according to a thermodynamic gradient (Odum 1983), whereas others are characterized by a negative entropic regime (Schrödinger 1944), dramatically increasing the flow of information (Reza 1961) and distance from a high-probability state of entropic configurations (Ulanowicz 1997).

At first sight, a complex system is perceived as a complicated one, but this is a misleading effect largely caused by a lack of a full understanding of the rules that govern the relationships between its component parts. Such rules are sequences of ecological codes that are used to connect at least two interacting subjects.

Similar to all organic codes (Barbieri 2003), the ecological codes can be defined as mechanisms that establish an arbitrary set of connections between two or more components (organisms and/or their aggregations) of a complex system. The ecological codes are the tools that organisms use in everyday life to relate themselves to the external world. Ecological codes are visual, acoustic, tactile, chemical, and cultural and exist at every scale of living organization.

More functions require more ecological codes, which results in more possibilities for organisms to interact with their perception of the external environment, or Umwelt, sensu von Uexküll (1982 (1940), 1992 (1934)), and finally to benefit for resources (Farina 2012).

The ecological codes classify external objects and process analogue gradients as discrete meaningful digital units. In birds, for example, the distance from a safe place is transformed into discrete distance units, each associated with a specific habit (e.g., stay confident, quickly escape, alerting, etc.). Such transformations are performed by cognitive code makers and produce codes that are incorporated into the genetic or the cultural reservoir of every organism.

To understand an ecological code is necessary to start from a specific function that in turn is activated by a specific physiological need. For every function there is a dedicated code sequence that guides such function to tracking the resource requested to satisfy a specific need. For instance, the need to acquire food is satisfied by a predator by using a search image of the prey, as a result of which the right prey is selected and every other organism is excluded. An insectivore bird, for example, has a codified (cognitive template) image of its prey.

A food specialist has a restricted repertoire of “cognitive templates,” which allows it to conserve energy, whereas a generalist has a more broad collection of templates and wastes much energy in random attempts. The dance of honeybees is an example of codification about food source and abundance.

More complex codes are requested to optimize the search for resources. In addition to the codes for recognizing a prey, for example, there are codes for identifying the suitable environment in which to search for a prey. To every resource is associated a spatial configuration that is a carrier of meaning [see the theory of the eco-field (Farina and Belgrano 2006)] which must be recognized.

Today, the degradation of the ecological codes and the changes in the environmental conditions in which codes have evolved are crucial issues in the challenge to preserve biodiversity and the associated ecological processes.

The acoustic codes are units of informational acoustics carried out by special organs (e.g., syrinx in birds, vocal cords in humans) with specific sequences to produce meaning (alarm calls, song, contact calls, etc.).

The acoustic codes are particularly complex in songbirds and relatively simple in other birds and in frogs and fishes.

An acoustic code is composed of indivisible units that represent the alphabet in the sonic language. Although the majority of the communicative processes is devoted to intraspecific communication, several species have the capacity to understand the sonic communication of other associated species. This ability creates a favorable condition for the organization of interspecific indirect exchange of information, often invoked as proof of the existence of a community level over the population level.

Considering that every signal is species specific and that humans cannot understand the meaning of animals signals and vice versa, indirect methods must be used to explore this informative dimension.

For instance, when a signal is repeated under the same circumstances it is reasonable to find a meaning correlated with that circumstance after a reasonable accumulation of evidence. In this way we cannot determine the meaning of the signal but can perceive its effects on the receiver. In particular, signals that vary dichotomously are particularly adapted to study the semantics of these signals. In particular, the study of the acoustic repertoire of those species whose song repertoire can be divided in distinct categories is useful. Thus, species of the wood warbler genera *Vermivora*, *Dendroica*, *Parula*, *Mniotilta*, and *Setophaga* have a song repertoire that can be divided into two distinct categories.

An important study was conducted by Byers (1996) that has selected categorical variables for investigating the acoustic repertoire of the chestnut-sided warbler (*Dendroica pensylvanica*).

The following categorical variables were selected:

Stage of breeding
Song before male movement

Song following male movement
Direction of male movement after song, relative to border
Direction of male movement before song, relative to border
Direction of male movement after song, relative to female
Direction of male movement before song, relative to female
Song before female movement
Song after female movement
Direction of female movement after song, relative to male
Direction of female movement before song, relative to male
Direction of female movement after song, relative to border
Direction of female movement before song, relative to border
Direction of female movement before song, relative to nest
Song, before interaction
Song, after interaction
Type of interaction
Neighbor's song category.

For each category a specific code was coupled, for instance, at the category "Song before female movement" the possible codes are 1 = last song before female movement, 2 = second to last, 3 = third to last, 0 = not within three songs of female movement.

Accented-ending songs were heard in connection with unmated status, and unaccented-ending songs were associated with singing near territorial borders, but this author stated that "the predictive power of song category was not absolute," but each song category is associated with a set of behavioral possibilities. Further specific messages are included in the acoustic signal but encoded in a finer level of signal variation. This work demonstrates how it is possible to explore the complexity of the song repertoire when each type of song produces observable changes in the behavior of neighbors.

Song categories represent a type of language and can be considered at the top of a hierarchy of signal organization.

Other playback experiments conducted by Mathevon and Aubin (2001) have demonstrated how the blackcap (*Sylvia atricapilla*) is able to decode the sound signals from conspecifics as an adaptation to the degradation of signals in dense vegetation.

Manipulated songs to mimic the effect of song degradation by vegetation in a dense layer of undergrowth were used to examine the responses of blackcap males. Territory holders replied even to crudely manipulated artificial sound. Only when the frequency modulation was suppressed or the number of syllables strongly reduced there was a cessation of male reply to the playback. This species seems highly tolerant of song structure modification as an adaptation to the propagation of song in a dense forest environment.

4.11 How to Improve Signaling

The efficiency of a signal largely depends on the internal design and the capacity of an emitter to separate the signal from background noise.

Signals must be separated by background noise and must be minimized for their degradation by distance or by interference with the medium (soil, foliage, etc.). Signals must cope with the environmental conditions and at the same time overcome the environmental constraints to improve communication.

The signal must evolve to maximize the signal-to-noise ratio, and the simplest way is to increase the intensity or amplitude of the signal. For instance, males able to produce a louder song are preferred by females. The signal must contrast with the background noise, but when background noise is strong evolutionary pressure can produce a weak signal.

Noise reduction is an important issue in communication strategy; in particular, to reduce background noise behavior can evolve to send signals from the most favorable places, times, and seasons, when ambient noise is at the minimum (Fig. 4.9).

Signals produced throughout the day require more energy to mask background noise, but at night the signals may be weaker in a quieter sonic background. One main hypothesis for the dawn and dusk chorus (but see Chap. 7) resides in a quiet sonic environment that would improve the diffusion of acoustic performances.

A strategy to increase the signal-to-noise ratio consists of filtering out continuous/ monotonous signals such as background noise. This process is known as habituation.

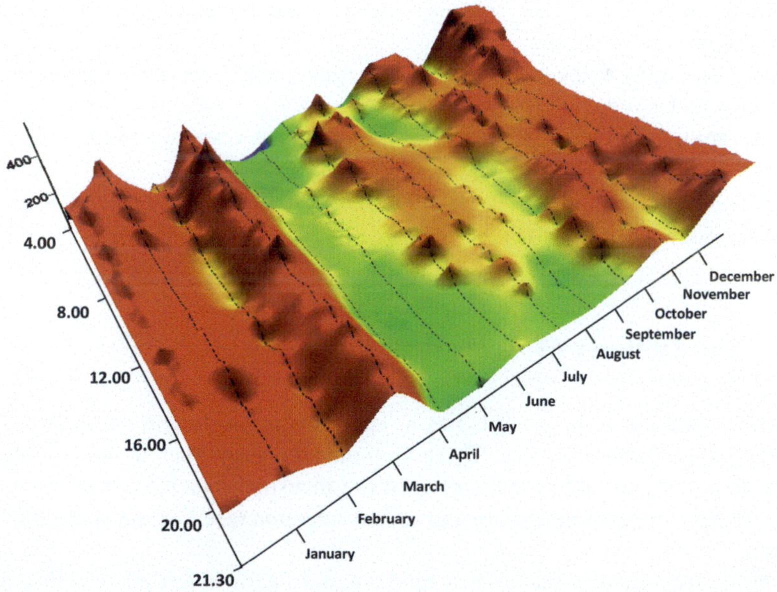

Fig. 4.9 Distribution of background noise ($<1,500$ Hz) during daytime hours and months resulting from application of acoustic complexity index (ACI) to sound files recorded at a site in a Mediterranean maqui (Reproduced from Farina et al. 2013)

Endler (1992) has defined sensory driver as the process that integrates environmental physics, biophysics, and neurobiology and has depicted some general rules for signaling with sound.

1. Use lower frequencies (<2 kHz) in order to minimize reverberation, attenuation, and scattering by the atmosphere, objects, and "shadows" resulting from temperature gradients.
   Do not use frequencies that are to low (>0.5-1 kHz) to minimize destructive interference and to avoid energetically costly coupling between low-frequency sound generation and the atmosphere.
2. Send signals from a position that is greater than 1 m above the ground and preferably upwind of the receiver, in order to minimize destructive interference from ground reflection and temperature gradient effects.
3. Avoid rapid syllable repetition rates in places (such as forests) with strong reverberation.
4. Use frequency modulation rather than amplitude modulation because reverberation and turbulence modify sound amplitude more than frequency.
5. Use redundant signal structure in order to average out background noise.
6. Use greater amplitude and/or physical structure to beam the signal more effectively.
7. Use higher frequencies if much information must be transmitted in a short time, because receptors have a faster frequency response at higher frequencies and there will be less degradation by turbulence.
8. Use species-specific frequency bands and tuned receptors in order to minimize noise at other frequencies.
9. Choose frequency bands, places, seasons, and times of day that minimize turbulence and/or background noise.
10. Avoid signaling at the same time as immediate neighbors (unless jamming their signals is desired).
11. Use rapidly degrading signals for short-distance communication and slowly degrading signals for long-distance communication.
12. Use simpler, more effective alerting signals to attract the receiver's attention before sending the main signal.

4.12 The Honesty of Acoustical Signals

Acoustic communication in birds has at least two major functions: defending the territory and attracting females. There are other functions such as social cohesion, cooperation, but generally the majority of the investigations have been carried out exploring the competition domain and sexual selection (see Gil and Gahr 2002 for a review).

In birds, song quality becomes a fundamental factor in sexual selection and in mating success. The physical condition of a male is connected to the quality of song

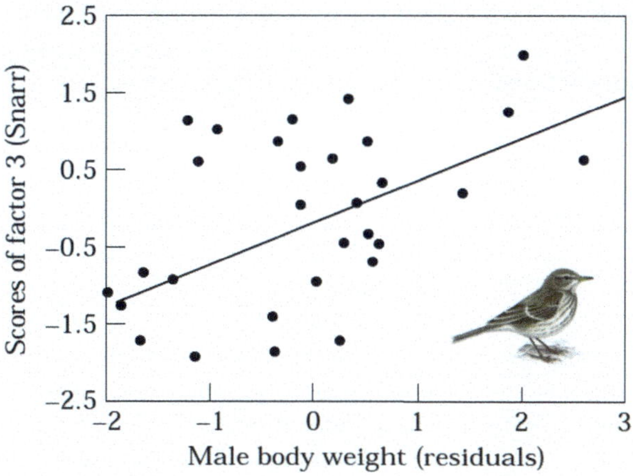

Fig. 4.10 Relationship between the amount of "snarr" in the song of the water pipit (*Anthus spinoletta*) and male body weight (Reproduced with permission from Rehsteiner et al. 1998)

being uttered, although acoustic performance is not sufficient to explain male reproductive success and other characteristics aside from song quality concur to this success, such as time spent singing, song rate, length and amplitude, etc.

In birds like starlings (*Sturnus vulgaris*) and in *Acrocephalus* warblers the females select the males with the broadest repertoire, where the repertoire size is defined as the number of different elements of song that a male can produce.

In sexual selection the content of a song repertoire seems more relevant than its size. In many birds close neighbors share a relevant percentage of their repertoire. In many species local conditions favor the birth of dialects, and female usually prefer males that are adopting such dialects. In some species, such as canaries (*Serinus canaria*), the presence of a fast trill of large bandwidth is very attractive for females, which means that special syllables can be involved in the process of sexual selection. In the case of the water pipit (*Anthus spinoletta*), males that present a special "snarr" were mated more often than males with low scores (Rehsteiner et al. 1998). Individuals that performed "snarr" had the highest body weight, although other phenotypic characters (length of the eighth primary, tarsus length) and age seem to not be related (Fig. 4.10). Finally, the "snarr" sounds seems to increase reproductive success with important results in male–male competition.

It is well documented that body size of species is an indicator of frequency range [*Regulus regulus* (8.5–9.5 cm) and *Turdus merula* (23–29 cm) have a dominant frequency of 8,000 and 4,000 Hz, respectively]. If this constraint is present also within a species, females should distinguish larger from smaller males according to their acoustic characters, and this character could be an honest signal in male competition.

Also, timing of singing may be an indicator of individual quality. For instance, in conditions of bad weather only high-quality males are singing. At dawn, there are dominant males that sing early, which should have an energetic cost, but the advantage is that a male sings a great deal when the female indulges in extra copulation and when mate guarding is not possible. Individual singing early than average times could represent an extra cost.

In some cases males are countersinging, and this behavior grades aggressiveness by singing the same song, by overlapping, or increasing the rate at which song types are switched.

Song reliability may be controlled by the cost paid by the signaler or by the behavior of the receiver. Signals can be controlled by individual performance or by physical constraints: both are important issues to drive signal design and cost type.

Song is the result of muscular contraction that requires much energy: singing has an energetic cost, and availability of energy affects singing performances. This cost has been proved by supplying food to singers, and observing higher acoustic activity during warm and calm days.

Singing is costly but singing activity is incompatible with feeding. Thus the time subtracted by feeding produces a reduction of energy input and conversely reduces singing performance for lack of energy budget. Direct physiological measurements have proved that the consumption of oxygen during singing performance increases 1.05- to 1.38-fold compared with the pre-singing metabolic level. A difference in energy consumption has been observed between loud and calm song, and species with a higher metabolic rate have lower song outputs.

Song quality could be modified by the physical conditions to which an individual was exposed during the first period of its life (nestling period) (Nowicki et al. 2000).

In some species there are syllables that represent a stimulus for the neighboring males to challenge this individual more frequently. During singing activities males are exposed to aerial predation if singing from exposed perches. When a singer localizes a predator, it immediately switches the song to an inconspicuous alarm.

In all oscine passerines song is the result of a learning process: when no tutor is present during the period between hatching and the time young males establish a territory, the lack of tutoring reduces song complexity of the song, which appears to be an honest age-related signal.

The theory of the honest signal has been investigated by Laiolo et al. (2004) on a population of short-toed larks (*Calandrella rufescens*). This species emits distress calls in the presence of a predator, and the harshness of the call has been demonstrated to be connected with body condition and T-cell-mediated immunological conditions. The harshness of the distress calls seems an honest signal of the health status of the prey, and as a consequence a signal of ability to escape predators and a way to inform the predator about the level of difficulty should the individual be captured.

In a population of short-toed larks of Fuerteventura (Canary Islands), of which 55 % was infected with poxvirus, modifications were observed of the spectro-temporal structure of the lark distress calls. Infected birds had shorter and lower-pitched calls when compared with virus-free individuals. Call harshness was independent of virus infection and was correlated with bird immune and body condition, but in stressed birds the call pulse rate decreased. Infection in this case produces two types of effects: a reduction of fitness directly by the disease and indirectly an increase predatory pressure caused by a lower antipredatory behavior.

The rich song repertoire of male birds is an important attribute not only for male choice but also for parental effort during the breeding period, as shown by the investigation of Buchanan and Catchpole (2000) in the sedge warbler (*Acrocephalus schoenobaenus*). They found that chick weight, when controlling for chick age, was correlated with the repertoire complexity (size) of the male. Repertoire size is recognized by the female of this species as an important factor of mating selection. Thus, the selection of a male with a complex sound repertoire assures indirect and direct benefits such as increase of parental effort.

The honest signal hypothesis has found a confirmation in a population of song sparrow (*Melospiza melodia*) living on Mandarte Island, British Columbia, Canada. Reid et al. (2005) have observed that males with a larger song repertoire contributed more independent and recruited offspring and independent and recruited grand-offspring. This result was possible because these males were living longer and reared a greater portion of hatched chicks. Independent offspring of males with a larger song repertoire had a higher probability to recruit and then to have more grand-offspring than males with a lesser song repertoire.

After a male acoustic performance, the females decide whether to mate or to ignore the male. The choice may depend on several factors. Food seems one of the most important factors acting as an evolutionary force, but recently Ritschard and Brumm (2011) tested this hypothesis on the zebra finch (*Taeniopygia guttata*) by manipulating food availability in males. The results did not demonstrate a relationship between food availability and syllable repertoire, proportion of sound versus silence within a song, and mean song frequency. Treated birds showed a reduced song rate, an increased latency to sing, and a lowered song amplitude and fundamental frequency.

4.13 Dialects as a Divergent Communication

Dialects are important variations in communication sets among birds (and also among humans). Dialects represents variation of the acoustic theme performed by a local population of a species under specific environmental constraints. Acoustic divergence seems the first step for a more pronounced and definitive divergence for speciation. Recently, Slabbekoorn and Smith (2002) discussed the role and the signification of dialects in birds, arguing that many studies on dialects have not considered the role of ecology; in other words, the consequences on fitness for female mate preferences have not been considered, or only divergences in realized acoustic variation without a consideration of the potential acoustic variation have been considered. These authors stress the importance of the learning characteristics of an individual in song variation at the population level.

Acoustic mechanisms contribute to species differentiation and isolation. The subspecies represents the first degree of speciation, and acoustic cues can contribute to this process. Experiments conducted in *Emberiza schoeniclus* and in the Mediterranean subspecies (*E. s. intermedia*) have not produced definitive results in terms of acoustic differentiation (Matessi et al. 2000, 2001), but this work represents an attempt to investigate the role of acoustic performance as initiator of an isolating process that in the long term should produce speciation.

4.14 Acoustic Partitioning in Birds and Frogs

The involuntary overlap of heterospecific acoustic performances reduces the efficiency of the signals. To reduce interspecific acoustic interference, species can adopt two main strategies at least: choice of time and space for singing, or evolving different typology of songs.

Often a signal is produced in a noisy environment, especially in environments with a high diversity of vocal species, and detectability and discriminability are the two properties that a signal should conserve to be useful, reducing the interference with irrelevant signals.

Considering that the environmental noise rarely is continuous, as during a heavy rain or close to the sea during a tempest, or near a waterfall, there are possible gaps in which a signaler can place its signal, reducing overlap and interference with other signals.

This process is called acoustic partitioning and consists of changing the time at which to start an acoustic performance in the presence of other concurrent singers.

During a dawn chorus many species and individuals sing with a temporal turnover to minimize the interference effect.

Time partitioning was observed by Luther (2008): among 106 species of birds that sang at dawn in Alta Forest, Mato Grosso, Brazil, 34 species sang in a 30-min or 1-h time. Of these species, 2 sang within 30 min of sunrise: the white-browed antbird (*Myrmoborus leucophrys*) and warbling antbird (*Hypocnemis cantator striata*). Two other species, the chestnut-backed antshrike (*Thamnophilus palliatus palliatus*) and the black-faced antthrush (*Formicarius analis*), sang more frequently during 1–2 h after sunrise (Fig. 4.11). Playback experiments were conducted with these 4 species, receiving replies from each species during the typical song time. Response outside the typical song period was observed but was less intense and less immediate. This experiment has proved that some species use temporal partitioning to reduce acoustic interference.

This author has confirmed in another location a clear dispersion of the song of 82 sedentary species that sang at dawn (Luther 2009). In fact, song of congener and family members was more dispersed than the song of species selected at random. This research has demonstrated that species present in the same place and time partition signal space, a phenomenon observed also by Malavasi and Farina (2013) in dusk choruses of birds in a Mediterranean woodland.

In a neotropical highland-frog community, Lüddecke et al. (2000) have found species partitioning the acoustic space although such a community is composed of only five species (*Hyla labialis*, *Eleutherodactylus elegans*, *Eleutherodactylus ebogotensis*, *Eleutherodactylus enervicus*, and *Colostethus subunctatus*) that have distinct ecological traits. These species have distinct call patterns and are calling at different periods of the day. Despite the low frog diversity and the use of land sites from which to call, only one species calls in ponds (*Hyla labialis*); the other are calling in different types of plants and at different heights from the soil, so acoustic partitioning remains an important behavioral trait for reliable vocal communication.

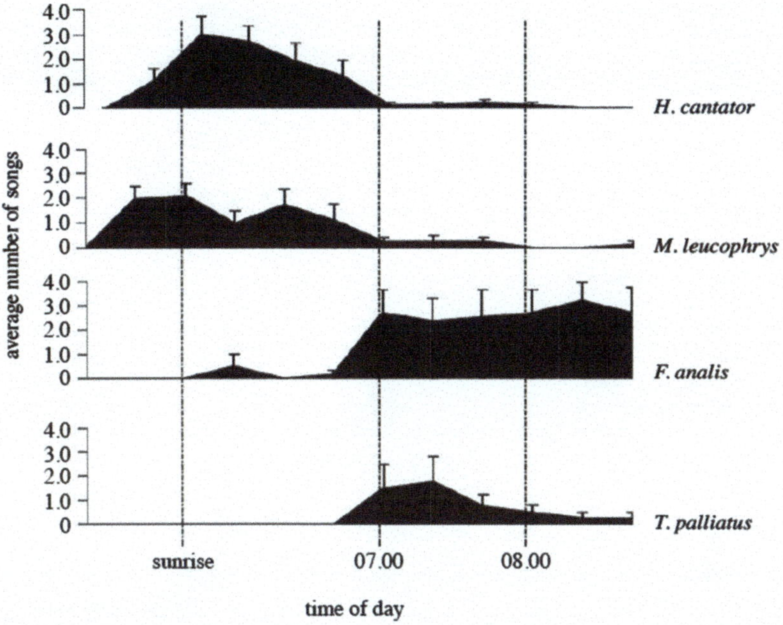

Fig. 4.11 Amount of song activity during morning hours of warbling antbird (*Hypocnemis cantator striata*), white-browed antbird (*Myrmoborus leucophrys*), black-faced antthrush (*Formicarius analis*), and chestnut-backed antshrike (*Thamnophilus palliatus* palliatus) (Reproduced with permission from Luther 2008)

4.15 Communication on Steep Soils

Generally acoustic communication is observed and studied in flat areas or on gentle slopes, and when moving along the steep sides of a mountain range the topographic constraint immediately appears that modifies direction and intensity of the sonic ambient. In fact, the sonic ambient is strongly influenced by the steepness and by a nonuniform distribution of the acoustics waves that are not easily perceived. So, steepness not only affects abundance, distribution, and movement of organisms but also the communication of vocal animals. Birds that sing upslope have a greater far distribution of sonic energy, and in contrast birds singing downslope have a very narrow distribution of sound (Fig. 4.12) Hunter (1989) reports an interesting observation made on 40 birds, including 21 passerine birds, during their singing behavior along a steep slope in the Kulu and Ravi valleys of the Western Himalayas, India. Of 48 observations 37 were upslope, 9 parallel to the slope, and only 2 downslope, confirming the strong orientation of birds on steep slopes. However, factors such as steepness of the slopes, tree height, and perch height were not significantly correlated with body orientation of the birds during their singing performance. Further research would clarify the effect of topography on the communication of birds and on their distribution in a landscape.

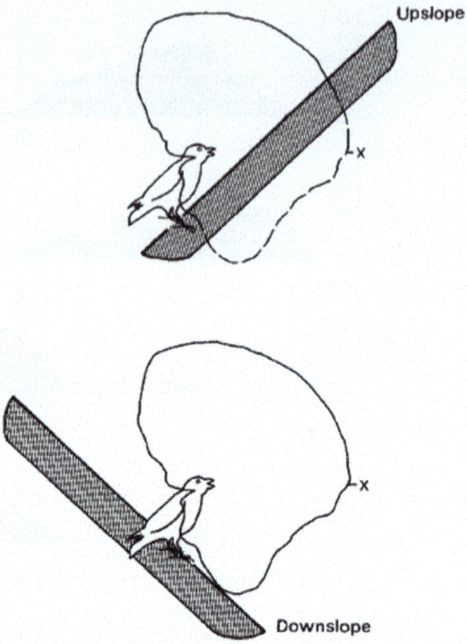

Fig. 4.12 The orientation of a singing bird along a steep slope makes a difference in terms of acoustic propagation. In upslope orientation, the acoustic impact on the surroundings is more efficient than from a downslope orientation. The sonic envelope around the bird silhouette has been generated using in an anechoic chamber a 8-kHz pure tone (Reproduced with permission from Hunter 1989)

4.16 Signaling Systems and Climatic Proxies

Signaling systems may diverge along an ecological gradient, and this has great implications in speciation and diversification. Several studies have focused on environmental drivers such as vegetation but little attention has been devoted, as already stressed, on the effects of climatic changes on signal transmission (Snell-Rood 2012). In fact, climate is important for sound propagation because humidity, temperature, and air turbulence largely depend on climatic conditions. It is well known that a signal decreases in intensity because of spherical spreading, but atmospheric absorption represents an important proxy. This sound absorption increases with the frequency of sound but it varies also for temperature and relative humidity. So, signaling systems are conditioned by climate. Absorption is not high, indeed, but from a biological point of view this can represent an important element in an evolutionary dimension. Snell-Rood (2012) reports that although an avian vocalization may have an absorption of only 0.1 dB/m, this effect may be 10 dB on a surface of 1 ha of territory. It remains unclear if the value of absorption between species is the same within the species, and the effect of climate change on the evolution of acoustic signals remains to be investigated. This point is of extreme importance. We have often observed how seasonality favors or can affect signal transmission, expressed as a useful time to sing. It is well documented that during bad weather conditions, song activity is significantly depressed.

To study in depth this interesting point, Snell-Rood (2012) has investigated North American wood warblers (50 species), which represent a diverse and well-studied monophyletic clade of birds. This group is geographically distributed across a wide range of climates. The song repertoire ranges from 3 to 9 kHz. Divergence in song is connected to the speciation process, and the climatic conditions can affect this process. In the same study Snell-Rood investigated the bats of the American Southwest that represent the richest collection of all North America north of Mexico. This region has a great variety of elevation gradients characterized by different temperatures and humidity.

In bats, the signal process is largely based on a self-hearing process, although eavesdropping in some species of bats has been observed (see in this chapter the section devoted to eavesdropping). Considering that bats emit sounds at a frequency from 12 to 100 kHz and considering also the strong influence of absorption of the high frequencies, bats remain good candidates to study the effects of climatic changes on signaling systems.

At least two hypotheses can be presented on this matter:

1. Frequency characteristics of sounds should be linked to sound absorption.
2. Signal duration should increase with the increase of signal absorption because detectability is connected to duration.

A significant variation in sound absorption has been found across the geographic range of the warblers and a negative correlation has been found between frequency and atmospheric absorption. This result has been confirmed also in the signals of bats in which the echolocation call frequency is inversely connected with sound absorption.

The Snell-Rood work suggests that organisms under a variable environment are able to change the signal structure to optimize the communication. In particular, in bats has been recognized the capacity to change the duration and frequency according to wet and dry seasons. As illustrated in Fig. 4.13, during the wet season the frequency is lower and the duration is higher. Considering that the absorption is higher during the wet season of 0.58 dB/m and that bats use echolocation in a range of 20 m away, during the wet season the signal will be affected by an absorption of 24 dB. These mammals have the ability to change call structure also in relationship to a change in atmospheric pressure.

4.17 Song and Reproductive Timing

The song repertoire is an adaptive character, and the richness of such repertoire is a more effective signal. Considering that male parental ability, age, genetic quality, and physiological condition may be correlated with song complexity, a female will achieve benefits by mating with a male that expresses high complexity in singing repertoire. During the time of female fertility the song appears extremely important for the choice that females will make. Males will benefit for female choice and also

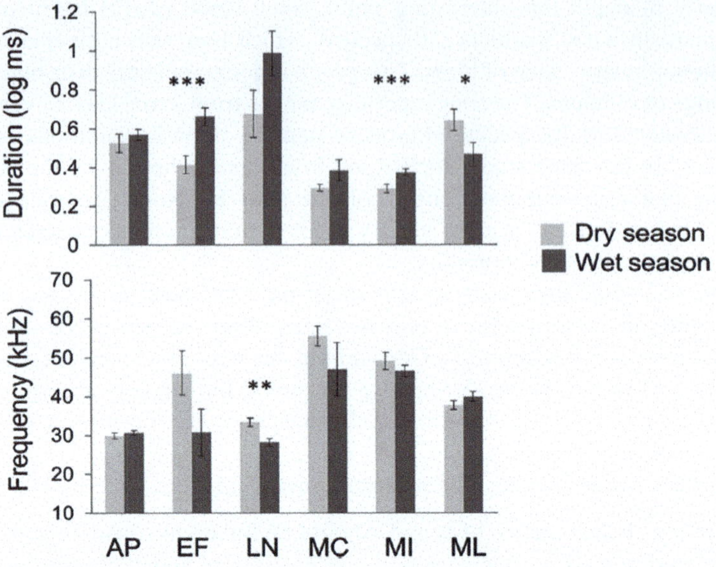

Fig. 4.13 Change in frequency and duration of calls of six species of bats during dry and wet season: AP *Antrozous pallidus*, EF *Eptesicus fuscus*, LN *Lasionycteris noctivagans*, MC *Myotis californicus*, MI *Myotis ciliolabrum*, ML *Myotis lucifigus*, *one asterisk* $P < 0.05$, *two asterisks* $P < 0.01$, *three asterisks* $P < 0.001$ (Reproduced with permission from Snell-Rood 2012)

for extra-pair fertilization. It is known that in many species song complexity increases during the season, but the change in the repertoire is not easily distinguished from other factors. The grosbeak (*Guiraca caerulea*) sings a single song type but rearranges the elements of its repertoire when the females are fertile, as demonstrated by Ballentine et al. (2003), who found that male grosbeaks present a more varied, more versatile arrangement of elements and have more syntax consistency during the fertile period of the females than outside this period (Fig. 4.14).

Conspecific competition usually is observed between males during the selection and the maintenance of a territory. It seems strategic for competing males to enter into a communicative contact to prevent trespassing in the neighbor territory. For this reason, the acoustic communication is one of the best (and peaceful and relatively inexpensive) methods to control a border, to respect the hierarchy, and finally to have the necessary resources to raise a clutch. The rules by which two or more males set the borders of each territory are quite complex. In the nightingale (*Luscinia megarhynchos*) the singers can alternate their song, can overlap, or can be neutral to the song of other males. Playback experiments carried out by Naguib (1999) (Fig. 4.15) have confirmed that during the overlap treatment males were singing at a higher rate and interrupted the singing performance more frequently than in alternating experiments. When males were exposed to alternating treatment, however, they were singing at higher rates when they were exposed to overlap treatment than when the alternating treatment was the first treatment they received. This experiment confirms the importance of the time at which an individual receives some conspecific stimuli.

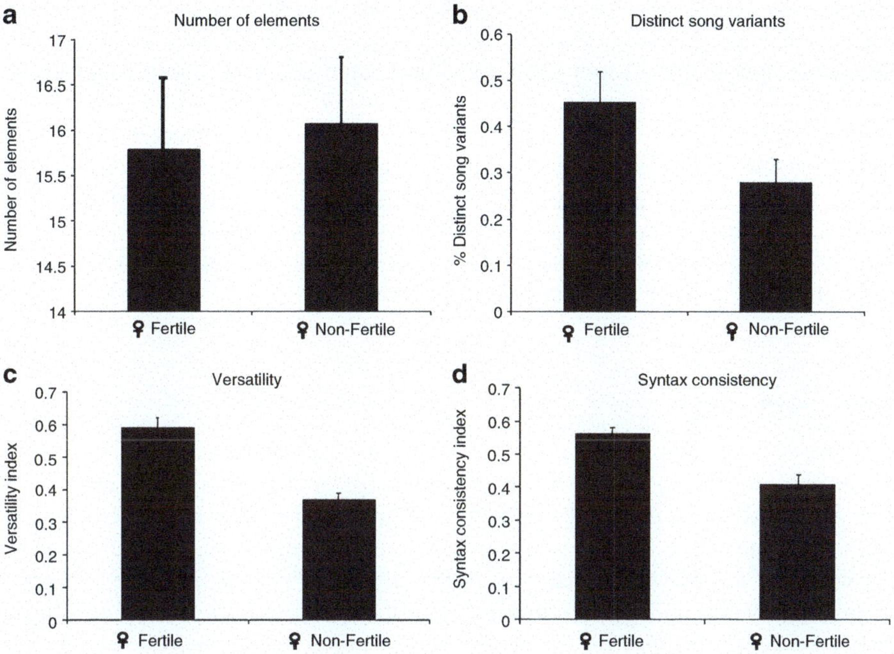

Fig. 4.14 Characteristic of male song in blue grosbeaks (*Guiraca caerulea*) during female fertile and nonfertile periods: number of elements in male repertoire (**a**); percent distinct song variation (**b**); song versatility (**c**); syntax consistency for each period (**d**) (Reproduced with permission from Ballentine et al. 2003)

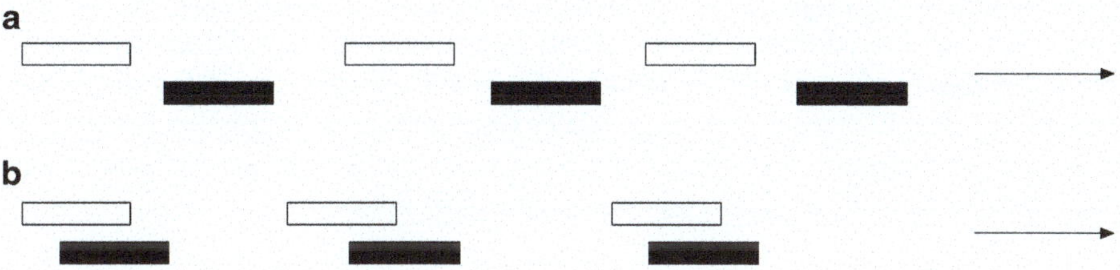

Fig. 4.15 Design of two models of experimental playback treatment for nightingale (*Luscinia megarhynchos*): alternating treatment (**a**) and overlapping treatment (**b**). *Black rectangles* playback songs, *open rectangles* subject songs (Reproduced with permission from Naguib 1999)

4.18 Acoustic Cues and Antipredatory Performance

Krams (2001) observed how singing chaffinches (*Fringilla coelebs*) changed their perch position when exposed to a sparrowhawk (*Accipiter nisus*) model in mature pine forest and in pine saplings, which represent two habitats with different cover from predatory exposition. In fact, without a specific "threat" chaffinches were singing inside the canopy, but after the exposure to the sparrowhawk model all the singing birds selected the conspicuous perch. The males that before the experiment were in the upper part of the canopy moved to the lower canopy, whereas males that were singing below the canopy did not change their position. The perches in the lower canopy exposes the chaffinches to a higher risk of predation but at the same time the visibility of an oncoming predator is very good and this at the end could prevent attacks. This is a case in which the spatial component of a song is adapted to current conditions. The capacity of the chaffinch to modify perch position under the presence of a sparrowhawk demonstrates the flexibility adopted to reduce the risk of predation to a minimum. The acoustic implication of the change of perch position that improves vigilance has to be considered carefully.

4.19 Soft Song

A low-amplitude song, more known as "soft song" or "quiet song," is common in male songbirds and a different significance is attributed to this in the literature. At least two categories of soft song are located according the context. In pied flycatchers (*Ficedula hypoleuca*) and in the great tit (*Parus major*), the soft song is uttered only during courtship. In white-throated sparrows (*Zonotrichia albicollis*) and in redwings (*Turdus iliacus*), males were observed to emit the soft song during aggressive contexts.

In common blackbirds (*Turdus merula*) and in dunnocks (*Prunella modularis*), this song is heard both in courtship and in aggressive encounters.

In the song sparrow (*Melospiza melodia*), the soft song is present only during aggressive behavior; to test if this soft song is a predictor of attack, playback experiments were conducted by Anderson et al. (2007). Two categories of soft song were considered during the playback experiment: a "crystallized soft song" and a "warbled soft song." The females responded with less courtship display to crystallized than to warbled soft song. Males responded with aggressive behavior to both types of playback soft song. From this experiment results showed that in the song sparrow the soft song is a signal of aggressive intent.

4.20 Bioacoustics Interactions Between Plants and Insects

Climate changes are favoring animal displacement, invasion, and rapid pest spreading. In the forest system the balance between soil, plants, and animals is disrupted by climatic change that modifies precipitation abundance and temporal distribution, air temperature, and wind regime. Drought stress that recently has affected many boreal and temperate regions has favored the outbreak of deforesting insects at a growing intensity and capacity to cover large areas.

In conclusion, higher deforestation is driven by insects and modulated by climate (Dunn and Crutchfield 2009). In particular, boreal forests could face a reduction of half their size if the actual climatic trend in rising temperature is confirmed. One risk of deforestation depends on fires, which have doubled in recent years. Forests damaged by fires are easily attacked by species of opportunistic insects. This concern recently has been extended also to temperate forests that under drought stress are attacked by bark beetles.

It is evident that some damages are the result of just a unique species of insect; in Alaska has been reported that the spruce bark beetle (*Dendroctonus rufipennis*) has killed 4.4 million acres of forests. We do not enter into the details of the damages caused also by other species in the American subcontinent or in Siberia, but the situation is in continuous evolution and forests are facing an unprecedented risk of radical modification of their assets with consequences on the carbon storage balance.

Among the several hypotheses of a modified relationship between insects and trees, the bioacoustic ecology hypothesis represents a novel field of investigation.

The changes in temperature and moisture create conditions for insect outbreaks, changing the relationship between plants and insects. Insects are poikilothermic organisms and a change in temperature, for instance in winter, can favor the active presence of insects all year around.

Trees have no capacity to react at the same speed of climatic changes. At the southern range plants under an increased temperature reduce the reproductive cycle and die; on the northern border they migrate at a low speed, and the total surface occupied by a species is progressively reduced.

Between the different aspects that favor the spreading of insect outbreaks in many northern biomes, little is known about the role of bioacoustics in this process because the insects responsible for the decline of forests use intensively acoustic mechanisms that are not well understood at the present.

In many bark beetles, there is a *pars strident* organ; at least 14 types of stridulating organs have been described in this group of insects for 30 different families. In *Ips confusus* beetles this organ is located on the back of the head and is stroked by a plectrum on the underside of the dorsal anterior edge of the prothorax. In other species the *pars strident* has a different anatomical position. The sound produced by these organs consists of a single-impulse click or a range of different multi-impulse chirps. All these sounds are considered stridulations.

In some species, as in the genus *Dendroctonus*, only the male has a stridulation organ but the female is the pioneer. The opposite is observable in the *Ips* genus where the male is the pioneer and without *pars stridens* and it is the female that has such an organ.

It is evident that bark beetles use either chemical and acoustic signals to regulate aggression, attachment on host trees, courtship, mating behavior, and population density regulation. The chemical signals seem to be used for communicating at a distance and the acoustic signal for short-distance communication. In the complex behavior of this species chemical and acoustic cues are intensively utilized, although little information is yet available on the hearing organ mechanisms.

On the other hand plants emit, when under hydric stress, acoustic signals produced by cavitation phenomena. Cavitation is the breaking of the water column in the conducting xylem tissue of leaves, stems, and trunks. The sound represents the vibration produced by the single-cell collapse caused by gradual dehydration and prolonged water deficit. This acoustic emission is audible only in part; the majority pertains to ultrasonic frequencies. It is commonly used by botanists to measure such acoustic emissions to evaluate the level of drought stress in living plants and more precisely to determine the threshold water potential at which the water-conducting system initiates damage (Jackson and Grace 1996). This approach can only determine the proportion of embolism in a conducting system and is not able to determine the percentage of reduction in hydraulic conductivity.

Acoustic utilization seems more important than the amount of research could speculate. In 1987 Mattson and Haack (1987) advanced the hypothesis that drought-stressed trees could produce a cavitation sound audible to plant-eating insects. It is not clear if the bark beetles are able to listen to the sounds produced by trees, but in many insects the capacity to capture ultrasonic signals has been found. For instance, the capacity of some butterflies and moths to escape the echoes emitted by bats has been documented.

Summary
Communication is an important component of animal life, contributing to reproduction and to survival life traits. In communication there is a sender and a receiver connected by an informative signal. A third subject is the eavesdropper or the involuntary receiver that utilizes the information which circulates between a sender and a receiver.

A communication theory represents a general framework used to describe the plethora of observations, rules, and exceptions described in the variegated context of animal communication.

Communication is costly in term of energy budget allocated by vocal individuals, and a well-designed communicative framework should be the result of a selective pressure.

Signal generation and emission is the first step of the communication process, followed by other important steps/constraints such as transmission through the medium, signal reception and processing, and finally the decisions based on the perceived signal.

In the communication process the indirect intercept of a message/signal directed to others is called eavesdropping, and this process represent an important evolutionary phenomenon.

Echolocation, so common in bats, is a mechanism by which a communicative channel is entirely used to evaluate the consistency and distance of a physical context and potential prey and predators.

Heterospecific attraction may be defined as the capacity by a species to copy the information shared by other species in the same location. For instance, resident species can offer important information to transient species in terms of refuges or foraging opportunities. Conspecific attraction remains a fundamental mechanism that guides in territory selection.

In the communication process, the male–female song recognition seems an important element influencing the way resources are shared within the pair, assuring food and predatory avoidance.

The capacity to evaluate the distance at which a potential competitor is during a precise moment is essential to save energy in territorial defense and in antipredatory behavior. Ranging is defined as the ensemble of mechanisms that an individual enacts to evaluate the distance to another individual. Several pieces of evidence have confirmed the diffusion of this capacity among vocal animals, although amplitude fluctuation, reverberation, blurring, attenuation, and signal-to-noise ratio may reduce the efficiency of ranging procedure.

Among the communication strategies adopted by species the distress call represent a way to inform conspecifics and heterospecifics on the presence of a potential predator. To be efficient these calls must warn other species about the menace and at the same time to not inform predators about the location of the emitter.

Acoustic signals are used in several groups of species, humans included, to find resources, to avoid predators, and to find or adapt locations for reproduction. In other terms the acoustic cues are used in the orienting processing.

In the analysis of the complexity of the acoustic communication codes, the acoustic codes specifically represent an interesting perspective of analysis. Codes in biology are numerous and well documented. Every acoustic message is the result of a sequence of elementary parts whose syntax represents a message. Acoustic codes operate at intraspecific level but also at interspecific level, improving the efficiency of communicative behavior. Code sequence may have a so robust structure that despite the loss of a relevant part of the codes the message is not degraded. An example of this aspect is represented by the acoustic structure of blackcap song, where experimentally it has been verified that only when the frequency modulation was suppressed or the number of syllables strongly reduced was there a cessation of male reply to the playback.

Improving the efficacy of a signal is important to reduce the loss of energy and on the other hand to optimize the information contained in the signal. The efficiency of a signal largely depends on internal design and the capacity of an emitter to separate the signal from the background noise.

A central principle in the communication theory states that biological signals are honest: this means for an acoustic signal such as a song that the complexity of the repertoire and all the secondary associated characters are expressions of natural selection. Thus, the theory of honest signals becomes a cornerstone in the ecology of acoustic communication. There are several types of evidence that confirm this hypothesis for many animals.

Dialects, which are local variations of acoustic performance, are an important aspect of adaptation to local conditions and represent the initial stages of future acoustic divergence.

To reduce interspecific competition, birds and frogs can use acoustic partitioning that consists of changing the time at which to start an acoustic performance in the presence of other concurrent singers. This process allows species to share the same sonic context, reducing the risk of interspecific masking or having equivocal acoustic signals.

Among the environmental factors that can have some influence on acoustic communication, the steepness of soils may have an important role. Field observation have confirmed that in birds bodily orientation is a fundamental factor responsible for appropriate transmission of acoustic signals.

Climatic conditions are recognized as important proxies in acoustic communication, especially in birds and bats. In particular in bats has been recognized the capacity to change the duration and frequency of the signals according to wet and dry seasons.

Considering that most of the acoustic repertoires are used by vocalizing animals for reproductive purposes, the optimization of song performance during the reproductive time confirms this hypothesis, as demonstrated for instance by the grosbeak (*Guiraca caerulea*) song that presents more variants, more versatile arrangement of elements, and more syntax consistency during the fertile period of the females than outside this period.

A strict relationship emerges between acoustic cues and antipredatory strategies, producing changes in behavioral traits and singing performance.

The soft song appears as a secondary acoustic manifestation of some species of birds to which has been attributed a different significance that ranges from a courtship to an aggressive behavior. Soft song represents a further categorization of a great variety of acoustic cues that vocal animals (in this case birds) utilize during the reproductive time.

Climatic changes have produced dramatic modification in the physiology of many trees that suffer hydric stress and fires. Insects are important proxies of forest declines, especially in boreal areas. It has been proven that plants and insects are deeply connected by an acoustic mechanism. Plants under hydric stress produce a cavitation sound audible to plant-eating insects that select the more exposed plants for pest attack.

References

Anderson RC, Nowicki S, Searcy WA (2007) Soft song in song sparrows: response of males and females to an enigmatic signal. Behav Ecol Sociobiol 61:1267–1274

Ballentine B, Badyaev A, Hill GE (2003) Changes in song complexity correspond to periods of female fertility in blue grosbeaks (*Guiraca caerulea*). Ethology 109:55–66

Balsby TJS, Dabelsteen T, Pedersen SB (2003) Degradation of whitethroat vocalizations: implications for song flight and communication network activities. Behaviour 140:695–719

Barbieri M (2003) The organic codes. Cambridge University Press, Cambridge

Bee MA (2007) Sound source segregation in grey treefrogs: spatial release from masking by the sound of a chorus. Anim Behav 74:549–558

Buchanan KL, Catchpole CK (2000) Song as an indicator of male parental effort in the sedge warbler. Proc Biol Sci 267(1441):321–326

Byers BE (1996) Messages encoded in the songs of chestnut-sided warblers. Anim Behav 52:691–705

Dabelsteen T, Larsen ON, Pedersen SB (1993) Habitat-induced degradation of sound signals: quantifying the effects of communication sounds and bird location on blur ratio, excess attenuation, and signal-to-noise ratio in blackbird song. J Acoust Soc Am 93(4):2206–2220

De La Torre S, Snowdon CT (2002) Environmental correlates of vocal communication of wild pygmy marmosets, *Cebuella pygmaea*. Anim Behav 63:847–856

Doutrelant C, Leitao A, Otter K, Lambrechts MM (2000) Effect of blue tit song syntax on great tit territorial responsiveness—an experimental test of the character shift hypothesis. Behav Ecol Sociobiol 48:119–124

Dunn D, Crutchfield JP (2009) Entomogenic climate change: insect bioacoustics and future forest ecology. Leonardo 42(3):239–244

Eggers S, Griesser M, Nystrand M, Ekman J (2006) Predation risk induces changes in nest-site selection and clutch size in the Siberian jay. Proc R Soc B 273:701–706

Emmering QC, Schmidt KA (2011) Nesting songbirds assess spatial heterogeneity of predatory chipmunk by eavesdropping on their vocalizations. J Anim Ecol 80(6):1305–1312

Endler JA (1992) Signals, signal conditions, and the direction of evolution. Am Nat 139: S125–S153

Endler JA (1993) Some general comments on the evolution and design of animal communication systems. Philos Trans R Soc Lond B 340:215–225

Farina A (2012) A biosemiotic perspective of the resource criterion: toward a general theory of resources. Biosemiotics 5:17–32

Farina A, Belgrano A (2006) The eco-field hypothesis: toward a cognitive landscape. Landsc Ecol 21:5–17

Farina A, Pieretti N, Morganti N (2013) Acoustic patterns of an invasive species: the red-billed *Leiothrix* (*Leiothrix lutea* Scopoli 1786) in a Mediterranean shrubland. Bioacoustics 22(3):175–194. doi:10.1080/09524622.2012.761571

Fay R (2009) Soundscapes and the sense of hearing of fishes. Integr Zool 4:26–32

Forrest TG (1994) From sender to receiver: propagation and environmental effects on acoustic signals. Am Zool 34:644–654

Fotheringham JR, Martin PR, Ratcliffe L (1997) Song transmission and auditory perception of distance in wood warblers (Parulinae). Anim Behav 53:1271–1285

Gil D, Gahr M (2002) The honesty of bird song: multiple constraints for multiple traits. Trends Ecol Evol 17:133–141

Hoffmeyer J (2008) The semiotic niche. J Mediterr Ecol 9:5–30

Hunter ML Jr (1989) Himalayan birds face uphill while singing. Auk 106:728–729

Jackson GE, Grace J (1996) Field measurement of xylem cavitation: are acoustic emissions useful? J Exp Bot 47(304):1643–1650

Jones PL, Page RA, Hartbauer M, Siemers BM (2011) Behavioural evidence for eavesdropping on prey song in two Palearctic sibling bat species. Behav Ecol Sociobiol 65:333–340

Krams I (2001) Perch selection by singing chaffinches: a better view of surroundings and the risk of predation. Behav Ecol 12(3):295–300

Laiolo P, Tella JL, Carrete M, Serrano D, Lòpez G (2004) Distress calls may honestly signal bird quality to predators. Proc R Soc Lond B (Suppl) 271:S513–S515

Lampe HM, Larsen ON, Pedersen SB, Dabelsteen T (2007) Song degradation in the hole-nesting pied flycatcher *Ficedula hypoleuca*: implications for polyterritorial behaviour in contrasting habitat-types. Behaviour 144:1161–1178

Lind H, Dabelsteen T, McGregor PK (1996) Female great tits can identify mates by song. Anim Behav 52:667–671

Lüddecke H, Amézquita A, Ximena B, Guzmàn F (2000) Partitioning of vocal activity in a Neotropical highland-frog community. Stud Neotrop Fauna Environ 35:185–194

Luther DA (2008) Signaller:receiver coordination and the timing of communication in Amazonian birds. Biol Lett 4:651–654

Luther DA (2009) The influence of the acoustic community on songs of birds in a neotropical rain forest. Behav Ecol 20(4):864–871

Malavasi R, Farina A (2013) Neighbours' talk: interspecific choruses among songbirds. Biosemiotics 22:33–48
Margoliash D (1983) Acoustic parameters underlying the response of song-specific neurons in the white-crowned sparrow. J Neurosci 6:1643–1661
Margoliash D, Hale ME (2008) Vertebrate vocalizations. Science 231:347–348
Margoliash D, Konishi M (1985) Auditory representation of autogenous song in the song-system of white-crowned sparrows. Proc Natl Acad Sci USA 82:5997–6000
Matessi G, Debelsteen T, Pilastro A (2000) Responses to playback of different subspecies songs in the reed bunting *Emberiza schoeniclus*. J Avian Biol 31(1):96–101
Matessi G, Debelsteen T, Pilastro A (2001) Subspecies on discrimination in a Mediterranean population of the reed bunting *Emberiza schoeniclus intermedia*. Ital J Zool 68:311–314
Mathevon N, Aubin T (1997) Reaction to conspecific degraded song by the wren *Troglodytes troglodytes*: territorial response and choice of song post. Behav Processes 39:77–84
Mathevon N, Aubin T (2001) Sound-based species-specific recognition in the blackcap *Sylvia atricapilla* shows high tolerance to signal modifications. Behaviour 138:511–524
Mathevon N, Aubin T, Brémond JC (1997) Propagation of bird acoustic signals: comparative study of starling and blackbird distress calls. C R Acad Sci Paris 320:869–876
Mathevon N, Dabelsteen T, Blumenrath SH (2005) Are high perches in the blackcap *Sylvia atricapilla* song or listening posts? A sound transmission study. J Acoust Soc Am 117 (1):442–449
Mathevon N, Aubin T, Viellieard J, da Silva ML, Sebe F (2008) Singing in the rain forest: how a tropical bird song transfers information. PLoS One 3(2):e1580
Mattson WJ, Haack RA (1987) The role of drought in outbreaks of plant-eating insects. Bioscience 37(2):110–118
McGregor PK (2005) Introduction. In: McGregor PK (ed) Animal communication network. Cambridge University Press, Cambridge, pp 1–6
McGregor PK, Krebs JR, Ratcliffe LM (1983) The reaction of great tit (*Parus major*) to playback of degraded and undegraded songs: the effect of familiarity with the stimulus song type. Auk 100:898–906
Mönkkönen M, Forsman T (2002) Heterospecific attraction among forest birds: a review. Ornithol Sci 1:4–51
Mönkkönen M, Helle P, Soppela K (1990) Numerical and behavioral responses of migrant passerines to experimental manipulation of resident tits (*Parus* spp.): heterospecific attraction in northern breeding bird communities? Oecologia (Berl) 85:218–225
Morton ES (1987) The effects of distance and isolation on song-type sharing in the Carolina Wren. Wilson Bull 99(4):601–610
Morton ES, Gish SL, van der Voort M (1986) On the learning of degraded and undegraded songs in the Carolina Wren. Anim Behav 34:815–820
Naguib M (1995) Auditory distance assessment of singing conspecifics in Carolina wrens: the role of reverberation and frequency-dependent attenuation. Anim Behav 50:1297–1307
Naguib M (1996) Auditory distance estimation in song birds: implications, methodologies and perspectives. Behav Processes 38:163–168
Naguib M (1999) Effects of song overlapping and alternating on nocturnally singing nightingales. Anim Behav 58:1061–1067
Naguib M (2005) Singing interactions in songbirds: implications for social relations and territorial settlement. In: McGregor PK (ed) Animal communication network. Cambridge University Press, New York, pp 300–319
Naguib M, Klump GM, Hillmann E, Griessmann B, Teige T (2000) Assessment of auditory distance in a territorial songbird: accurate feat or rule of thumb? Anim Behav 59:715–721
Nowicki S, Hasselquist D, Bensch S, Peters S (2000) Nestling growth and song repertoire size in great reed warblers: evidence for song learning as an indicator mechanism in mate choice. Proc R Soc Lond B Biol Sci 267:2419–2424
Odum HT (1983) Systems ecology: an introduction. Wiley, New York

Peake T (2005) Eavesdropping in communication networks. In: McGregor PK (ed) Animal communication networks. Cambridge University Press, Cambridge, pp 13–37
Penna M, Llusia D, Marquez R (2012) Propagation of natural toad calls in a Mediterranean terrestrial environment. J Acoust Soc Am 132(6):4025–4031
Redmond LJ, Murphy MT, Dolan AC, Sexton K (2009) Public information facilitates habitat selection of a territorial species: the eastern kingbird. Anim Behav 77:457–463
Rehsteiner U, Geisser H, Reyer HU (1998) Singing and mating success in water pipits: one specific song element makes all the difference. Anim Behav 55:1471–1481
Reid JM, Arcese P, Cassidy ALEV, Hiebert SM, Smith JNM, Stoddard P, Marr AB, Keller LF (2005) Fitness correlates of song repertoire size in free-living song sparrows (*Melospiza melodia*). Am Nat 165:299–310
Reza FM (1961) An introduction to information theory. Dover, New York
Richards DG (1981) Estimation of distance of singing conspecifics by Carolina Wren. Auk 98:127–133
Richards DG, Wiley RH (1980) Reverberations and amplitude fluctuations in the propagation of sound in a forest: implication for animal communication. Am Nat 115(3):381–399
Ritschard M, Brumm H (2011) Zebra finch song reflects current food availability. Evol Ecol 26 (4):801–812. doi:10.1007/s10682-011-9541-3
Schrödinger E (1944) What is life? Cambridge University Press, Cambridge
Slabbekoorn H, Bouton N (2008) Soundscape orientation: a new field in need of sound investigation. Anim Behav 76:e5–e8
Slabbekoorn H, Smith TB (2002) Bird song, ecology and speciation. Philos Trans R Soc Lond B 357:493–503
Snell-Rood E (2012) The effect of climate on acoustic signals: does atmospheric sound absorption matter for bird song and bat echolocation? J Acoust Soc Am 131(2):1650–1658
Stilz WP, Schnitzler HU (2012) Estimation of the acoustic range of bat echolocation for extended targets. J Acoust Soc Am 132(3):1765–1775
Thomas RJ, Drewitt EJA, Kelly DJ, Marples NM, Semple S (2003) Nocturnal playbacks reveal hidden differences in singing behaviour between populations of robin *Erithacus rubecula*. Bird Study 50:84–87
Übernickel K, Tschapka M, Valko EKV (2012) Selective eavesdropping behaviour in three neotropical bat species. Ethology 119:66–76
Ulanowicz RE (1997) Ecology, the ascendant perspective. Columbia University Press, New York
von Uexküll J (1982 (1940)) The theory of meaning. Semiotica 42(1):25–82
von Uexküll J (1992 (1934)) A stroll through the worlds of animals and men. Semiotica 89 (4):319–391
Wiley RH (1998) Ranging reconsidered. Behav Ecol Sociobiol 42:143–146
Wiley RH, Richards DG (1978) Physical constraints on acoustic communication in the atmosphere: implications for the evolution of animal vocalizations. Behav Ecol Sociobiol 3:69–94

Chapter 5
Human Dimension of the Soundscape: From Individuals to Society

The sonic ambience associated with visual aesthetics is important to human well-being, and the search for a quiet area becomes a priority strategy in human societies surrounded by growing anthropogenic noise. The study of the soundscape from the human perspective requires a psychological approach that allows a synthesis between visual and acoustic stimuli. A semantic approach able to describe the feeling that sounds produce in individuals allows categorization of the cultural and social effects of a subjective sonic environment.

The quality of the sonic ambience increases the acoustic capacity in human hearing and alters the relationship between acoustic cues and their cultural and social interpretation.

Noise sensitivity is defined as an attitude to rating sound according to an individual-based criterion; it represents a major antecedent of individual noise annoyance, and the results are independent of the predisposition to perceive a sound event more intensely or the capacity to better discriminate sounds.

Direct and indirect effects of noise exposure on human health produce hearing impairment, reduction of speech communication, cardiovascular risk, sleep disturbance, psychological and performance reduction effects, and a subjective feeling of annoyance. Exposure to outdoor noise represents the highest risk to human health. Insomnia is one of the most common effects, but in the long term ischemic heart disease and increase in blood pressure have to be considered equally probable.

The soundscape of urban spaces is a complex system in which physical, psychological, and social components concur to human sonic assessment. A sonic environment of high quality in urban areas means not only a low level of sonic energy but also the presence of natural elements such as green spaces and a tangible biodiversity

A. Farina, *Soundscape Ecology: Principles, Patterns, Methods and Applications*,
DOI 10.1007/978-94-007-7374-5_5,

5.1 Introduction

The recent changes in the global economy, human migratory fluxes, and occupying new land for development has modified the traditional landscape structure, increasing humanity to a new level of noise exposure that is considered a form of invasive environmental pollution. Quiet areas are associated with relevant ecological values, and in national parks and protected areas their value is compared to clean air quality.

The perception of the sonic context requires a specific recognition that has its origin either in the genetic code or in cultural individual experience. In other words, sonic perception requires a specific capacity by every perceiving agent (humans and nonhuman animals) that must be trained. For this, measurements of sound magnitude [e.g., A-weighted sound pressure level (L_{Aeq})] is not sufficient to characterize the sonic ambient, and other factors have to be considered.

The sonic environment, as a neutrality-based soundscape (sensu Farina 2006), exists in terms of energy and information but to be perceived, which requires a species-specific cognitive approach that converts such sonic energy to meaningful information.

In fact, the sonic environment can be described in at least three levels of growing complexity.

The first level is dominated by the physical properties of the sounds, ranging from infrasound to ultrasound frequencies.

The second level is the level of the physiological perception of every single species. For instance, in this level human perception is limited to 20–20,000 Hz, but whales can heard infrasound ranging from 20 Hz down to 0.001 Hz, and bats in contrast can perceive ultrasonic frequencies up to values of 210 kHz.

A third level pertains to the interpretation of the sounds according to genetic or cultural/social decodification of the sonic signals (Fig. 5.1). At this last level pertains the concept of the sound ambient environment (SAE) where human experience and acoustic characters are considered at the same time. In conclusion, perception of the sonic ambient requires, from the receiver, action dependent on genetic and cultural mechanisms.

The sonic environment is associated to aesthetics agreeing with human well-being in outdoor and indoor situations. To recover the sense of well-being is increasingly important for our stressed everyday life. Reflection and relaxation are two functions that are necessary for our psychological comfort and equilibrium and which can be achieved only in special places we recognize as tranquil, quiet, or in other terms, "beautiful." Such places are not common around us. In many cases such places have vanished with the spread of infrastructures and urbanization, and developing new tranquil spaces combining acoustic and visual components of the environment has become a priority.

The state of "tense arousal" was essential for our ancestors to escape predation, enabling them to cope constantly with environmental variables and every type of event. Of this heritage, modern people have retained only a few components. Of

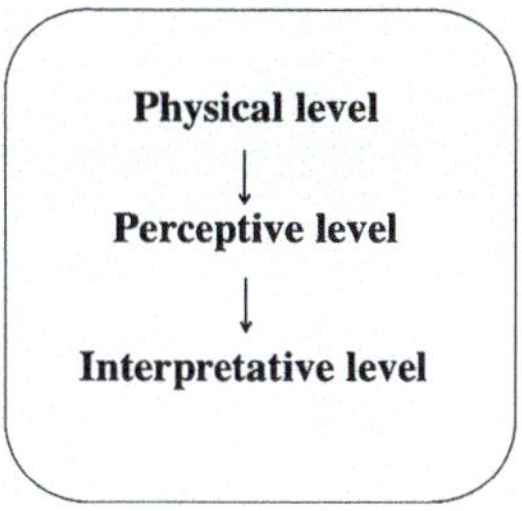

Fig. 5.1 Schematic representation of the three levels at which humans interact with the soundscape

these the sense of tranquility, that means one is safe from enemies, is probably just one of the major remnants.

The "prospect and refuge" theory of Appleton (1975) is an important step, although its reductionistic vision can limit the application to a more general problem. However, this theory well explains the attitude of the first hominids to select favorable places from which to scan the surroundings and to maintain a safe refuge from predators or hostile competitors.

To a soundscape evaluation concur not only the sonic characters of the environment but also the visual (aesthetic) components. For instance, a bimodal stimulus has been proved important for characterizing tranquil spaces, and in fact it has been observed (Watts et al. 2009) that "the medial prefrontal cortex receives significantly enhanced contribution from the auditory cortex under tranquil visual conditions when compared to non-tranquil visual conditions." When a soundscape is appreciated through a visual scene, this definitely modifies the auditory perception.

Truax (1994) has distinguished three levels of attention toward sonic ambience:

Listening research, which is an active conscious activity for the listener, tuned in to a specific sound. Examples include a birdwatcher walking in the fields or the soundwalking of a researcher in urban parks.

Listening in readiness when the listener's attention is directed elsewhere but is ready to receive meaningful information.

Background listening or distracted listening, when the listener is engaged in other activities and does not pay attention to a specific sound.

In reality, these three categories of attentiveness can be better explained in terms of active functions: in the first case an active function activates contemporarily a sonic eco-field, and the second case and the third case can be associated with the same situation based on a nonactive function that could be coupled with a sonic eco-field. For instance, reading a book is a function that does not require a sonic eco-field, but considering that acoustic signals enter in our ears independently of our will, if the signal is strong this creates an instinctive reaction and if the signal is weak it can be neglected by our central nervous system.

Alarm or alert sounds modify the perception scenario decreasing the rating also in the presence of highly valuable visual components.

Another important element to be considered when a sound is analyzed is represented by the physical structure of sound that can generate alert or alarm. In this case it seems reasonable to imagine that primitive instinctive components are activated. When an alerting or alarming acoustic signal is perceived, the mobility of eye movements associated with the exploration of the surroundings increases.

The internal status of a listener functions as a filter and activates the specific sonic eco-field.

Several factors influence the assessment of sound annoyance, and this largely depends on nonacoustical features, or the visibility of an unwanted noise source. For example, the position in the city makes further differences: living along a pretty street produces a reaction to noise less stressful than to live in a noisy periphery. It is necessary to approach the soundscape using a semantic distinction of the sound sources and of the context in which a sound is perceived and not only the noise by using level measurements. We have to introduce the concept of meaning by using a semiotic approach of interpretation, as recommended by Raimbault and Dubois (2005).

When the sound and in particular noise is considered, it seems important to distinguish between voluntary exposure to noise, as in a disco, or involuntary exposure to a noisy ambience, such as at the side of a busy road. The social consequences are different, and also the regulation tools are different. Probably, however, direct consequences on human health are more severe when hearing full-volume music than when walking in a urban park considered acoustically polluted.

5.2 Psychological Approach to Soundscape Interpretation

Soundscape evolves and develops according to the character of landscapes, according to the time and the technology, and is strongly influenced by cultural and political issues as well (Botteldooren et al. 2006). Consequently, to be efficient soundscape studies must rapidly develop toward a more integrated holistic approach in which the visual component is blended with psychological feelings such as the sense of safety, or with the air quality. In fact, vision and hearing are strictly connected in human perception of the surroundings. For instance, the presence of vegetation in a noisy ambience creates the conditions for a lower perceived noise, although the vegetation does not function as an efficient barrier, as argued by Anderson et al. (1984).

Plants along roads are regarded as true barrier to noise diffusion, and a study conducted by Watts et al. (1999) has demonstrated experimentally that noise rating was depressed in people to whom noise was supplied in concomitance with the vision of green barriers. In other words, the noise rating was higher when the source of noise was visually screened. This experiment demonstrates how the psychological effect of visual cues reduces the effect of acoustic exposition. When a metal barrier was submitted to noise rating versus a barrier of willow shrubs, no differences were found in noise reduction between the two type of barriers. Of course this study was not conducted to demonstrate the efficiency of green barriers, which can be provided by direct measurements, but to indicate how the visual source of noise affects the final results.

In natural areas far from a significant source of noise pollution where the A-weighted sound pressure level is <20 dB it is possible to hear a snowmobile at 3–5 km and an aircraft at 15–20 km. In this case, the assessment of the sonic environment is modified significantly in respect to a sonic environment in which the background noise is higher and able to mask key sounds.

The sonic quality of an environment has absolute values in terms of sound pressure level but subjective interpretation and effects according to the area, the people, and the circumstance in which people are engaged. Annoyance and quality of the sonic environment are strictly connected, and annoyance (see Sect. 5.4) becomes a relative parameter of noise exposure.

Sophisticated acoustical and psycho-acoustical measurable and quantifiable parameters are today available. Soundscape characters can be described by words or measured with numbers, but the best result is obtained when both methods are combined. A soundscape is not only words or numbers but rather their meaningful combination.

5.3 Noise Sensitivity

People have different reactions to different noise levels for many reasons, principally connected to personal experience, culture, and context. In 1978 Weinstein proposed a noise sensitive index (Weinstein 1978) (Table 5.1), an approach able to determine a score between people and noise annoyance. In fact, there is interindividual sensitivity of people at different noise levels.

Noise sensitivity is defined as an attitude to rating sound according to an individual-based criterion and represents a major antecedent of individual noise annoyance. Job (1999) defines noise sensitivity (NS) as "*Noise sensitivity refers to the internal states (be they physiological, psychological [including attitudinal], or related to life style or activities conducted) of any individual which increase their degree of reactivity to noise in general.*"

Noise sensitivity has indirect effects on health, creating a stressful psychological condition, increasing psychological reactivity of the cardiovascular system and covarying with the degree of psychopathology. Noise-sensitive people (NSP) have significantly greater increases in heart rate and systolic and diastolic blood pressure when compared with non-sensitive noise people (NNSP). Non-sensitive noise people seem more connected to psychiatric problems such as phobic disorders and depression. Neurotic people seem more sensitive to noise level when by the term neurotic we mean a person who is too nervous or unstable.

This difference depends not only on a different hearing capacity but by other causes. For instance, when exposed to 80 dB of an helicopter there are no significant differences between noise-sensitive people and non-noise sensitive people.

Noise sensitivity represents a personal trait that is stable over time.

Noise exposure (NE) is the first factor of noise annoyance but after this the major factor is represented by individual noise sensitivity (INS). According to Job (1988),

Table 5.1 The Weinstein noise sensitivity scale

1. I wouldn't mind living on a noisy street if the apartment I had was nice.	AGREE	1 2 3 4 5 6	DISAGREE
2. I am more aware of noise than I used to be.	AGREE	6 5 4 3 2 1	DISAGREE
3. No one should mind much if someone turns up his stereo full blast once in a while.	AGREE	1 2 3 4 5 6	DISAGREE
4. At movies, whispering and crinkling candy wrappers disturb me.	AGREE	6 5 4 3 2 1	DISAGREE
5. I am easily awakened by noise.	AGREE	6 5 4 3 2 1	DISAGREE
6. If it's noisy where I'm studying, I try to close the door or window or move someplace else.	AGREE	6 5 4 3 2 1	DISAGREE
7. I get annoyed when my neighbors are noisy.	AGREE	6 5 4 3 2 1	DISAGREE
8. I get used to most noises without much difficulty.	AGREE	1 2 3 4 5 6	DISAGREE
9. How much would it matter to you if an apartment you were interested in renting was located across from a fire station.	A LOT	6 5 4 3 2 1	NOT MUCH
10. Sometimes noises get on my nerves and get me irritated.	AGREE	6 5 4 3 2 1	DISAGREE
11. Even music I normally like will bother me if I'm trying to concentrate.	AGREE	6 5 4 3 2 1	DISAGREE
12. It wouldn't bother me to hear the sounds of everyday living from my neighbors (footsteps, running water, etc).	AGREE	1 2 3 4 5 6	DISAGREE
13. When I want to be alone, it disturbs me to hear outside noises.	AGREE	6 5 4 3 2 1	DISAGREE
14. I'm good at concentrating no matter what is going on around me.	AGREE	1 2 3 4 5 6	DISAGREE
15. In a library, I don't mind of people carry on a conversation if they do it quietly.	AGREE	1 2 3 4 5 6	DISAGREE
16. There are often times when I want complete silence.	AGREE	6 5 4 3 2 1	DISAGREE
17. Motorcycles ought to be required to have bigger mufflers.	AGREE	6 5 4 3 2 1	DISAGREE
18. I find it hard to relax in a place that's noisy.	AGREE	6 5 4 3 2 1	DISAGREE
19. I get mad at people who make noise that keeps me from falling asleep or getting work done.	AGREE	6 5 4 3 2 1	DISAGREE
20. I wouldn't mind living in an apartment with thin walls.	AGREE	1 2 3 4 5 6	DISAGREE
21. I am sensitive to noise.	AGREE	6 5 4 3 2 1	DISAGREE
	TOTAL SCORE________		

Source: Reproduced with permission from Weinstein (1978)

the INS explains 10.2 % of variation in noise annoyance (NA) whereas noise exposure (NE) explains 17.6 %.

Perception and evaluation are two separate processes and NS is part of the first process. A further hypothesis argues that NS can be the result of greater hearing acuity (Job 1999), but few studies have demonstrated that NS is related to some physiologically specific performance.

The difficulties in applying psychometric measures directly to NS create a problem connecting such factors to physical attributes of sounds. Ellermeier et al. (2001) have investigated a new approach to "measure" noise sensitivity, and they concluded that people with a higher NS have no predisposition to perceive a sound event more intensely nor the capacity to better discriminate, but attitudinal/evaluative components should be more active than sensory components. From this study the hypothesis of a higher auditory acuity is refused.

The experience of a noisy neighbor can create the condition for a NS. The effects disappear when these people are relocated in a quiet environment but their characteristic of NS remain.

Noise sensitivity can be discussed according to two points. For noise-sensitive people noise becomes an important component of their well-being, because these people consider noise as threatening element and more out of their control than do NNS people. NS people react to noise in a negative way and they can adapt to noises more slowly. Their annoyance threshold to noise is lower despite the level of noise. These two elements are connected to psychiatric disorder in terms of potential noise threat and in terms of annoyance susceptibility.

An investigation carried out on 3,445 persons exposed to noise in Amsterdam by Meijer et al. (1985) demonstrated that the people with higher sensitivity to noise belonged to a higher socioeconomic status.

Another psychological reaction to a new stimulus is called orienting response (OR) or orienting reflex. This reflex is reduced after a stimulus is repeated several times, and this reaction is called habituation. NS have a harder time with habituation than NNS subjects. Women subjected to repeated sounds of 50, 75, and 100 dB (A-weighted) had a slower habituation threshold when the sounds were repeated.

In NS the distractibility is higher than in NNS. When moderately loud sound are submitted to NS people the annoyance and unpleasantness are considered greater. NS People exposed to road traffic noise declared a higher annoyance than NNS people at a low level but converged when the intensity was increased.

The NS becomes a physiological status that may have important consequences on the choice for such individuals of a place in which to live (see Luz 2005). Considering that 22 % of a urban population may be affected by the NS condition, diffuse noise in most urban metropolitan areas represents a relevant problems for many people.

Noise sensitivity is not related to noise exposure: NS alters the effect of noise exposure on noise annoyance and has not simply an additive effect. In an experiment Miedema and Vos (2003) investigated the effect of aircraft traffic noise exposure. Noise sensitivity was found to change the influence of noise exposure on noise annoyance. Subjects with NS have a predisposition to discriminate environmental conditions and to evaluate them, but this attitude is weak in NNS.

Age seems a discriminant for noise annoyance. An investigation conducted by Van Gerven et al. (2009) in a Dutch sample and in a specific data set (Amsterdam Airport Schiphol) of total 62,893 individuals aged between 15 and 102 years demonstrated that noise sensitivity had a curvilinear distribution with a maximum around 45 years, and a minimum was found between youngest and oldest age segment (Fig. 5.2). This patterns is independent from noise exposure and self-reported noise sensitivity. The large data set and its internationality assures the significance of the results and the intercultural effects. It seems a surprise that the people who should be more resilient to noise are at the same time the most vulnerable.

5.4 Noise Annoyance: Definition and Evaluation Procedures

Annoyance is defined as a feeling of displeasure evoked by noise (Berglund et al. 2002). This parameter is used not only to evaluate the effect of noise on sleeping function but also for other adverse effects of noise. Many researches have confirmed that annoyance is correlated with the (perceived) loudness and the sound pressure level.

For instance, personal and situational variables were considered by Fields (1993) in the evaluation of noise annoyance in residential areas. According to this author noise annoyance is not caused by ambient noise levels or by the amount of time residents are at home, the type of interviewing method, or the demographic variables (age, sex, social status, income, education, home ownership, type of

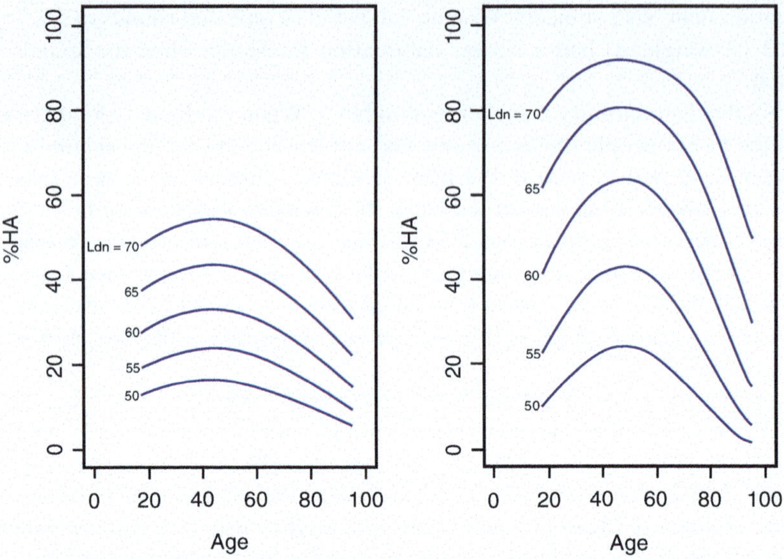

Fig. 5.2 Predicted percentage of highly annoyed persons (%HA) in relationship to noise level [Ldn in dB(A)] of international pooled data set (*left image*) and the Amsterdam Airport Schiphol data set (*right image*). *Ldn* average equivalent noise level during daytime (*d*) and nighttime (*n*) (Reproduced with permission from Van Gerven et al. 2009)

dwelling, length of residence, receipt of benefits from the noise source). The annoyance is related to amount of isolation from sound at home and to five attitudes (fear of danger from noise source, noise prevention beliefs, general noise sensitivity, beliefs about the importance of the noise source, and annoyance with non-noise impact of the noise source).

It is quite clear that annoyance assessment based on decibel measurement is not sufficient to evaluate the quality of the sonic urban ambience. The soundscape concept seems a more integrated and efficient way to assess such quality.

Annoyance is a quite contentious argument because a noise for a people can be a kind of lifestyle for others. The perception of a sound is a complex process in which sound source is identified, the direction and distance evaluated, the recall to a memorized similar or dissimilar sound, and contemporarily the situation during which that sound was heard at that time. Sound memory produces the evocation of feelings and visual scenes. In the analysis of sonic annoyance it is necessary to separate analytically the structure of the sound from personal subjective characterization (e.g., loudness, unpleasantness) as suggested by Schulte-Fortkamp and Nitsch (1999). Finally, different groups of people (residents, visitors, workers, etc.) may have a different judgment of a specific sound.

Annoyance is the result of long exposure to unwanted sounds and can be rated using questionnaires graded as “none/a little/rather much/very much.”

The level of annoyance expressed by a subject reflects their personality and noise sensitivity. Defeat reaction, increase of blood pressure, and disturbance of hormonal secretions are effects of massive exposure to noise. In populations living near an airport, serious effects can be observed such as apathy, decreased secretion of hormones, depression of the immune system, and also decreased social contact. The increase of blood pressure is an important effect that represents some uncertainty because other factors such as diet, smoking, and heredity have a great influence.

The effects of noise exposure on the urban society are of growing concern, but the major problem to control and possibly to reduce such exposure largely depends on methods that can measure in efficient and realistic ways the noise dose important for our physiological and social system.

The effect of annoyance is only in part caused by noise exposure, but other factors such as landscape appreciation may play an important role, as already described. A positive evaluation of the landscape can improve or can reduce the effects of annoyance. Three components are important in soundscape evaluations: the context, the focus of attention, and personal knowledge/experience; this means that both the environscape and psychscape are important. Finally, annoyance is reduced or is perceived as more important according to the level of appreciation of the surroundings.

5.5 A Soundscape Semantic

Soundscape categorization has been obtained using objective acoustical parameters such as sound pressure level, roughness and fluctuation strength, sharpness, and binaural parameters (Rychtarikova and Vermeir 2013).

When the noise is considered simply as a quantity of energy (and measured as sound pressure level), we exclude the psychological and social factors. But a more qualitative approach to describe how noise is interpreted by the human mind in its semantic components seems more appropriate. In fact, the semantic methods allow connecting people's feelings with a linguistic and psychophysical approach and allows us to consider the soundscape an emotional perceptual space. As a direct consequence, the difference between noise and sound becomes of an emotional type.

A complex multidisciplinary approach to a semantic of soundscape, recently proposed by Davies et al. (2013), recognized at least three semantic levels: sound sources, sound descriptors, and soundscape descriptors.

Source sources refer to physical entities that can be originated by multiple agents (e.g., traffic) or components (e.g., brakes). The sound descriptors refer to the description of sounds either as nouns (roar, crash) or as adjectives (bipping, flapping) and phrases (different language).

Finally, soundscape descriptors refer to the emergent sound from indistinct agents (cacophony, hubbub, constant, temporal) (Fig. 5.3).

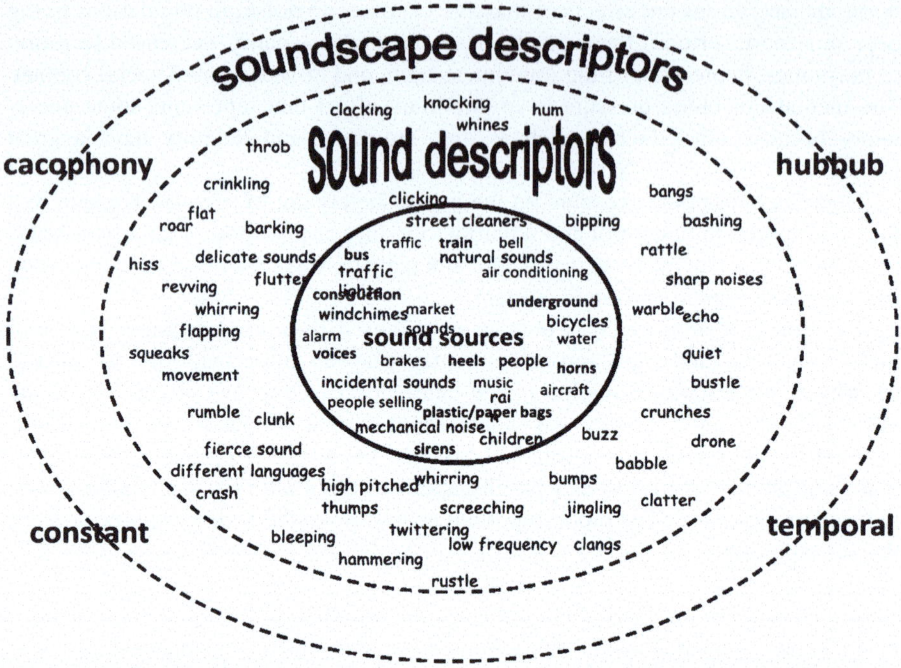

Fig. 5.3 The different components of a soundscape as the result of individual reaction to the sonic environment after a soundwalking (Reproduced with permission from Davies et al. 2013)

Cacophony is a term used to describe a soundscape that is perceived in a negative way and therefore is associated with a negative listening experience.

Hubbub is a term used to describe a mix of sounds that are perceived positively and are associated with a positive listening experience.

Constant refers to a constant sound, as is typical of a stream water flow noise that is constant and masks every other sound. In some cases this category can produce a negative feeling (for instance, when a birdwatcher cannot heard the song of birds close to a stream border) but could be positive for a backcountry visitor.

Temporal refer to the sounds that are repeated at intervals from a few seconds to minutes, such as the sea waves.

These four categories can be sorted two by two: cacophonies–hubbub, and constant–temporal.

The first two describe the way sounds are blended, which can produce pleasantness or unpleasantness. The second two describe the rhythm and the level of monotony of the soundscape and contain few elements of judgment by people.

For instance, the semantic differential technique is proposed to address complex sounds. This technique rates a number of different soundscapes and sounds (Kerrick et al. 1969). According to this technique, Cain et al. (2013) have isolated two principal components of sound evaluation: calmness and vibrancy, and with the combination of these two axes other attributes of the soundscape can be distributed.

Payne (2013) has presented a perceived restorativeness soundscape scale (PRSS) designed to assess perceptions of the potential of a soundscape to offer psychological restoration.

5.6 Elements of Psychoacoustics

Psychoacoustics is a field of research that focuses on the use of the psychological measurements of sound, meaning what people perceive from a sound. The psychoacoustic studies consider either the human reaction to noise or the physic character of noise. This approach presents more advantages than the psychological methods because measures are collected that are independent by visitor self-reports of noise exposure.

This sound quality evaluation finds many applications in everyday life, because our life (mostly spent in urban contexts) is immersed in a growing plethora of artificial and technological sounds.

The perceptual quality of sounds are considered auditory attributes: psychoacoustical research aims to find the quantitative relationships between the physical attributes of sounds and their perceived properties.

The physical attributes are evaluated by analyzing the sound signals, and the psychoacoustical research focuses on auditory attributes such as loudness, pitch, duration, and sharpness.

Measuring auditory attributes means "providing quantities representative of what a user perceives" (Susini et al. 2012). Among the different approaches, the judgment on a multiple semantic scale represents a fruitful scale that uses a single semantic descriptor (unipolar scale) or a pair of antonymic descriptors (bipolar scale).

Psychoacoustic techniques use the semantic differential (SD), which is represented by a scale of judgment labeled with two opposite descriptors (good–bad, poor–rich); such labels are called semantic descriptors. A multidimensional scaling technique (MDS) is an innovative approach (Susini et al. 2012) based on dissimilarity ratings.

The use of different semantic scales concurs to create a semantic profile. In sound evaluation the use of different perceptional dimensions allows exploring in detail the relationship between sound attributes and their effects in different aspects of human life, and this can range from perceptual dimension to cognitive and emotional aspects.

Psychoacoustical models provide tools for measuring the sensations that sounds produce in the listeners. Loudness, sharpness, roughness, and fluctuation strength are four models used to estimate the quality of sound (Ferguson et al. 2006), although the complexity of a soundscape reduces the potentiality of the psychoacoustic approach as illustrated by Jennings and Cain (2013) in Fig. 5.4.

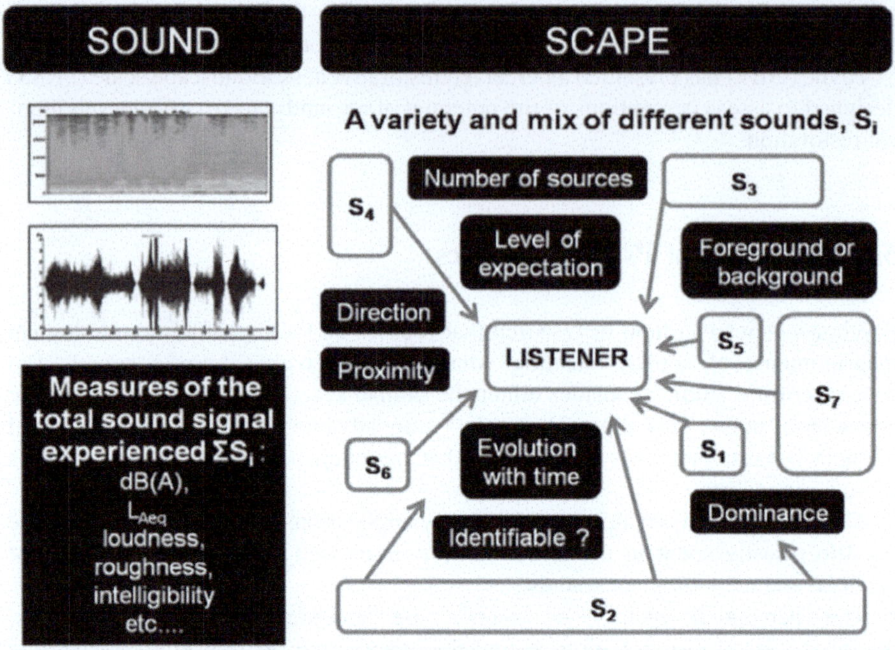

Fig. 5.4 Framework representing the complexity of the soundscape where the acoustic and psychoacoustic approach is constrained by a variety of sonic sources, their direction, proximity, evolution in time, level of dominance, and position in foreground or background from the listener (Reproduced with permission from Jennings and Cain 2013)

Loudness is the most fundamental of auditory sensations.

The unit of loudness is the *sone*: the ratio scale referenced against the sensation produced by 1 kHz sine tone with a sound pressure level of 40 dB. This parameter is not considered as a robust auditory scale or data representation. Loudness has an inherent dependency on frequency content.

Loudness is affected by the dominant frequency of the signal, the signal bandwidth, and the signal duration.

Sharpness (or brightness) is one of the most important features of timbre. This model describes how a sound is "sharp," "harsh," or has "an edge."

The unit of sharpness is the *acum*, referenced to a band of noise centered on 1 kHz at a level of 60 dB.

Roughness is the sensation caused by a quite rapid amplitude modulation within auditory filters.

The unit of roughness is the *asper*, which is referenced to a 1-kHz tone at 60 dB with 100 % amplitude modulation at 70 Hz.

Fluctuation strength describes the sensation caused by relatively slow amplitude modulation within auditory filters with a maximum sensitivity at 4 Hz.

The unit of fluctuation strength is the *vacil*, referenced to 60 dB 1-kHz pure tone 100 % amplitude modulated at 4 Hz.

Pitch is defined as that auditory attribute of sound according to which sounds can be ordered on a scale from low to high. To date this definition still is a useful basis, although it must be complemented by considering certain additional aspects.

For instance, in a first experiment (Gygi et al. 2007) 145 sounds were rated in 20 semantic dimensions, and the rating suggests that 90 % of variance was associated with four factors: harshness, size, complexity, and appeal. In a second experiment the sounds were submitted to a low- and high-pass filter to understand what spectral region was selected in the sound identification of a set of 70 environmental sounds (Table 5.2) using limited spectral regions(Gygi et al. 2004). Subjects were able to identify sounds also when they were severely filtered (70 % or more of correct identification). The spectral region requested for identification ranges from 1,200 to 2,400 Hz, similar to the spectrum related to speech recognition. This type of approach investigates the minimum code for sonic identification.

A psychological space or a cognitive template probably is when an acoustic stimulus is captured by our hearing system and processed in our brain. The minimum structure sound probably increases the speed and the efficiency of the identification. A three-dimensional multidimensional scaling (MDS), a set of related statistical techniques often used in information ordination, was adopted to identify similarity in sounds and to reduce the high-dimensional acoustic space to few perceptual dimensions.

People perceived sounds by attributing categories according to the acoustic features of those sounds, including harmonicity, spectral spread, continuity, periodicity, and envelope modulation. An MDS approach can create similarity between the different types of sources, for instance, vocalizations, impacts, and water sounds.

Birds and waterfalls are considered pleasant sounds but technological sounds are associated with negative feelings of a site. Children at play and talking people are considered positive. To discover the relationship between these psychoacoustic processes and the physical composition of a soundscape remains an important research issue. Axelsson et al. (2010) have utilized 50 binaural soundscape excerpts of 30 s in duration selected from a large set of urban recordings in London and Stockholm among a great variability of soundscapes and sound events with a broad range of sound pressure level (43–79 dB $L_{\mathrm{Aeq,\ 30s}}$) (Fig. 5.5). These 50 excerpts were from ten kinds of location: urban courtyard, motorways, pedestrian street, school yards, suburban parks, suburban recreational areas, suburban residential areas, urban parks, urban square market, urban street.

The technological sounds were from airplanes, individual cars, motorcycles, road traffic, car alarms, car horns, chainsaw, rock drill, street sweeper, construction work, train ventilation fans, and sirens of emergency vehicles.

Table 5.2 Seventy sounds used in experiment by Gygi et al. (2004)

Sound	Code	Duration (ms)	Sound	Code	Duration (ms)
Airplane flying	Air	3,426	Harp being strummed	HAR	2,839
Baby crying	BAB	3,119	Helicopter flying	HCP	3,100
Basketball bouncing	BBA	3,184	Horse neighing	HEN	1,464
Beer can opening	BER	431	Horse running	HRR	2,046
Bells chiming	BLS	3,100	Ice dropping into glass	ICE	738
Billiard balls colliding	BIL	1,634	Laughing	LAF	1,836
Bird calling	BRD	1,283	Match being lit	MAT	1,366
Bowling	BOW	3,062	Paper being crumpled	PAP	1,697
Bubbling	BLB	2,212	Phone ringing	PHN	3,100
Buzzer sounding	BZZ	2,523	Ping-pong ball bouncing	PNG	3,945
Camera shutter clicking	CAM	1,398	Printing paper	PRN	3,154
Car accelerating	CRA	3,209	Projector running	PRJ	2,909
Car starting	CRS	3,551	Rain	RAI	3,362
Cars crashing	CRC	3,110	Rocking chair	RCK	3,234
Cars honking	CRH	3,367	Rooster crowing	ROO	2,004
Cash register closing	REG	2,841	Scissors cutting paper	SCI	2,807
Cat meowing	CAT	1,054	Screen door closing	SCR	3,292
Chimp calling	CHP	2,315	Sheep baaing	SHP	1,365
Chopping wood	AXE	2,624	Shoveling	SHV	1,585
Clapping	CLP	1,035	Siren blaring	SIR	1,534
Clock ticking	CLK	2,882	Sneezing	SNZ	844
Coughing	COF	1,043	Splash	SPL	2,338
Cow mooing	COW	1,290	Stapling	STA	822
Cymbals being hit	CYM	1,204	Tennis ball being hit	TEN	2,761
Dog barking	DOG	1,007	Thunder rolling	THU	2,895
Door opening and closing	DOR	1,802	Toilet flushing	TOI	2,494
Drumming	DRM	2,747	Train moving	TRN	3,333
Electric guitar (strum)	GTR	2,087	Tree falling	TRE	3,424
Electric saw cutting	ESW	2,600	Typing on keyboard	TYK	1,033
Flute playing	FLU	2,566	Typing on typewriter	TYT	2,755
Footsteps	FST	2,700	Water pouring	POU	3,302
Gargling	GRG	2,523	Waves crashing	WAV	2,876
Glass breaking	GLS	1,018	Whistle blowing	WHI	2,047
Gunshot	GUN	902	Windshield wipers	WWP	3,023
Hammering a nail	HAM	1,833	Zipper	ZIP	1,733

Source: Reproduced with permission from Gygi et al. (2004)

The natural sounds included sounds from birds, wind whispering, rustling leaves, moving water (rain), fountain jets, waterfalls, and a purling water spring. Sounds from human activity included children at play, footsteps, and human voices.

The binaural soundscape excerpts were listened to by volunteers in an anechoic chamber, and their assessments were collected for rating. The results of this

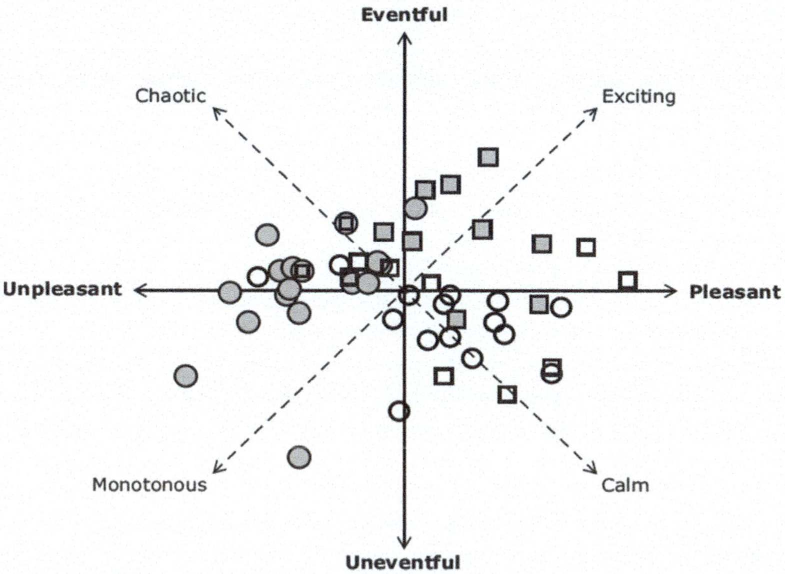

Fig. 5.5 Ordination of 50 soundscape excerpts of binaural recordings of urban outdoor soundscapes on 116 attribute scales according to pleasantness and eventfulness. *Filled squares* human sounds, *filled circles* technological sounds, *open squares* natural sounds, *open circles* no dominant sound-category (Reproduced with permission from Axelsson et al. 2010)

complex experiments confirm that soundscape perception can be described in terms of three basic components: pleasantness, eventfulness, and familiarity.

Pleasantness explains the 50 % of variance in soundscape measures creating an ordination between pleasant and unpleasant on a continuum. Eventfulness explains 16 % of the variance, ordering the soundscape in an eventful–uneventful continuum. Familiarity expresses 8 % of the variance according to a familiar–unfamiliar continuum. This indicator is considered less significant in the sonic assessment and was perceived as similar among the different categories of respondents. However, this factor could be important in cases of intercultural comparison and could be more efficiently applied in touristic cities.

Other possible classifications of the soundscape regard the emergent attributes of exciting and calming. Exciting mixes together high arousal with pleasure, and calming mixes low arousal with pleasure. Technological sounds were found affecting the pleasantness negatively after the control of the overall intensity. Analysis of the natural sounds does not reflect a direct relationship with pleasantness. For instance, a far background noise that has a low-frequency spectrum is difficult to be masked by natural sounds that are generally, as for bird songs, acting at higher frequencies. Human sounds are considered more eventful than natural sounds and this fact should be considered for future soundscape planning. Human sounds such as from playgrounds or a cafeteria increase the eventfulness of a pleasant sonic environment.

In conclusion, an extended set of soundscape attributes (116) were well integrated in the three components: pleasantness, eventfulness, and familiarity. Pleasantness and eventfulness aggregate the soundscape attribute in a circumplex pattern. An exciting soundscape is the combination of pleasant and eventful, a calm soundscape is the result of the combination of a pleasant and uneventful condition. A calm soundscape would be pleasant and uneventful, but a chaotic soundscape would be unpleasant and eventful, and a monotonous soundscape would be the result of unpleasant and uneventful.

Soundscape excerpts dominated by technological sounds were considered unpleasant, those dominated by human sound were considered eventful, and these relationships survived after the control of loudness.

Definitively, the outdoor urban soundscape can be represented by two dimensions: pleasantness and eventfulness. This model confirms the simplicity of soundscape perception, the few basic dimensions related to the informational properties of the categories of sounds. The high number of excerpts and their variety should assure a good set of sonic combinations and, conversely, stability, representativeness, and generalizability to the proposed model.

In conclusion these authors propose a model in which a sonic environment is divided into two-dimensional spaces defined by the following attributes: pleasant, exciting, eventful, chaotic, unpleasant, monotonous, uneventful, and calm. These spaces represent eight vectors separated by 45° of a "circumplex model of soundscape perception."

5.7 Noise Pollution: A Growing Problem for Human Societies

Noise is related to human activities such as road traffic, air traffic, industry, civil engineering, recreational activities, and outdoor equipment. The exposure of people to noise effects is dramatically high in the world. It is estimated that at least 500 million inhabitants are exposed to hazardous levels of noise.

More than half the European population is living in an uncomfortable sonic environment: 30 % of the people are exposed during the night to a sound pressure level >55 dB(A) that creates sleep disturbances. In developing urban areas, the situation is worst with people exposed during 24 h to a noise dose of 75–80 dB(A). In total, 80 million Europeans are affected by an unacceptable dose of noise, 30 million are exposed to noise in the work environment, and 170 million are living in "grey areas" where the noise produces severe annoyance during the day, and 40 % of the European population is exposed to road traffic noise exceeding 55 dB(A) in the daytime and 20 % is exposed to levels >65 dB(A). For instance, in Germany it is reported that road traffic and air traffic are the two major source of annoyance, but 6.5 % of German citizens are highly annoyed by the noise produced by their neighbors. In Sweden more than 2 million people (25 % of the population) are exposed to a traffic noise level $L_{Aeq,\ 24\ h}$ >55 dB and $L_{Amax} = 70$ dB; this trend is growing with time (Skanberg and Ohrstrom 2002; Ohrstrom et al. (2006).

Noise is considered in urban areas a real problem that is connected also to topography, meteorology, land use patterns, and the lifestyle of the residents.

In residential areas, the soundscape shows variation in time and space connected with the rhythms of the people. Buildings and artificial barriers create shields for sound propagation that modify the sonic environment.

People locate the quiet areas that have a higher commercial value and concentrate their leisure time in open greenery spaces in which the noise level is lower: 30 % of variance associated with annoyance is linked to noise exposure. The weak relationship between dose exposure and annoyance largely depends on the fact that people move from one place to another and that noise exposure changes in space and time.

5.8 Consequences of Noise for Human Physiology and Health

In humans the hearing mechanism is complex and has implications for different organs (middle ear, acoustic nerve, cortex, and deep brain). Sound produces reflexes in the muscular system and in the endocrine system also. Sound also elicits also emotional reactions. Thus, the complexity of the acoustic performance depends on several factors such as previous experience of the sounds, the simultaneous presence of other environmental stimuli, the health status of the subject, etc. (Rylander 2006).

The physiological reaction to different levels of sonic saturation reflects the mechanism of the *hormesis*. Hormesis is defined as a dose response wherein a small amount of the agent is beneficial and a large dose may have negative effects, and this hormesis principle (Fig. 5.6) can be applied to the sonic ambience. People relegated to an environment without acoustic stimuli, as in an anechoic chamber, feel isolation and insecurity. People require a background of sounds to maintain their physiological functions.

A sound is transformed into a noise after a subjective interpretation; definitively, noise does not exist, but in some cases the mixing of different sounds creates a white noise effect that does not allow distinguishing the component sounds. For instance, a bird chorus at the maximum level during springtime prevents distinguishing individual species and the sensation is of a diffuse "pleasant" noise. Some types of music are considered unpleasant by some people and classified as noise; for truck drivers, the sound of a truck engine can represent a familiar and attractive sound while for other people this is just an unpleasant noise.

When exposed to a loud sound the subject tries instinctively to defend some parts of the body with an involuntary contraction of the muscles around the eyes, in the abdomen, and in the legs and the arms, preparing to escape subjecting the body to

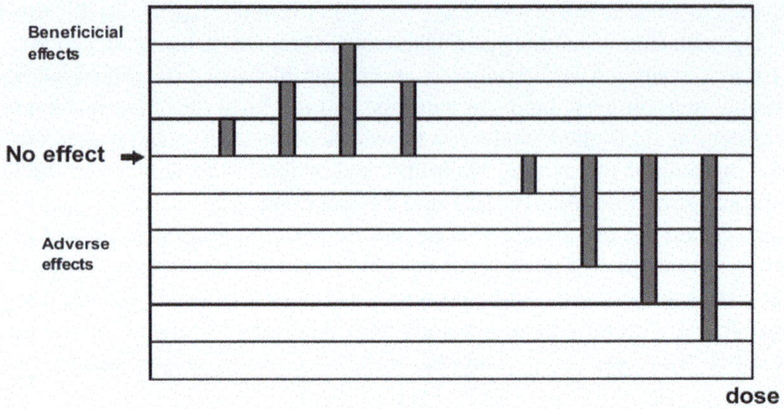

Fig. 5.6 Hormesis dose–response relationship (Reproduced with permission from Rylander 2006)

this stress. This reaction is particularly evident when the stimulus is sudden and unexpected. In this way, a gunshot also except for people such as policemen who are used to this sound, causes a contractive reaction. This reaction is created to escape, to defend oneself from possible danger, because a strong sound is the result of great energy that is released (e.g., strong sounds are produced by landslides, volcano eruptions, earthquakes, tempests, etc.).

The loss of concentration in reading or thinking is a common reaction after a sudden sound.

To measure exposure to traffic, usually the noise dose is averaged over 24 h, and this contrasts with the principle that the highest noise level and the noisiest event are the proxies of human pathologies. It seems apparent that the annoyance increases with the increase of noise peaks. When traffic barriers are erected we do not always observe improvement in the annoyance reaction because the background noise is significantly reduced, but not the noise peaks, which persist although sonic barriers are present. For example, when the traffic noise is continuous this has less effect on the concentration of schoolchildren than sudden noises such as those produced by aircraft. Peaks are more difficult to shield than continuous sounds.

Noise pollution is increasing as urban growth spreads. The negative effect of noise on health produces hearing impairment, reduction of speech communication, cardiovascular risk, sleep disturbance, psychological and performance reduction effects, and a subjective feeling of annoyance.

Cardiovascular risk and hearing loss are effects caused by high noise exposure in the occupational field. Sleep disturbance provoked by high noise exposure have many effects on human health. Subjective sensitivity, sex, age, health status, and socioeconomic status are related conditions.

Subjective illness and chronic diseases are factors that make the difference in noise effects; in fact, sound intolerance can occur in subjects who are under working or health stress.

The auditory system, differing from the visual system, is permanently active also during sleeping times, and even a low-level sound during the night can cause the subject to awaken, creating, if this situation is repeated, fatigue and depression.

The effect of noise on sleep can be tested in the laboratory or in the field. The two approaches have positive and negative elements, and integration between the two approaches is required because large discrepancies have been found. People exposed to heavy noise as in the proximity of an airport respond with sleep disturbance, but when the noise dose is lower there are not evident differences from the effects of road traffic noise.

If it is possible to insulate houses and sleeping rooms, reducing the effect of noise on dwellers, the consequences of outdoor noise on sleep rhythms remain obscure. People exposed during the day to heavy noise have difficulty sleeping. Habituation to traffic noise reduces insomnia but is still connected to interindividual variation in sensitivity to outdoor noise.

Prolongation of the awake state and shallow sleep reduces the useful time for deep sleep and rapid eye movements (REM). Residents exposed to noise have poorer sleep and at the morning show higher irritability. REM status decreases in length according to increase of noise exposure.

Insomnia increases in inhabitants living close to busy highways, as has been found in populations exposed to a traffic noise above 65 L_{eq} dBA when compared with inhabitants living in quieter areas. Physiological, psychological, and social characteristics are further elements to be considered. Sleep disturbances become evident when the subjects complain of difficulties falling asleep (delay) or of awakening from sleep in the middle of the night or early morning.

The quality of noise makes the difference in evaluating the effect on people, so intermittent noise has more adverse effects than a continuous noise level. Noise peaks have more effects than the averaged weighted L_{eq}. Metabolic syndromes, sleeping apnea, job stress, and shift work are all components that negatively affect sleep performance, so noise exposure alone is not a sufficient factor to characterize sleep disturbance (Kawada 2011).

Health effects are expected in long-term periods that can create ischemic heart disease and increased blood pressure. Epidemiologic studies on the relationship between transportation noise and ischemic heart disease have demonstrated a higher risk of myocardial infarction in subjects exposed to high levels of traffic noise (Babisch et al. 2005). An investigation carried out on 1,881 subjects aged 20–69 years affected by myocardial infarction (MI), compared with 2,234 controls, has found the adjusted odds ratio for men exposed to sound levels of more than 70 dB(A) (adopting the acoustic map of the Berlin city). These results confirm the hypothesis that people exposed to high levels of traffic noise are more at risk for cardiovascular diseases than people living in quiet areas.

In conclusion, several studies have focused on the effect of noise on human health, especially in the outdoor and indoor urban environment. Urban sounds are not only noises but a complex communication system that affects in different ways the human social network, well-being, and ultimately human health.

Acoustics, physics, psychology, medicine, ecology, bioacoustics, and sociology are all disciplines that have common roots in the soundscape. The future challenge is to create a procedure that incorporate the different aspects of soundscape, human societies, and the environment.

5.9 The Characters of the Urban Soundscape

The soundscape in an urban space is really a complex system that has relationships with physical, psychological, and social factors. The urban soundscape represents a complex system in which environmental factors are shared with psychological, social, and cultural attributes of the living societies. Physical noise has been for long time an important component of the quality of the urban spaces, but a reduction of the noise does not automatically create better comfort, as recently argued by Kang and Zhang (2010).

Studying the urban soundscape means to investigate at least in two distinct directions: (1) assuming an holist hearing without semantics, where the background noise is evaluated in terms of annoyance without distinguishing the composing events; and (2) distinguishing the different sources of sound/noise and their rating according to a detailed categorization after a recognition of meaningful sounds.

Using the semantic differential analysis first introduced by Osgood et al. (1957) to identify the emotional meaning of the words, Kang and Zhang (2010) have evaluated the different sounds in open urban public spaces in which an important role is assumed by acoustic comfort and noise (Table 5.3). In the design used to evaluate the soundscape quality (students) respondents have demonstrated a clear preferences for natural sounds and green spaces, confirming that the sonic ambient is of primary importance between the different environmental conditions.

Sources of urban noise seems a controversial issue for the variety of urban sounds and for the generation of new sounds according to technological improvement.

Leobon proposed in 1995 a theoretical classification reported in Table 5.4 where sounds were classified as "unpleasant," "pleasant," and "unusual."

The noise in urban areas is produced not only by transport but also by all types of human activity and by nature too.

The total noise (L_{cn}) of an urbanized area is the result of superimposition of the acoustic background (L_{bn}) (all sounds produced by human activity, excluding transport noise) and the noise generated by transport (L_{tn}). For more details see Lebiedowska (2005).

Table 5.3 Categories of sounds used to classified soundscape walks in four urban open public spaces in Sheffield

	Very	Fairly	Little	Neutral	Little	Fairly	Very	
Agitating	3	2	1	0	−1	−2	−3	*Calming*
Comfort	3	2	1	0	−1	−2	−3	*Discomfort*
Directional	3	2	1	0	−1	−2	−3	*Everywhere*
Echoed	3	2	1	0	−1	−2	−3	*Deadly*
Far	3	2	1	0	−1	−2	−3	*Close*
Fast	3	2	1	0	−1	−2	−3	*Slow*
Gentle	3	2	1	0	−1	−2	−3	*Harsh*
Hard	3	2	1	0	−1	−2	−3	*Soft*
Interesting	3	2	1	0	−1	−2	−3	*Boring*
Like	3	2	1	0	−1	−2	−3	*Dislike*
Meaningful	3	2	1	0	−1	−2	−3	*Meaningless*
Natural	3	2	1	0	−1	−2	−3	*Artificial*
Pleasant	3	2	1	0	−1	−2	−3	*Unpleasant*
Quiet	3	2	1	0	−1	−2	−3	*Noisy*
Rough	3	2	1	0	−1	−2	−3	*Smooth*
Sharp	3	2	1	0	−1	−2	−3	*Flat*
Social	3	2	1	0	−1	−2	−3	*Unsocial*
Varied	3	2	1	0	−1	−2	−3	*Simple*
Beautiful	3	2	1	0	−1	−2	−3	*Ugly*
Bright	3	2	1	0	−1	−2	−3	*Dark*
Friendly	3	2	1	0	−1	−2	−3	*Unfriendly*
Happy	3	2	1	0	−1	−2	−3	*Sad*
High	3	2	1	0	−1	−2	−3	*Low*
Impure	3	2	1	0	−1	−2	−3	*Pure*
Light	3	2	1	0	−1	−2	−3	*Heavy*
Safe	3	2	1	0	−1	−2	−3	*Unsafe*
Steady	3	2	1	0	−1	−2	−3	*Unsteady*
Strong	3	2	1	0	−1	−2	−3	*Weak*

Source: Reproduced with permission from Kang and Zhang (2010)

5.10 The Sonic Environment of Urban Parks

Urban parks are areas in which people find physical and psychological advantages. In every city green spaces are distributed according a criterion to offer to the residents a "green lung" in which to spend free time, to walk, to meet other people for recreational purposes, and to benefit from a sonic ambience of high quality. The ecosystem services offered by a urban park are several and range from peacefulness to quietness and the contact with natural objects such as trees, flowers, and animals. These (therapeutic) values of urban parks are inherent to the planning and realization of parks in urban areas. But if the visual aspects of parks can be easily ascertained, the opposite happens for its sonic environment. Recent investigations addressed to 400 people, split 50:50 male/female, by Payne (2008) in two urban parks of Sheffield, UK have demonstrated that participants perceived the

Table 5.4 Leobon's theoretical classification of urban noises as unpleasant, pleasant, or unusual

	Perception		
	"Unpleasant"	"Pleasant"	"Unusual"
	High-pitched, piercing, strong and continuous or irregular and intermittent	Regular sound that evokes personal memories, entertainment	Unexpected, unknown, strange or foreign; has emotional valences
Background noise	Hum, urban noise	Combination of different sounds, simultaneous	Remote, not identified clearly
Mechanical equipment	Road, railway and aircraft noise. Equipment in large industrial plants and construction sites	Small industrial plants	Alarm sirens, abrupt braking
Human activity	Demonstrations, bells, sirens, trade, household noise (vacuum cleaner, drill)	Noise of small economic activity (workshops, markets), celebrations. State and religious holidays, entertainment	Doorbell, alarm, firecrackers
Noises of nature	Storms, thunder, wind, creaking	Birds, rain	Cracking, implosion, falling down
Human presence	Neighbors (telephone, barking)	Games, crunching, children	Screaming
Speech and communication	Effects of the human body	Voice, music	Discussion

Source: Reproduced with permission from Lebiedowska (2005)

soundscape in terms of amount of time they heard a certain type of sound and the volume at which they heard such sound.

In this research a questionnaire was created according to a list of seven types of perceived sounds: background city, happy people sound, sounds from surrounding buildings, natural sounds, object sound from people in the park, individual vehicles, or aircraft sounds, and sad and angry people. For every sound type was requested the percentage (0 %, 50 %, 100 %) of the time each sound was heard. Also for each sound type was requested the "average volume" they heard according a scale ranging from quiet (1) to loud (7). Sound predominance was obtained by multiplying the two scores.

The perceived restoration was evaluated by eight items articulated in seven semantic differential scales from disagree to agree.

Four items concerned the recovery:

(1) Renew your energy
(2) Regain the ability to concentrate
(3) Reduce any tension
(4) Become yourself again

Four items concerned the reflections:

(1) Ponder your daily experiences
(2) Think about your relationships with others
(3) Think about important issues
(4) See things in a new perspective

This study has confirmed that the urban park users' soundscape perception plays an important role in the restorative experience. The perception of the sound levels was found to be correlated with the measured sound level. Although the two parks studied were surrounded by traffic and by other technical sounds, the "natural and happy people sounds" were perceived predominant. This finding suggests that such sounds are distinctively perceived over the traffic noise or city background noise. This study demonstrates the great capacity for people to create a private perceptive and cognitive Umwelt that results strongly independent of objective information.

5.11 Pleasant Sounds, Sound Pressure Level, and Acoustic Comfort

The physical quality of a sound does not match the pleasantness, and such pleasantness, as it results from a listening function, is quite more complicated than the physical quality of a sound.

Listening is a psychological function by which people relate to the surrounding world, and this function depends on the noise sensitivity that is a subjective individual property and which varies according the personality of the subject. Definitively objective parameters not always are enough to evaluate the quality of the sonic ambience. In fact, people when interviewed about the level of sonic ambience generally are in accordance with the mean L_{eq}, especially when the sound level is below the level of 73 dBA, as argued by Yang and Kang (2005). These authors have found individual differences from a set of 9,200 interviews across 14 urban open public spaces in five European countries. From this investigation, people seemed more tolerant in terms of the sonic comfort evaluation where the sound source type affects greatly the acoustic comfort evaluation. In this case a pleasant sound can improve acoustic comfort also in high sound levels. To create a comfortable sonic ambience in an urban open public space it is necessary to reduce the background sound level.

Pleasant sound when introduced in a sonic ambience influences the relationship between acoustic comfort and sonic level that becomes weaker. Age differences separate teenagers from the older groups when interviewed. The teenager were more unsatisfied with the acoustic comfort.

5.12 The Value of Soundmarks in the Urban Landscape

As in the aesthetic field, there is primary landscape quality, in the same way, in the soundscape there are important soundmarks that make the difference.

Soundmarks are important elements in the design of a soundscape. Soundmarks are passive and active. Passive soundmarks are created by specific objects such as fountains or sonic sculptures; active soundmarks are the result of interesting activities.

For instance, water and twittering birds were the sounds most preferred by the subjects involved in a study by Yang and Kang (2005) when these elements are posed in a pool. Music from passengers cars is less appreciated and a source of annoyance. According to the geometric organization of a square and the presence of passive objects such as trees or facilities for the young such as skateboarding, the soundscape changes accordingly. Music is a further important component of the sonic ambience in a urban square independent of type. Loud music is always annoying and quiet music may receive more attention.

5.13 The Interactions Between Visual and Acoustic Stimuli

It is recognized that sound–image congruence reinforces environmental preferences and that incongruities interfere. The scenes that increase the acoustic rating are those with abundant water and vegetation, but an acoustic disturbance can dramatically reduce this appreciation. On the other hand, natural sound can improve the negative impact of a built-up environment.

It is well known that the visual landscape associated with sonic ambience produces an informative space in which characteristics change according to space and time. In a human-dominated landscape especially the combination of visual and acoustic cues creates an informative space that can be useful to manage and to reduce anthropogenic noise, can improve the quality of the soundscape, may better explain the relationship between sound and human behavior, can improve the efficacy of planning and designing rural and urban areas and natural parks, and can offer new audiovisual tools for public awareness and environmental education (Matsinos et al. 2008).

The interactions between visual and acoustic stimuli have received much attention by researchers.

To evaluate the role of aesthetics in the assessment of the quality of a landscape, Hetherington et al. (1993) performed an experiment by tests consisting of the representation of a combination between video/sound, video/no sound, or static/no sound, demonstrating the importance of motion when associated with sounds in rating the quality of a proposed landscape, especially in the presence of significant dominant elements.

Table 5.5 Presentation of the six categories of sounds that are combined with six categories of images proposed for a questionnaire

Sounds	Images	Combinations
V–Village	v–village	Vv Vw Vc Ve Vr
S–Stream	s–stream	Ww Wc Ws We Wr Wv
B–Busy park	b–busy park	Bc Bs Be Br Bv Bw
T–Thunderstorm	t–thunderstorm	Ts Te Tr Tv Tw Tc
Q–Quiet park	q–quiet park	Qe Qr Qv Qw Qc Qs
R–Residential neighborhood	r–residential neighborhood	Rr Rv Rw Rc Rs Re

Source: Reproduced with permission from Carles et al. (1999)

In particular, Carles et al. (1999) investigated the relationships between visual and acoustic stimuli for landscape perception. In Table 5.5 is summarized the sequence of 36 sound–image combinations proposed in a questionnaire in which 75 interviewers (43 women, 32 men) were requested to rate, from 1 = very unpleasant to 5 = very pleasant, the sound stimuli presented first separately, then in combination with visual stimuli. The main result of this investigation can be summarized as follows.

Natural sounds are considered positive and contribute to the appreciation of natural and artificial settings. Some natural sounds such as water contribute to improving the appreciation of natural and manmade spaces. It is well known that water and bird songs are sounds that produce relaxation in humans more than do human sounds, as has been proved physiologically by measuring heart rate, skin conductance, and electromyographic responses. The sound of the human voice increases the relationship with natural sound but the combination of the human voice and technological sounds is widely rejected. This is not the case of the human voice in a quiet park, where a human voice is considered a disturbance.

Experiments regarding a combination of sounds and images in the assessment of the quality of a courtyard were conducted by Gidlof-Gunnarsson et al. (2007) in a laboratory on 24 subjects to whom were shown photographs combined with auditory stimuli (road traffic and bird song) at different levels of L_{Aeq}. The images of the courtyard were manipulated to create attractive and unattractive images. Road traffic noise affected soundscape quality and the noise annoyance. With the increase of traffic noise, the rating was rapidly degraded toward annoyance, but this trend was reduced by the presence of bird song. The visual effect of a nice courtyard reduced the annoyance effect but when the road traffic noise was raised to 55 dB this effect was less evident and was judged more stressful in the attractive courtyard than in the unattractive courtyard setting. It seems reasonable to admit that for an attractive courtyard a peaceful sonic ambience is expected, meaning that the more attractive a courtyard is, the more sonic expectations are required, and that the increase of road traffic noise can degrade the attractive courtyard more than the unattractive ones. Low road traffic noise in combination with bird song increases the rating of attractiveness of the experimental courtyard.

An attempt to measure tranquility has been proposed by Pheasant et al. (2008) for 11 English rural and urban landscapes. Visual and acoustic data were supplied to

44 volunteers to make a subjective assessment. The maximum sound pressure (L_{Amax}) and the percentage of natural features present at a location were found to be key factors that influence tranquility. From empirical data these authors have validated a regression analysis to extract the tranquility rating (TR), which can be calculated as follows:

$$TR = 13.93 - 0.165L_{Amax} + 0.027NF(49\ \text{dB} < L_{Amax} < 85\ \text{dB})$$

where NF is the proportion (%) of the natural features.

TR can be calculated also starting from L_{Aeq}:

$$TR = 8.57 + 0.0036NF - 0.11L_{Aeq}$$

Soundscape evaluation is strongly affected by nonauditory factors such as visual images, daylight, and olfactory perception.

The soundscape analysis requires the collection of many concurrent factors that express a subjective assessment of the annoyance rating. A study conducted in Seoul metropolitan area by Jeon et al. (2010, 2011) have ascertained the quality of different urban soundscapes by an integrated approach. Ten different sites were selected and the SPL was measured in each (Table 5.6).

A questionnaire was administered, divided into four sections:

(1) perception of urban soundscape, (2) soundmark, (3) semantic differential, and (4) individual information.

1. Perception of urban soundscape was divided into three sections:

   Overall impression
   Acoustic comfort
   Other contexts

2. Soundmarks
   A general description was requested

3. Semantic differential was composed of 12 statements:

   Pleasant–unpleasant
   Comfortable–uncomfortable
   Disturbing–not disturbing
   Quiet–noisy
   Calm–loud
   Natural–unnatural
   Distinct–usual
   Steady–unsteady
   Pulsating–not pulsating
   Nearby–far
   Wide–narrow
   Dry–reverberant

Table 5.6 Urban sites in Seoul City and sound pressure levels at each site (L_{eq}, dBA)

Site	Main function	Sound sources	SPL[a]	Site	Main function	Sound sources	SPL[a]
1	Commercial, office	Traffic, footsteps, music from buildings, surrounding speech	72.1 (3.1)	6	Office, cultural (historic building)	Traffic, footsteps, speech	66.9 (2.5)
2	Commercial, office	Traffic, footsteps, construction, music from buildings, surrounding speech	68.7 (2.8)	7	Relaxation, office	Traffic, footsteps, speech, water (stream)	65.2 (2.0)
3	Square, office	Traffic, water (fountain), construction, children	68.9 (3.0)	8	Commercial, office	Traffic, footsteps, water (stream), music from buildings	69.0 (2.0)
4	Square, office, cultural	Traffic, water (fountain), surrounding speech	75.5 (4.2)	9	Relaxation	Wind, birds, children, traffic	52.2 (1.2)
5	Commercial	Traffic, footsteps, music from buildings	63.3 (1.8)	10	Relaxation, recreation	Wind, birds, speech, traffic	61.0 (2.2)

Source: Reproduced with permission from Jeon et al. (2011)
[a]Numbers in parentheses represent standard deviation of sound levels

For every couple of statements seven degrees were scored: extremely, quite, slightly, neither, lightly, quite, extremely

4. Individual information was articulated as

   Gender and age
   House type: apartment/row house/detached house/other
   Number of visits to urban spaces (0–10)
   Your sensitivity to noise (0–19 scores).

The use of a questionnaire and of soundwalks allowed collecting important details about the composition of a soundscape. The urban soundscape can be characterized by soundmarks where acoustic comfort, visual images, and day lighting dominate the soundscape survey. Reverberation does not discriminate, but this factor can be easily replaced by physical measurements. Openness and density seem important visual characters.

An experiment was conducted by Viollon et al (2002) on a group of subjects who were requested to rate eight urban sound environments (recordings) at the same time with five visual settings (four color slides showing a different degree of urbanization and a control without slides, with two sound scales

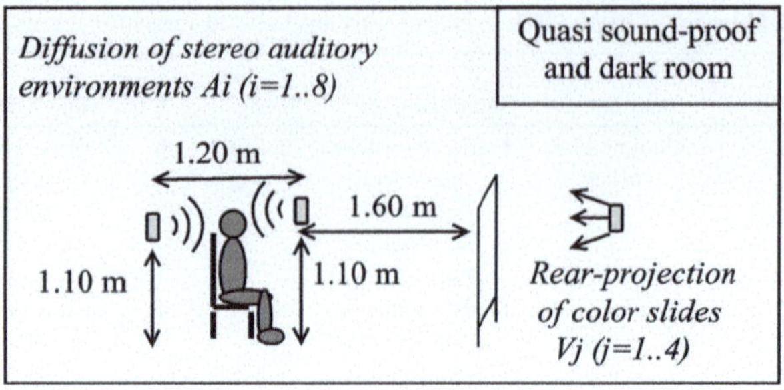

Fig. 5.7 Experimental design for an audiovisual environment in which to experiment with sonic ratings (Reproduced with permission from Viollon et al. 2002)

(unpleasant–pleasant, stressful–relaxing). In Fig. 5.7 is shown the experimental design to investigate the reaction of a subject to the contemporary projection of a slide and one typology of sound. In particular, the effect of visual scenes have been constrained by the type of sound. The subjects associated the increasing urban setting with a more negative sound rating. For instance, birdsong and traffic noise were judged more negatively when associated with more urban scenes. In contrast, all the anthropogenic sounds were indifferent by co-occurring visual stimuli and independent of the degree of urbanization contained inside the slides.

Very stressful sounds such as road traffic and very relaxing sounds such as bird choruses were significantly influenced by the visual degree of urbanization.

These authors have found a negative correlation to be increasing between a more urban visual environment and the rating at which the sound is interpreted.

The acoustic assessment in green areas inside urban cores is particularly important because of the extensive use of these areas for recreational and therapeutic use. Often the traffic noise of urban areas is not buffered from the green areas and noise invades such areas, nullifying the recreational effect expected. Evidences of this widespread phenomenon have been collected in a recent study by Szeremeta and Zannin (2009) in a Brazilian city where from the acoustic survey of four urban parks more than 65 % of sampled points had a sound level above the 55 dB(A) that is the local limit fixed by the municipal law. A contemporary interview of a sampled population in the investigated parks has demonstrated how important are the esthetic components of the landscape in mitigating the negative effect of the soundscape. Several factors have to be considered: environmental and urban zoning, land use, traffic routes, residential streets, vegetation, type of public transportation, park typology, and sound sources coming from outside. The unpleasant reaction of the interviewees places the soundscape in secondary importance compared with water pollution and social problems (prostitution and vagrancy). The low level of reaction from people to heavy acoustic pollution, shifting the attention on other "visual" factors, exposes the park visitor to a high level of sound with negative health effects.

5.14 Quiet Areas as Solution to Decrease People's Annoyance

To efficiently combat the invasion of noise pollution, one possible solution is to offer quiet areas to inhabitants. In fact Ohrstrom et al. (2006) have proved that the percentage of annoyed subjects when exposed at different sound levels ($L_{\text{Aeq, 24 h}}$) with access to a quiet site their level of annoyance decreases compared with people without access to a quiet site (Fig. 5.8).

In this study was demonstrated that dwellers who have access to quiet indoor and outdoor areas are less sensitive to annoyance effect by noise dose. The reduction of sound level perceived is of 5 dB at the most exposed sites. Of course, to protect people the level of the sound from traffic should not exceed 60 dB ($L_{\text{Aeq, 24 h}}$) even if the dwellers have access to a quiet side.

This research confirms how important, from a psychological point of view, is the availability of a quiet site. This provision does not reduce the physical effect of noise but contributes significantly to the psychological effects that in turn modifies the physiological reaction of the human body.

5.15 The Aesthetic Component of Soundscape Interpretation

Acoustic comfort, defined as the state of mind that expresses satisfaction with the acoustic environment, should be the main outcome expected by urban park visitors. However, there is scant and fragmented information about the processes that create this type of comfort. A multivariate model has been proposed by Tse et al. (2012) approaching the study of soundscape in four parks of Hong Kong. They have included in a questionnaire, administrated to 595 respondents, natural rather than

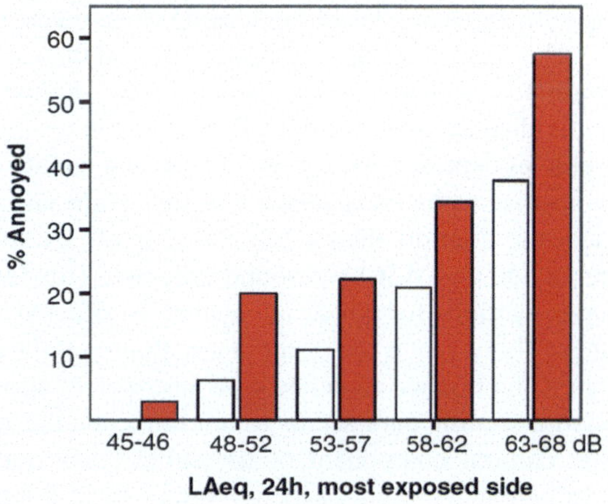

Fig. 5.8 Percentage of annoyed subjects at different sound levels ($L_{\text{Aeq, 24 h}}$) for individual without (*black bars*) and with (*white bars*) access to a quiet site in their dwelling areas (Reproduced with permission from Öhrström et al. 2006)

Table 5.7 Types of sounds investigated in a park in Hong Kong by Tse et al. (2012)

Natural sounds		
Insects	Fountain	Tree murmur
Twittering of birds	Sea wave	Running water
Calling crows	Barking dogs	Waterfall
Wind	Rain	Thunder
Anthropogenic and mechanical sounds		
Car horn	Car reverse horn	Heavy vehicles
Bicycle ring	Ship siren	Talking
Siren from ambulance/fire	Train	Screaming
engine/police vehicles	Airplane	Footsteps
Music	Light vehicles	Cell phone ring

Source: Reproduced with permission from Tse et al. (2012)

anthropogenic and mechanical sounds (Table 5.7). The interviews were carried out contemporarily to the measurement of the sound pressure level and at A-weighted percentile level. Personal information were also collected during the interview:

Gender (male: female); age (years) (<15, 16–25, 26–40, 41–60, >60); purpose of visiting the park (resting, undertaking physical activities, raising kids, other purposes); duration of stay (min) (<30, 30–59, 60–89, 90–119, >120); local resident (yes, no); self-rated auditory capability (very poor, poor, average, good, very good).

Specific statistics were presented to characterize the acoustic comfort (very uncomfortable, uncomfortable, neutral, comfortable, very comfortable), acceptability of the environment (very unacceptable, unacceptable, neutral, acceptable, very acceptable), preference to stay because of soundscape quality (leave right away, considering leaving, neutral, considering staying, continue to stay). From a multivariate analysis resulted an interrelationship among acoustic comfort evaluation, acceptability of environment, and preference to stay in a park. The acoustic-related factors, park environmental factors, and individual receptor characteristics contribute to the acoustic comfort of urban parks. The comfort factor is influenced not only by the sound pressure level (L_{eq}) and by subjective sound evaluation but also by specific types of sounds such as breezes, bikes, or heavy vehicles.

The natural sounds are more preferred than anthropogenic and mechanical ones. Natural sounds (birds and falling water), although preferred, do not contribute to individual acoustic comfort. Breeze and bikes increase the comfort score but heavy vehicles depress such a score. The visual quality of a park contributes to formatting a positive rating of acoustic comfort. High visual comfort attracts 2.2 times more acoustic comfort than a low subjectively evaluated sound level and 7.6 times a reduction of 1 dB(A) of sound pressure. This finding confirms the results of other investigations for which annoyance is affected by visual quality of the surroundings. This effect is the result of a spatiality of the cognitive maps that we have in our brain and that we associate with an expected sonic environment. The motivation to visit a specific park seems to not influence the final acoustic comfort rating. Also, the cultural component of the subject has not a clear trend. Acoustic comfort evaluation represents a better proxy of individual preferences to stay in a park than sound pressure level (L_{eq}) or the subjectively evaluated sound level, but is the visual comfort that has a greater influence. In conclusion, improving the acoustic comfort increases the acceptability of a park.

Semantic differential scaling is often used to evaluate the quality of the sonic environment, and several field studies have demonstrated that the informative properties of sound (i.e., categories of sounds) are better predictors of the perceived sonic quality than acoustic measures of the soundscape (i.e., L_{eq}).

Aesthetic landscapes and acoustic comfort are important components of overall physical comfort in urban areas. For this reason planners and architects must reserve to the soundscape aesthetic the same role attributed to other visual components of the urban landscape. Modeling is required to understand the sonic environment complexity in which perception and interpretation are both fundamental proxies to assess the sonic quality of a site.

The comfort acoustic model seems to work better than soundscape quality maps, as observed experimentally by Yu and Kang (2009) when an artificial neural network is utilized.

5.16 Conclusions

Noise represents, especially in urban areas, one of the major sources of citizen dissatisfaction and for this reason policies to develop efficient control tools and methods of assessment have to be urgently provided by policymakers. Noise exposure affects people's well-being. The evaluation of noise depends on noise type and on psychological and social characteristics.

The psychological perspective assumes that different people perceive differently the sound of the environment according to status, gender, culture, attitude, etc. In fact, according to the level of development that an area experiences, the noise is evaluated differently. For instance, in underdeveloped areas the mechanical noise are evaluated more negatively than in developed areas where people expect a mix of mechanical and nonmechanical noise. The tolerance to noise intrusion is higher in developed areas. Foreground task can change the level of perception of the noise. A wilderness hiker can tolerate the noise produced by a rescue helicopter more than that by a tourist airplane that is considered less necessary.

Noise requires tools to be recorded with reliability and a way to represent its distribution in space and time, in other words, to transform point-recording data in spatially distributed surfaces. For this the use of GIS represent the natural evolution of sound representation.

Public and private behavior can strongly improve the sonic ambience. Planning new buildings in quiet areas far from very noisy surroundings is a reasonable measure. In the private domain, the maintenance of a indoor low volume of radio and TV, and a lower voice during outdoor conversation on mobile telephones, are some possible "micro" actions highly beneficial for a common (public) sonic environment.

The dramatic effects of unwanted noises on human health recall the urgency to improve the sonic environment with structural and behavioral actions. Quiet areas in urban spaces become strategic for social well-being that must be coupled to the maximum protection of quality of the sonic environment in remote wild areas as well.

In urban areas, the abatement of noise appears as an effective action to attract animal biodiversity, which contributes, when present, to enhancing the sonic quality of open spaces.

Summary

The sonic environment is the result of the physical properties of the sounds, of the physiological species-specific perception of every single species, and of cultural/social interpretation. The sonic ambience associated with visual aesthetics concurs to human well-being, and the search for a quiet area becomes a prior strategy in human societies surrounded by increasing anthropogenic noise.

The study of the soundscape from a human perspective requires a psychological approach that allows a synthesis between visual and acoustic stimuli. The quality of a sonic ambience increases the acoustic capacity in human hearing and alters the relationship between acoustic cues and their cultural and social interpretation.

Annoyance is defined as a feeling of displeasure evoked by noise and is used to evaluate all the negative effects produced by noise on human feelings.

Noise sensitivity is defined as an attitude to rating sound according to an individual-based criterion and represents a major antecedent of individual noise annoyance; results are independent of the predisposition to perceive a sound event more intensely or to the capacity to better discriminate. Noise-sensitive people (NS) are exposed to a major stressful psychological condition, to an increasing psychological reactivity of the cardiovascular system, related to psychopathology. Noise becomes an important component of the well-being for noise-sensitive people because these people consider noise as threatening element and out of their control. Noise is discriminated at the same rating when high in NS and non-sensitive people (NNS) but this evaluation diverges between the two categories when the noise is administered at low amplitude.

Soundscape description based on physical characteristics seems insufficient to explain the complex relationship between humans and the sonic environment. A semantic approach able to describe the feeling that sounds produce in individuals allows categorizing the cultural and social effects of a subjective sonic environment. Cacophony, hubbub, constant, and temporal are some of the semantic categories in which the different sounds/noise are recognized.

Psychoacoustics research aims to find the quantitative relationships between the physical attributes of sounds and their perceived properties and provides tools to measure the sensation that sounds produce in the listeners. The most common auditory attributes used in the psychoacoustic approach are loudness, pitch, duration, sharpness, roughness, fluctuation, and strength.

Pleasantness, eventfulness, and familiarity are further sound categories or emergent characters of a soundscape; their combination allows the classification of several sonic environments.

Noise pollution is a growing worldwide phenomenon affecting millions of people. In Europe alone 80 million people are exposed to an unacceptable dose of noise. In an undeveloped country, noise exposure has a major impact.

Direct and indirect effects of noise exposure on human health produces hearing impairment, reduction of speech communication, cardiovascular risk, sleep disturbance, psychological and performance reduction effects, and a subjective feeling of annoyance. Exposition to outdoor noise represents the highest risk to human health where insomnia is one of the most common effects, but in the long term ischemic heart disease and rising blood pressure have to be considered equally probable.

The soundscape of urban spaces is a complex system in which physical, psychological, and social components concur to human sonic assessment. A sonic environment of high quality in urban areas means not only a low level of sonic energy but also the presence of natural elements such as green spaces and a tangible biodiversity.

Urban parks are areas where people find physical and psychological advantages, thanks to several positive cognitive objects such as peacefulness and quietness and natural physical natural objects such as trees, flowers, and animals. In urban parks visitors have been demonstrated able to extract positive acoustic cues such as natural sounds and happy people sounds.

Pleasant sounds reduce the relationship between sound pressure level and the acoustic comfort. The age of subjects is important: teenagers are more unsatisfied with acoustic comfort when compared with older groups.

Sonic and visual objects are major actors in the urban landscape, but the sound has distinctive characters compared to the visual objects. In fact, sound has no precise border and is ubiquitous; the aural harmonization is temporal and is a field of energy. A sound could be less rich in information but full of emotion. Rating a soundscape largely depends on the individual skill, experience, attentiveness, cultural model, and social and economic stratum.

Passive soundmarks are defined as every sonic object that is perceived as a priority by a listener such as fountains and sonic sculptures, whereas active soundmarks are the result of interesting activities (concerts, talks, local markets).

The combination of visual and acoustic cues creates an informative space that can be useful to manage and to reduce anthropogenic noise, and definitively to improve the quality of the soundscape. Some natural sounds such as water contribute to improve the appreciation of natural and manmade spaces, and the combination of visual and acoustic stimuli represents efficient strategies to assess the value of a landscape and the associated soundscape. Soundscape evaluation is generally performed by administrating questionnaires that have specific sections, such as (1) perception of urban soundscape; (2) soundmark; (3) semantic differential, and (4) individual information.

The possibility to benefit from quiet areas reduces in people the annoyance effect with physiological benefits, but to achieve this result the level of noise must be at least below the 60 dB $L_{\mathrm{Aeq,\ 24\ h}}$.

The assessment of the human-perceived soundscape requires diverse fields of practices and competencies such as sound quality, human acoustic comfort in buildings, music and relationships with wilderness and recreation management, urban and house design, landscape planning, and management.

Acoustic comfort, defined as the state of mind that expresses satisfaction with the acoustic environment, is the major outcome expected during the visit to a quiet urban area. A semantic differential scale represents a more efficient tool to assess the acoustic comfort for the major informative properties of sound (i.e., categories of sounds) than the physical acoustic measures of the soundscape (i.e., L_{eq}). The visual comfort contributes to the formation of the acoustic comfort.

Soundscape customer satisfaction represent a promising model in soundscape assessment in which attractive, one-dimensional, must-be, indifferent, and reverse are the quality categories that can be applied.

References

Anderson LM, Mulligan BE, Goodman LS (1984) Effects of vegetation on human response to sound. J Arboricult 10(2):45–49

Appleton J (1975) The experience of landscape. Wiley, London

Axelsson O, Nilsson ME, Berglund B (2010) A principal components model of soundscape perception. J Acoust Soc Am 128(5):2836–2846

Babisch W, Beule B, Schust M, Kersyen N, Ising H (2005) Traffic noise and risk of myocardial infarction. Epidemiology 16:33–40

Berglund B, Hassmen P, Preis A (2002) Annoyance and spectral contrast are cues for similarity and preference of sounds. J Sound Vib 250(1):53–64

Botteldooren D, De Coensel B, De Muer T (2006) The temporal structure of urban soundscapes. J Sound Vib 292:105–123

Cain R, Jennings P, Poxon J (2013) The development and application of the emotional dimensions of a soundscape. Appl Acoust 74:232–239

Carles JL, Barrio IL, de Lucio JV (1999) Sound influence on landscape values. Landsc Urban Plan 43:191–200

Davies WJ, Adams MD, Bruce NS, Cair Cusack R, Hall DA, Hume KI, Irwin A, Jennings P, Plack MJ, Poxon J (2013) Perception of soundscapes: an interdisciplinary approach. Appl Acoust 74:224–231

Ellermeier W, Eigenstetter M, Zimmer K (2001) Psychoacoustic correlates of individual noise sensitivity. J Acoust Soc Am 109(4):1464–1473

Farina A (2006) Principles and methods in landscape ecology. Toward a science of landscape. Springer, Dordrecht

Ferguson S, Cabrera D, Beillharz K, Song H-J (2006) Using psychoacoustical models for information sonification. In: Proceedings of the 12th international conference on auditory display, London, 20–23 June, 2006

Fields JM (1993) Effect of personal and situational variables on noise annoyance in residential areas. J Acoust Soc Am 93(5):2753–2763

Gidlof-Gunnarsson A, Ohrstrom E, Ogren M (2007) Noise annoyance and restoration in different courtyard settings: laboratory experiment on audio-visual interactions. Inter-Noise 2007, Istanbul

Gygi B, Kidd GR, Watson CS (2004) Spectral-temporal factors in the identification of environmental sounds. J Acoust Soc Am 115:1252–1265

Gygi B, Kidd GR, Watson CS (2007) Similarity and categorization of environmental sounds. Percept Psychophys 69:839–855

Hetherington J, Daniel TC, Brown TC (1993) Is motion more important that it sounds? the medium of presentation in environment perception research. J Environ Psychol 13:283–291

Jennings P, Cain R (2013) A framework for improving urban soundscapes. Appl Acoust 74(2):293–299

Jeon JY, Lee PJ, You J, Kang J (2010) Perceptual assessment of quality of urban soundscapes with combined noise sources and water sounds. J Acoust Soc Am 127(3):1357–1366

Jeon JY, Lee PJ, You J, Hong JY (2011) Non-auditory factors affecting urban soundscape evaluation. J Acoustic Soc Am 130(6):3761–3770

Job RFS (1988) Community response to noise: a review of factors influencing the relationship between noise exposure and reaction. J Acoust Soc Am 83:991–1001

Job RFS (1999) Noise sensitivity as a factor influencing human reaction to noise. Noise Health 3:57–68

Kang J, Zhang M (2010) Semantic differential analysis of the soundscape in urban open public spaces. Build Environ 45:150–157

Kawada T (2011) Noise and health – sleep disturbance in adults. J Occup Health 53:413–416

Kerrick JS, Nagel DC, Bennett RL (1969) Multiple ratings of sound stimuli. J Acoust Soc Am 45(4):1014–1017

Lebiedowska B (2005) Acoustic background and transport noise in urbanized areas: a note on the relative classification of the city soundscape. Transp Res Part D 10:341–345

Leobon A (1995) La qualification des ambiance sonores urbaines. Nat Sci Soc 3:26–41

Luz GA (2005) Noise sensitivity rating of individuals. Sound and vibration, Baltimore, August 2005, pp 14–17

Matsinos YG, Mazaris AD, Papadimitriou KD, Mniestris A, Hatzgiannidis G, Maioglou D, Pantis JD (2008) Spatio-temporal variability in human and natural sounds in a rural landscape. Landsc Ecol 23:945–959

Meijer H, Knipschild P, Salle H (1985) Road traffic noise annoyance in Amsterdam. Int Arch Occup Environ Health 56:25–297

Miedema HME, Vos H (2003) Noise sensitivity and reactions to noise and other environmental conditions. J Acoust Soc Am 113(3):1492–1504

Ohrstrom E, Skanberg A, Svensson H, Gidlof-Gunnarsson A (2006) Effects of road traffic noise and the benefit of access to quietness. J Sound Vib 295:40–59

Osgood CE, Suci GJ, Tannenbaum PH (1957) The measurement of meaning. University of Illinois Press, Urbana

Payne SR (2008) Are perceived soundscapes within urban parks restorative? Acoustic 2008, Paris, pp 5521–5526

Payne SR (2013) The production of a perceived restorativeness soundscape scale. Appl Acoust 74:255–263

Pheasant R, Horoshenkov K, Watts G (2008) The acoustic and visual factors influencing the construction of tranquil space in urban and rural environments tranquil spaces-quiet places? J Acoust Soc Am 123(3):1446–1457

Raimbault M, Dubois D (2005) Urban soundscapes: experiences and knowledge. Cities 22(5):339–350

Rychtarikova M, Vermeir G (2013) Soundscape categorization on the basis of objective acoustical parameters. Appl Acoust 74:240–247

Rylander R (2006) Noise, stress and annoyance. Noise Vib Worldw 37(6):9–13

Schulte-Fortkamp B, Nitsch W (1999) On soundscapes and their meaning regarding noise annoyance measurements. Iter-Noise 99, Fort Lauderdale, FL, US

Skanberg A, Ohrstrom E (2002) Adverse health effects in relation to urban residential soundscapes. J Sound Vib 250(1):151–155

Susini P, Lemaitre G, McAdams S (2012) Psychological measurement for sound description and evaluation. In: Berglund B, Rossi GB, Townsend J, Pedrill LR (eds) Measurement with persons: theory, methods, and implementation areas. Psychology Press, Taylor & Francis Group, New York

Szeremeta B, Zannin PHT (2009) Analysis and evaluation of soundscapes in public parks through interviews and measurement noise. Sci Total Environ 407:6143–6149
Truax B (1994) Acoustic communication. Ablex Publishing, Nordwood
Tse MS, Chau CK, Choy YS, Tsui WK, Chan CN, Tang SK (2012) Perception of urban park soundscape. J Acoust Soc Am 13(4):2762–2771
Van Gerven PWM, Vos H, Van Boxtel MPJ, Janssen SA, Miedema HME (2009) Annoyance from environmental noise across the lifespan. J Acoust Soc Am 126(1):187–194
Viollon S, Lavandier C, Dake C (2002) Influence of visual setting on sound ratings in an urban environment. Appl Acoust 63:493–511
Watts G, Chinn L, Godfrey N (1999) The effects of vegetation on the perception of traffic noise. Appl Acoust 56:39–56
Watts GR, Pheasant RJ, Horoshenkov KV, Ragonesi L (2009) Measurement and subjective assessment of water generated sounds. Acta Acoust uw Acustica 95:1032–1039
Weinstein ND (1978) Individual differences in reactions to noise: a longitudinal study in a college dormitory. J Appl Psychol 63(4):458–466
Yang W, Kang J (2005) Acoustic comfort evaluation in urban open public spaces. Appl Acoust 66:211–229
Yu L, Kang J (2009) Modeling subjective evaluation of soundscape quality in urban open spaces. An artificial neural network approach. J Acoust Soc Am 126:1163–1174

Chapter 6
Sonic Patterns I: The Noise

6.1 Introduction

Noise is a phenomenon that has been observed in acoustic, in visual (image), and electronic and vibrational (thermal) fields as well. Acoustic noise, in a communication perspective, is the result of a sound that is not used to communicate and that in many occasions can reduce the efficiency of direct and remote communications.

White noise and pink noise are examples of two models of noise. The first is characterized by a constant amplitude of pitches along the frequencies; in the second, the amplitude is a function of $1/f$ (Fig. 6.1). In most cases, noise has a small amount of meaningful information and sophisticated procedures are required to extract such information.

There is an impressive literature on noise source and propagation in different media and on the effects of noise in animals and humans. An interesting and abundant literature considers the possibility to reduce ambient noise from urban areas but also in natural areas (parks, natural reserves, recreational areas; see Chap. 10: Applications).

6.2 A Definition of Noise

Noise is an unintentional (mainly from background) sound that masks acoustic signals and may be generated by geophonies (e.g., strong wind, volcano eruptions, earthquakes), biophonies (e.g., urban roosting of starlings, breeding colonial birds and mammals), antrophonies (e.g., urban traffic, oil spill, stationary machines such as air conditioners). Noise can be generated by a continuous source, such as running waters, or by a discrete event such as thunder.

The noise may be a random event, such as a car horn, or a regular event, such as the sound of a bell, repetitive like the jackhammer or unique like a rifle shot. "The noise of . . .. an engine, a horn, etc." means also a sound with a reduced structure in

A. Farina, *Soundscape Ecology: Principles, Patterns, Methods and Applications*,
DOI 10.1007/978-94-007-7374-5_6,

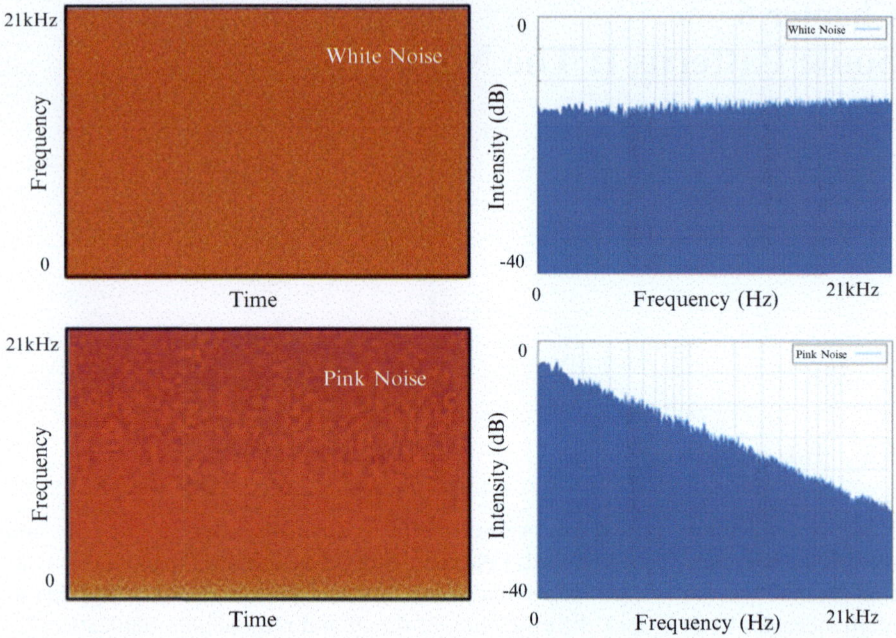

Fig. 6.1 Spectrogram and power spectrum of a *white* noise and a *pink* noise

frequency and amplitude modulation. The noise can be distinguished according to the bandwidth (broad or narrow frequency bands).

Despite the different definitions, a noise is a degraded sound or the summation of different sounds that create a confused pattern. In both cases it contains a low level of information. Adopting the three categories of geophonies, anthrophonies, and biophonies, we can say that the noise is acoustic energy that in turn masks and interferes one of these three sound source types.

The word noise derives from the latin *nausea* and has different definitions as stated in the book by Truax (1999): unwanted sound, unmusical sound, any loud sound or disturbance in any communication system.

An interesting definition of noise is proposed by Dodd (2001): "*noise = subjective sound which we would choose not to listen (i.e. it is subjective sound from which we would not choose to extract information).*"

When is a sound transformed to noise? Generally we refer to noise when we perceive indistinct and annoying background noise. But also a near-ground sound that is not identified or is unwanted may be considered a noise.

Noise is a voluntary or involuntary sound that when added to a signal reduces the capacity of that signal to be correctly perceived and interpreted. In an animal communication context, the noise is defined as "any factor that reduces the ability of receiver to detect a signal or to discriminate one signal from another" Brumm and Slabbekoorn (2005).

According to a soundscape perspective, noise can be defined as the portion of the acoustic spectrum that does not offer information per se, but the amount of noise and its distribution in time can be used to assess natural or anthropogenic events and to attribute a sonic quality to a landscape or to a site/location (see Fig. 4.9, page 87 as an example of environmental noise source).

Noise remains a not-clear concept that is used often to define every anthropogenic sound or and unwanted sound. We prefer such last definition because, for instance, if you wish to talk at a cell phone inside an exhibition of caged birds, immediately you realize it is not an appropriate location and it is not easy to communicate, although each bird produces a lovely song.

If we consider noise in the physical sense, in nature the noise is a rare event when there is no direct or indirect human intervention. But again, does a waterfall or a wind tempest produce sounds or noise? This question remain unanswered.

From a semiotic point of view, a waterfall can be considered a sonic landmark for some species and simply a noise for others.

When animals are engaged in dawn or dusk choruses, because of the contemporary song activity we perceive such choruses as an indistinct noise. The same is true for urban noise that is the product of road traffic, human voices, horns, bells, air conditioners, etc.

6.3 Characteristics of Anthropogenic Noise

Humans are the major factor responsible for increasing noise pollution. Noise has no boundaries, is difficult to be managed, and becomes chronic in urban areas and close to the great transportation infrastructures (Barber et al. 2009).

There is industrial noise, urban noise, ambient noise, aircraft noise, etc., and often noise is considered at the same level of importance of chemical substances as a source of pollution.

The transportation net is one the major sources of ambient noise, and air and road transport are so intense worldwide that entire ecosystems are exposed to growing chronic noise.

6.4 Active Space and Noise

Active space (AS) is defined as the distance from the sender at which a signal can be detected by a receiver. The AS depends on the quality of the signal, the level of noise that masks this signal, and the critical ratio of the receiver's sensory system. The critical ratio (CR) is the lowest level of a signal (in dB) at which the signal is detectable. The CR varies according to frequencies and across species.

Phase and time of signals are deeply modified by noise. Noises characterized by low frequency and high amplitude are good candidates to mask acoustic signals.

The active space for an equivalent signal is expected to be smaller in the urban context than in nonurban areas, although a recent investigation (Mokford et al. 2011) has found that urban habitat probably may degrade sound less than in rural settings.

Anthropogenic sound interferes with these parameters, but also the structure of urban spaces produces distortion in the sonic environment. For instance, large vertical surface of buildings reflect sound and produce reverberation.

Investigation on the brown-headed cowbird (*Molothrus ater*) by Gall et al. (2012) has proved that the active space of this species is affected by both habitat type and level of urbanization. The maximum active space was obtained at high frequency (4–6 kHz) and is larger in open urban spaces. The presence of reflecting surfaces probably favors the propagation of sounds.

6.5 The Noise Context

Noise is part of any environmental context and concurs to characterize a site assigning a well-being score or an economic value (see for instance, O'Connor 2008). According to its origin and level, noise can have a different impact on living organisms.

The substrate on which noise is produced can modify physical and perceptual noise. For instance, noise produced by the wind flowing in broadleaved trees is different from the noise produced by coniferous crowns.

Ambient noise is produced by a broad spectrum of different spectral frequencies. A large variety of frequencies is produced by wind passing through leaves (rustling) and twigs. Wind passing over substrates generally produces low-frequency sounds. Anthropogenic noises pertain to a wide spectrum of frequencies, but the anthrophonies that are real intruders at a growing rate in every environment are generally of low frequency. The spectral profile of every environment is peculiar for that ambient or, better, is habitat specific. It reflects geography, topography, botanical composition, and vegetational structure (clearings, core/edge areas, shape of patches, fragmentation level and typology, land use, level of human contamination, etc.).

In a nontechnological world, sounds are the manifestation of geophysical and biophysical dynamics. Thunder, wind, running water, and rain are common sounds from nature. But animals too contribute to such sounds, that Bernie Krause has called the "Great Animal Orchestra" (Krause 2012). In the modern world dominated by humans, the word "sound" is often neglected and substituted by the word "noise," that is defined as every sonic process far from quiet that is annoying and may be harmful.

When too loud (>100 dB), noise can produce temporary or permanent damage to the hearing system in humans and animals as well. Modern music when heard at the highest volume can create permanent deafness in youths. Noise is responsible for the so-called noise pollution, and mitigation or remediation actions are at the basis of modern planning strategies, especially in urban areas and protected wild areas.

The signal-to-noise ratio is an important factor that must be considered to understand the way organisms modify their acoustic patters to increase this factor.

6.6 Noise in Patchy Sonotopes

Ambient noise is one of the environmental factors that produces a selection pressure on acoustic signals. Signal-to-noise ratio (SNR) and the masked auditory threshold are the fundaments of the detectability of a signal.

The soundscape is heterogeneous, and this means that there is alternation of quiet and noisy sonotopes. Generalist species in terms of sonic tolerance can establish in both conditions and cope with local sonic conditions, but for specialist species with a narrow sonic tolerance only quiet conditions can be selected.

According to a strategy to reduce the masking effect of noise, chaffinches (*Fringilla coelebes*) living along torrents and close to waterfalls have been observed to modify their acoustic repertoire, repeating more often the same song type before changing to a new type (Brumm and Slater 2006) (Fig. 6.2). This observation could confirm the information theory, which states that in the presence of noise a signal must be repeated, creating a redundancy, assuring in this way the chance that a signal may be received. These authors have observed also that song types with faster trills were sung in shorter bouts, which opens the hypothesis that highly redundant song is limited by a trade-off with motor fatigue.

Many studies have demonstrated that song patterns diverge between a noisy or quiet sonic environment. However, it is reasonable to suppose that the changes occurring in the signalers have influence also on the receivers. Mockford and Marshall (2009) demonstrated, using playback experiments, a difference in spectral aspects of the song of the great tit (*Parus major*) living in an urban and rural context. These differences elicit the replay of the receiving males. Males living in quiet areas showed a significantly stronger response (e.g., faster to sing over the playback song, and approaching the speaker) as a reply to the male territory holder song with low background noise than to the song of males living in areas with high territory noise.

In this species the spectral differences between urban and rural song consists in a minimum frequency that was significantly lower in individuals living in quiet areas. Laboratory experiments carried out by Pohl et al. (2012) on the great tit (*Parus major*) have demonstrated that in conditions of urban noise this species benefits from high frequencies in detection and discrimination of intraspecific song.

If noise is important for animal communication, a study comparing ambient noise represents an important step to better understand how local animal communities aggregate and communicate. Slabbekoorn (2004) has conducted an investigation on the environmental noise in 14 sites in Cameroon, central Africa, 8 sites in rainforest and 6 in the ecotone region of dryer savanna. The results

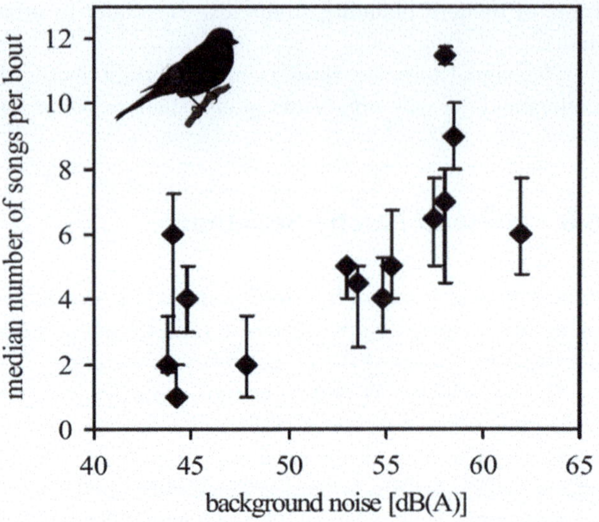

Fig. 6.2 Relationship between background noise level and median number of songs per bout in chaffinch (*Fringilla coelebs*) (Reproduced with permission from Brumm and Slater 2006)

demonstrated that the spectral profiles can be significantly and consistently different according to the habitat. This finding suggests that the ambient noise, because of its regularity, can drive the signal evolution of vocal animals. The masking of ambient noise can represent a significant adaptive mechanism to improve the signal-to-noise ratio with benefits for the species that can do this performance in an evolutive time. Considering the growing intrusion of anthropogenic noise and the rapid modification of habitats, the sonic environment can change abruptly without allowing species to adapt their communicative traits, as stressed by Slabbekoorn in the conclusive comment. This fact represents the "ecological debt" sensu Tilman et al. (1994) that can produce dramatic changes in biological diversity in future time.

6.7 Noise Pattern as a Source of Environmental Pollution

With the term noise pollution we intend every negative consequence produced by a noisy ambience on animals and human life as well.

Humans have dramatically changed many environments across the world, producing growing noise as one of the major consequences of economic development (Hanson 2008). As recently outlined by Frisk (2012), the increase of ambient noise levels in the sea is strictly related to the global economic trends. In the period 1950–2007, in the open oceans there has been an increase of 3.3 dB per decade (Fig. 6.3). The effects of such diffuse noise intrusion are not easy to detect because they influence the behavior of animals and cannot be monitored as can a chemical pollutant.

Among the consequences of noise on ecological and behavioral processes, noise can disrupt predatory–prey interactions, and it could facilitate the success of species

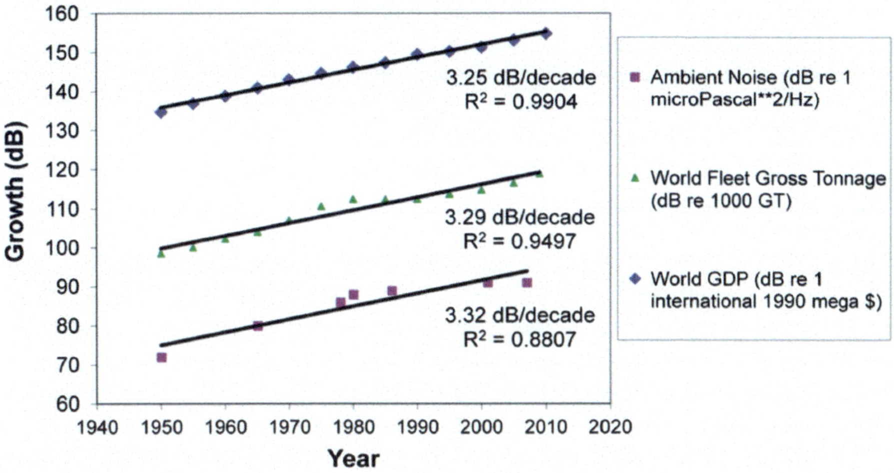

Fig. 6.3 Relationship between ambient noise level, gross tonnage of the world fleet, and world gross domestic product for the period 1950–2007 (Reproduced with permission from Frisk 2012)

that are more tolerant to noise, thus reshaping many communities and the relationships between species. For instance, Francis et al. (2009) have argued that urban-adapted species show an increase of breeding success facilitated by a decrease of predation (e.g., the decline of western scrub jay, *Aphelocoma californica*).

6.8 The Effects of Noise on Vocal Animals

Natural noise generally is handled by organisms with avoidance or acoustic adaptations, but human-generated noise affects animal physiology and behavior in different ways, producing annoyance, chronic stress, and hearing loss.

Anthropogenic noise creates disturbances in communicating animals, masking their frequencies.

As outlined by Slabbekoorn and Ripmeester (2008), ambient noise has many effects on animal populations: produces direct stress, masks the arrival of predators or the associated alarm calls, and interferes with the communication mechanisms in general.

To react to noise animals consume more energy, reduce the time for feeding, or, when vocal communications are masked, lose mating opportunities. The effect of noise can pertain indirectly to changes in habitat use, courtship and mating, reproduction, and parental care. Negative interactions occur also in the relationship between prey–predator signaling systems, altering the signals necessary to drive this complex relationship. In many cases the intrusion of noise produces a retreat from favorable habitat.

In social animals, anthropogenic noise can create serious difficulties in maintaining social aggregation or can reduce the exchange of strategic information such as the location of areas with abundant food.

Background noise and bird acoustic patterns seem strictly related. In fact, as emphasized by Slabbekoorn (2004), many animals use acoustic signals, and a low level of background noise is essential to attract mates, to defend territories, and to collect other information to track the necessary resources to maintain individual fitness

Many functions are connected to the active sonic ambience such as choruses and long-distance signals to inform individuals about the reciprocal presence or to signal special events such as water sources. The entire animal communication is based on the signal-to-noise ratio, and this largely depends on the amount of background noise that is present in an environment at a specific time of day or season, which interferes with the efficient of the signal emitted.

Rabin et al. (2003) has conducted research on the effect of highway noise on California ground squirrel vocalizations. The California ground squirrel (*Spermophilus beecheyi*) is a common vocal mammal that also lives in human-modified grassland landscapes under a different level of noise masking. This species has a broad repertoire of chatters, whistles, and single-note repetitive calls. Its vocalizations change according to the predator that is approaching. Chatter and repetitive call are emitted in presence of mammalian predators; whistles are emitted when a raptor appears. Variations in the structure of chatter calls are statistically different according to the presence of snakes, badgers, other mammalian predators, raptors, and conspecific competitors. These authors have observed that the California ground squirrel has the capacity to shift acoustic energy into those harmonics that do not overlap with highway noise. They argued that this is a case similar to the human Lombard effect that consists in change in amplitude to avoid the masking effect of noise. This species seems to have a high acoustic plasticity and online adjustment, demonstrated by a capacity to change the type of vocalization according to the type of menace.

6.9 Masking Effect on Animal Life

We know quite well the effect of a large dose of noise on psychology and in general on the health of exposed people, but little is known about the effects of noise on wild animals.

Vocal species produce sounds, and according to the acoustic niche hypothesis every species tries to accommodate itself in a way to reduce acoustic interspecific overlap.

Evaluating noise effects on an animal is a complicated issue because the effects vary between species and individuals, with time of the year, and by day.

In soundscape ecology noise is not simply an unwanted sound but a sound that reduces, by masking, the quality of every other acoustic signal. To reduce the masking effect of background noise, some species have modified their vocalizations, shifting the vocalizations to ultrasonic frequency bands. It is the case of the concave-eared torrent frog (*Amolops tormotus*), a species of arboreal frog of Central China that has ultrasonic harmonics used to communicate in a noisy environment (Narins et al. 2004). Ultrasonic components have been found also in the song of the black-faced warbler (*Abroscopus albogularis*), an oscine songbird that lives close to the torrents in a constant noisy environment (Narins et al. 2004).

To adjust to masking noise, some species increase the amplitude of their vocalization. As documented by Slabbekoorn and Peet (2003), in great tit (*Parus major*) populations a shift toward higher frequency bins has been observed for urban populations living in noisy conditions. The same effect has been observed from experiments on the zebra finch (*Taeniopygia guttata*). Birds subjected by Cynx et al. (1998) to various levels of white noise increased significantly the amplitude level of their vocalizations.

A background noise of 60–65 dB, such as that during the nights in tropical areas, can render the sound signal ambiguous, as studied by Hartbauer et al. (2012), who experimentally masked the acoustic mating signal of *Mecopoda elongate*, a tropical katydid. Synchronous entrainment was lost at a signal-to-noise ratio of −3 dB when a native noise was used.

Pytte et al. (2003) have observed that the male of the blue-throated hummingbird (*Lampornis clemencia*) has the capacity to modify the amplitude of their vocalizations when a natural noise occurs or when an experimenter-controlled change in ambient noise is created. This effect is known as the Lombard effect and has been found in many other animals. This species uses a serial chip territorial advertisement call that modulates in amplitude according to the level of environmental noise. With experiments of playback of a creek noise, these authors have observed a change in the amplitude of serial chip production. In this species this stereotyped simple call is emitted from a high perch for a long-distance transmission, and consequently the amplitude modulation seems an efficient mechanism to assure the long-distance broadcast.

The cacophony of sound perceived in urban contexts is considered a necessary price to be paid for living in the highly structured and social-service dispenser that is the urban environment.

Such sounds that are loud and intrusive become noise when perceived as indistinct and are accepted as a necessary compromise for logistic services offered by the urban environment. In natural areas, however, people consider every non-natural sound very disturbing and try to select remote areas with a quiet sonic ambience.

The effect of urban noise on animals (and especially on birds) can be distinguished at least in two major groups: amplitude shift and frequency shift (Fig. 6.4).

The amplitude shift or Lombard effect is the tendency for a speaker to increase the intensity of its vocal level to improve the intelligibility of a speech signal when in the presence of a loud background noise. This effect has been named after the French ear specialist, Etienne Lombard, who discovered this phenomenon in human speech (Lombard 1911).

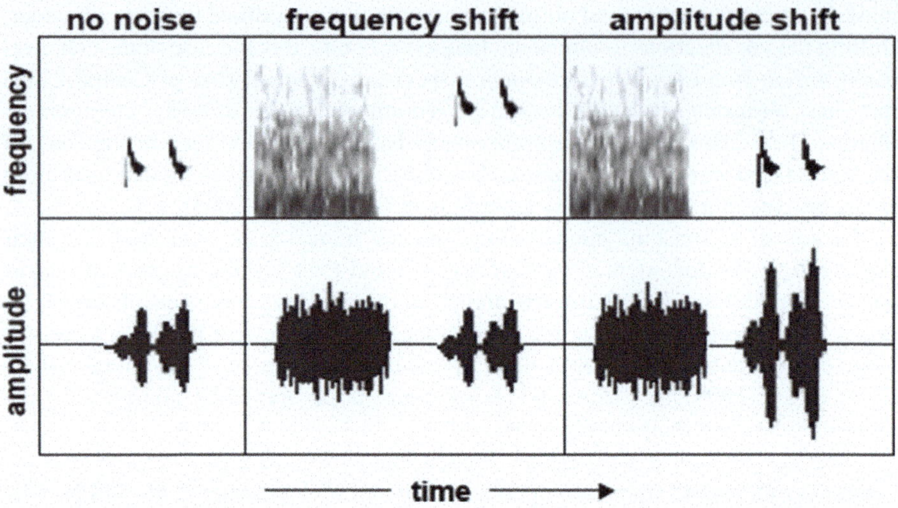

Fig. 6.4 Amplitude and frequency of the great tit (*Parus major*) call in a quiet area (**no noise**) and in a noisy area in which frequencies (**frequency shift**) or amplitude (**amplitude shift**) can be modified (Reproduced with permission from Katti and Warren 2004)

It seems a bit strange that this effect has been proved in animals 60 years later. One of the first researchers to investigate this effect on animals was Potash (1972), observing that the Japanese quail (*Coturnux coturnix japonica*) changes the amplitude of their vocalizations when experiencing a loud white noise.

Today several investigations have confirmed the Lombard effect, especially in urban areas, and some species such as the nightingale (*Luscinia megarhynchos*) are able to change the amplitude of their sound from working days as compared with weekend days (see Brumm and Todt 2003).

Also, frogs are able to change their amplitude when the background noise increases. In insects this phenomenon seems not be present. In fact, the stridulating mechanisms of an insect cannot be modulated.

Wind becomes noise when masking some biological signals, as studied by McNett et al. (2010) on the effects of wind-induced noise on the reproductive behavior of the treehopper *Enchenopoda binotata* ('ptelea').

This species utilizes plant-borne vibrations to communicate: the male and the female produce a duet before mating. These authors have observed that the male uses a relatively wind-free period to communicate. Females were observed to respond to male vibration in less measure in the presence of wind-borne vibrations

In the marine environment, Holt et al. (2008) have found that killer whales or orcas (*Orcinus orca*) increase their call source level by 1 dB for every 1-dB increase in background noise level (Fig. 6.5).

Laboratory experiments on hand-reared individuals of great tit (*Parus major*) (Langemann et al. 1998) have demonstrated that in this species in a noisy environment sensitivity was only in function of the noise level and not dependent on

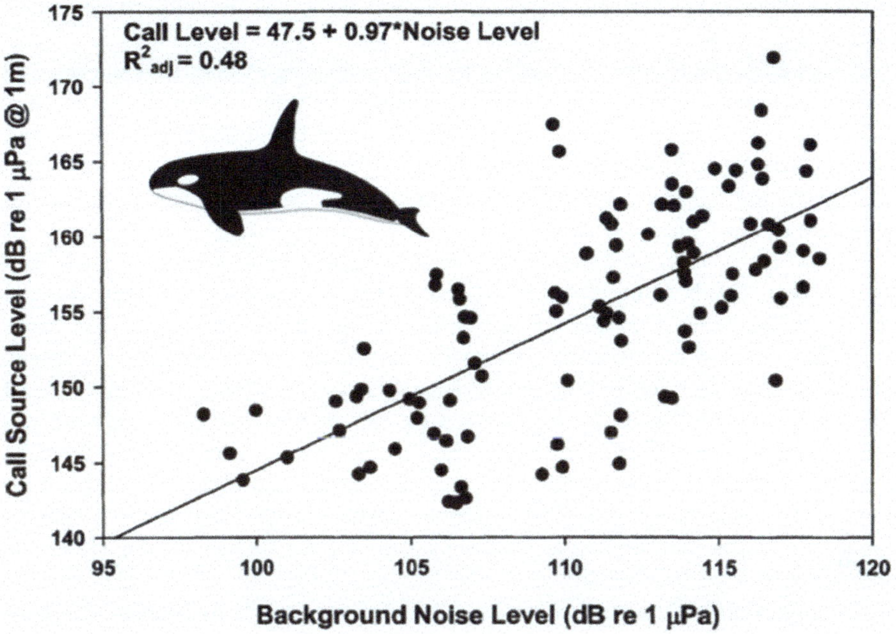

Fig. 6.5 Call source level as function of background noise level in the killer whale (orca, *Orcinus orca*) (Reproduced with permission from Holt et al. 2008)

frequency. In quiet conditions, the great tit was sensitive to frequencies between 2 and 4 kHz. In contrast to most other bird species, the great tit has low critical masking ratio at high frequencies. This finding further demonstrated that great tit communicates in a noisy environment using high-frequency vocalizations.

The airplane flyover is a source of unintentional noise that can affect animal communities according to intensity, duration, and period of the day. The increased level of such sound is strictly connected with a dramatic increase of air traffic favored by low-cost ticketing solutions.

The effect of airplane noise and playbacks of low-frequency motorcycle sounds has been examined by Sun and Narins (2005) on a frog chorus in central Thailand. Three acoustically active species (*Microhyla butleri*, *Rana nigrovittata*, *Kaloula pulchra*) reacted with a significant decrease of their calling notes, but *Rana taipehensis* had the opposite behavior, increasing its calling rate. This last species has been observed to emerge from the mixed chorus when the other species for natural reasons were decreasing their activity. These results illustrate the complexity of the effects of manmade sounds that produce a modulation in call rate and assign, involuntarily, advantages to species that are stimulated to increase their calls.

Studies conducted by Wood and Yezerinac (2006) have demonstrated that song sparrows (*Melospiza melodia*) modify their vocalizations to reduce the masking effect of urban noise and that males shifted more energy in the higher frequencies (4–9 kHz).

Although animals have evolved an acoustic signaling system very adaptive to background noise that minimizes the masking effect, the recent unprecedented diffusion of human-altered landscapes has modified the sonic environment, and animals have to face this relevant novelty.

In fact, anthropogenic noise often is louder and more persistent than natural sounds, and often the major ambient novelty.

The effect of anthropogenic noise was investigated by Francis et al. (2011) in two species of suboscine birds: ash-throated flycatcher (*Myiarchus cinerascens*) and grey flycatcher (*Empidonax wrightii*). At increasing noise exposure, grey flycatcher (*Empidonax wrightii*) occupancy declined but no frequency changes were observed. In contrast, the ash-throated flycatcher (*Myiarchus cinerascens*) does not show modification in density with the increase of noise dose but an increase of frequency of approximately 200 Hz was observed. This increase resulted in improving communication only slowly and could be the effect of an increase in vocal amplitude. The different behavior between these two species toward the same environmental changes opens a great debate on the effects of noise exposure for biodiversity.

The nightingale (*Luscinia megarhynchos*) has been proved by Brumm (2004) to react to urban noise by increasing song intensity (measured using the sound pressure levels parameter), demonstrating for the first time the noise-dependent vocal amplitude regulation (Fig. 6.6). This observation points out that birds try to reduce the masking effects produced by urban noise. However, this increases the energy cost of singing, which could create negative effects in the behavioral ecology of singing males.

The European robin (*Erithacus rubecula*), a highly territorial songbird, uses intensively vocal communication to defend territories. In the urban habitat, this species is also singing during the nighttime. This effect has been attributed first to light pollution (presence of artificial light all night), but recently Fuller et al. (2007) hypothesized that the nocturnal song may be a way to avoid noisy and masking daytime noise.

Noise playback experiments on the greater sage-grouse (*Centrocercus urophasianus*) conducted by Blickley et al. (2012) have demonstrated the sensitivity of this bird to noise, with a decrease of birds to leks, although this effect showed no "memory" in the second year after the noise playback experiment. These authors confirmed the threat that anthropogenic noise has on population viability on this sensitive species.

In fishes especially, hearing specialists are limited by the noise regime of their environment, as demonstrated by Wysochi and Ladich (2005) using audiograms obtained by two hearing specialist otophysine fishes (*Carassius auratus* and *Platydoras costatus*) and an hearing generalist (*Lepomis gibbosus*).

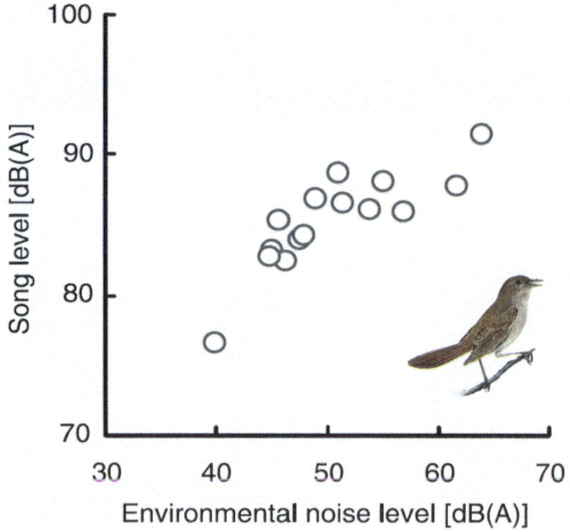

Fig. 6.6 Relationship between background noise and sound pressure of songs of residential nightingales (*Luscinia megarhynchos*) (Modified with permission from Brumm 2004)

6.10 Underwater Sound Sources

Sound propagate in the water much better than in air. In the aquatic environment (freshwater, seawater) the sound is not attenuated as is light or chemical substances, but can propagate for long distances, and for this reason sound is a carrier of information for fishes, other vertebrates, and arthropods.

Despite this evidence, the underwater world is considered from a human perspective as a quiet and silent world, but this largely depends on the fact that the hearing threshold of human ears in underwater conditions is relatively high, 84–100 dB, re 1 μPa. Many species living in water have a threshold of acoustic sensitivity higher than that of human ears (see for instance, Yan et al. 2010).

Marine mammal and fishes have no external or middle ears because the density of the medium is very similar to the internal body density and the pressure waves can be transmitted without significant loss through tissues and bones.

Marine mammals have sophisticated systems to capture sounds in a range of 3–50 kHz (hearing threshold of 60–70 dB), but some toothed whales have a sensitivity of 100 kHz and a hearing threshold of 30 dB.

Several species of fishes produce sounds that according to typology are distinct in two broad categories: stridulations and drumming sounds. The former are the result of rubbing hard parts of the body. Drumming results from contraction of the skeletal muscles along the body wall against a gas-filled structure.

The capacity to detect a signal depends not only on the ability of the receiver but also on environmental characteristics such as water depth, bottom structure (rocky, sandy, vegetated), temperature, and salinity, all conditions that produce distortion in the propagation of acoustic waves. As in the terrestrial ecosystem, noise in water depends on several factors. The hearing capacity of animals can be masked by ambient noise, which in streams and water currents can be quite high.

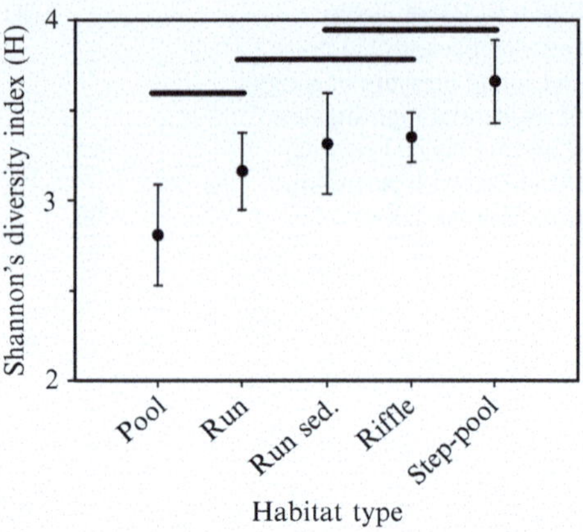

Fig. 6.7 Acoustic diversity (Shannon index, average + standard deviation) of five habitat types (pool, runs, runs with streambed sediments, and pools) in 12 rivers of the Swiss Confederation (Reproduced with permission from Tonolla et al. 2010)

The use of acoustic cues is very important for detecting predators, prey, and mates and for orientation (coastlines, torrents, water currents). For some fishes the acoustic signals must be 50 dB higher than background noise to communicate efficiently.

Twelve European freshwater habitats were selected to measure underwater background noise by Wysocki et al. (2007). In this investigation the linear equivalent sound pressure level (L_{Leq}) was measured where the quieter environment was a stagnant habitat (lakes and backwaters) with a noise level of 100 dB re 1 μPa (L_{Leq}) under absence of wind. The majority of noise was concentrated in the lower frequencies below 500 Hz. The noise in fast-flowing waters was greater than 110 dB with peaks at 135 dB. A large amount of sound energy was concentrated in a frequency range >1 kHz. Considering the acoustic capacity of fishes, noise was moderately masking the fishes' hearing in stagnant waters, but in a fast-flowing habitat this noise strongly affects the hearing capacity of fishes.

Recently Tonolla et al. (2010) investigated five common habitat types in 12 rivers of the Swiss Confederation using a couple of submerged hydrophones. Significant differences in acoustic signature or soundscape were found, confirming the importance of the aquatic soundscape for environmental characterization as perceived by aquatic life (Fig. 6.7). In a preceding experiment Tonolla et al. (2009) measured the noise in a flume, changing hydromorphic drivers such as water velocity, relative submergence, and flow obstructions. Noise emerged when the water velocity and submergence ratio of obstacles increased, and this effect resulted in a higher midrange of frequencies (125 Hz to 2 kHz). Thus, the obstruction of flow and air bubble formation produce a specific sound signature.

In the inshore Lake Jyväsjärvi (Finland), the most common underwater noise was that produced by motor vessels (ships, motorboats), especially along the waterways. Seppänen and Nieminen (2004) have measured such noise at 50–100 m from the waterways, recording a noise of 120–140 dB (linear SPL, re 1 μPa) where the majority of noise was concentrated at the frequencies 1–5 kHz.

6.11 Marine Sound Signature

The sonic sources in marine systems range between 1 Hz and 100 kHz: their origin is abiotic, biotic, and anthropogenic.

Tides and hydrostatic pressure changes produce large amplitudes and low frequencies in the spectrum. Seismic disturbance can generate sonic waves from 1 to 100 Hz. Turbulence in the oceans caused by water currents largely depends on the speed of the currents; for instance, current that moves at 1 knot generates noise around 106 dB (re 1 μPa). Surface waves generated by blowing wind have a frequency that ranges from 1 to 50 kHz. Precipitation (rain, hail, and snow) generate a spectral peak at 13.5 kHz for a wind speed of 1.5 m/s. Ship traffic generates frequencies from 50 to 500 Hz, and this noise can be heard at a distance of 1,000 miles (1,609 km).

The ocean soundscape is undergoing rapid change, and this signifies a great influence on animal life and its communication processes. Some changes have a positive direction, such as the reduction of naval sonar technology and regulations that have reduced the impact on pelagic fishes and marine mammals. On the other hand, seismic exploration in the deep seas has dramatically impacted animal life.

An increased merchant shipping activity has had a growing impact on the sea noise with an increase in the past four decades of 2.5–3 dB per decade in the frequency band 30–50 Hz. The implantation of extensive wind farms in coastal waters (<30 m) has increased the low-frequency sounds with impact on nearby marine mammals. Finally, the marine soundscape is under change caused by climatic change. Ocean acidification, from the reaction of CO_2 with the water and the formation of carbonic acid (H_2CO_3), has the effect of propagating sound farther, increasing the noise level in the oceans. For instance, a noise of 3 kHz would increase by 30 % if the pH decreases from 8.0 to 7.4 (Etter 2012).

Similar results have been found in the northeastern Pacific by McDonald et al. (2006). These researchers have found that the deep ambient noise at 30–50 Hz was 12–13 dB higher in 2003–2004 than in the 1964–1966 period, suggesting an increase of 2.5–3 dB per decade.

On the eastern Canadian continental shelf, Zakarauskas et al. (1990) have found frequencies from 30 to 900 Hz with a higher level of noise in wintertime and higher values on the Scotian Shelf than on the rest of the eastern Canadian continental shelf. Noise produced by the transit of ships contributes in relevant way for frequencies >900 Hz.

The main sources of the sonic environment in the ocean are as follows:

- Wind. This source produces noise between 150 and 200 Hz. This noise largely depends on local conditions and coast type. For instance, a net of 13 widely distributed receivers in the North Pacific confirmed a good correlation (0.56) between wind speed and marine sound (Curtis et al. 1999) (Fig. 6.8).
- Traffic. This source is produced by ship and boat engines and propellers of a frequency usually below 200 Hz. Cato (1976) reported the result of an

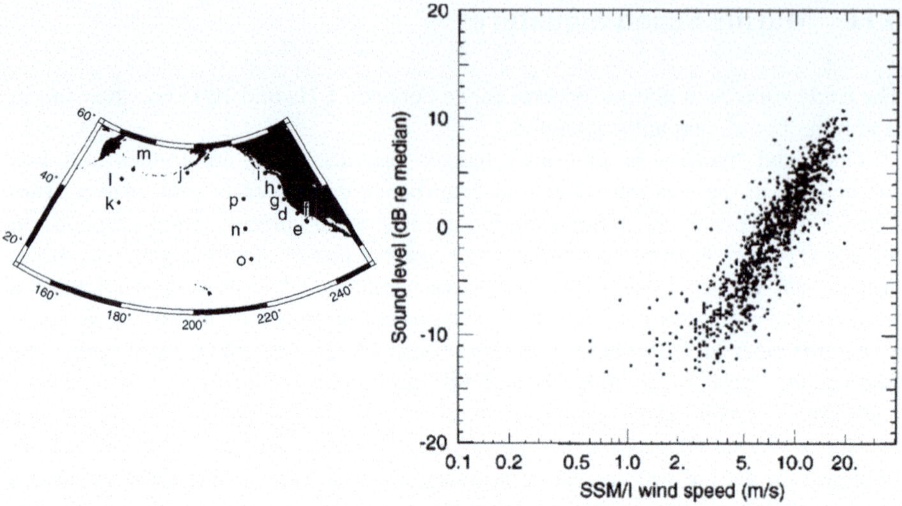

Fig. 6.8 Map of the study area locations and relationship between sound level received in the 200–400 Hz band and special sensor microwave imagers (SSM/Is) wind speed (Reproduced with permission from Curtis et al. 1999)

investigation in tropical waters near Australia, finding great variability caused by local conditions in terms of wind but also shipping traffic.

- Marine life. This source is extremely variable, from infrasound to ultrasound.

At the moment there is very little information about the marine soundscape, but every trait of the coastal or offshore parts of seas and oceans has a specific sound signature. Different soundscapes can have an important role for pelagic larval reef fishes and decapods to orient toward coastal ranges. The importance of this character is growing rapidly in research investigations.

For instance, Radford et al. (2010) have demonstrated that there are differences in the spectral and temporal composition of the sonic environment in three coastal habitats: a macroalgae-dominated reef, a sea urchin-dominated reef, and a sandy beach, all localized in northeastern New Zealand.

The sea urchin-dominated reef had an intense sound (800–2,500 Hz) when compared with the other two habitats (Fig. 6.9). The sound produced by snapping shrimp was higher on the sea urchin-dominated reef. The two coral reefs had higher activity at dusk than at noon time. These authors have stressed that the soundscape has an "immense value to the pelagic larval stage of a coastal organism."

Recently Staatermasn et al. (2013) have collected data about the distinct sonic signature between two coral reefs, one located in the Tropical Eastern Pacific (Panama) and one in the Caribbean (Florida Keys) which showed that in Panama the acoustic energy (biophonies) was higher than in Florida. In Panama, the snapping shrimp band was very active, indicating the greater activity of this species. Shrimp live in dead or fragmented corals, and we must use this acoustic evidence

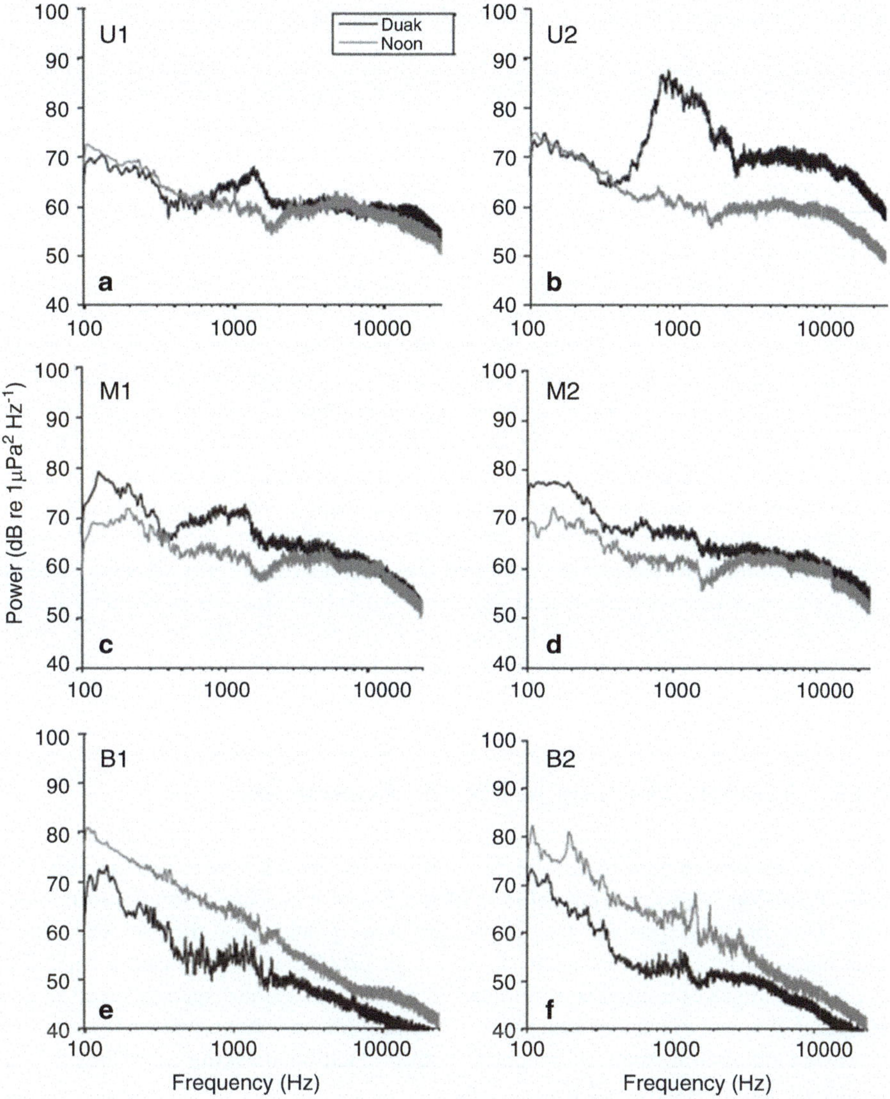

Fig. 6.9 Noon and dusk spectra collected at six locations along the northeastern coast of North Island of New Zealand: *U1, U2* urchin-dominated rocky reef, *M1, M2* macroalgae-dominated rocky reef, *B1, B2* open sandy beach (Reproduced with permission from Radford et al. 2010)

for a general coral reef assessment. Despite this the coral reef of Panama has 38–56 % of live coral compared with the Florida reef (23 % live corals). From this study it is seen that the soundscape may reveal important information about the status of the reef in terms of either complexity or level of degradation.

6.12 Acoustic Masking in Freshwater Ecosystems

In the aquatic environment, the acoustic cues are important characteristics for the survival of animals. Sound travels fast in the aquatic medium and more information is carried when compared with the terrestrial ecosystems. In shallow waters, the sound produced by aquatic insects has a range of approximately 1 m, which is quite different when compared with terrestrial insects that have a communication range of at least many meters (Aiken 1982).

In fish communities there are species considered specialists with enhanced auditory sensitivity and broad frequency ranges and non-specialists (or generalists) that do not have such characters. Specialists seem have evolved in quiet habitats such as the deep sea, lakes, backwaters of rivers, and slowly flowing streams. This hypothesis has been confirmed after an experimentation conducted by Amoser and Ladich (2005) in four different freshwater habitats near Vienna, Austria (Danube River, Triesting stream, Lake Neusiedl, backwaters of the Danube River). The sonic signature of these aquatic systems were played back to an hearing specialist, the common carp (*Cyprinus carpio*), and a hearing generalist, the European perch (*Perca fluviatilis*). Results have confirmed that carp hearing is only partially masked by a quiet sonic environment (mean threshold, 9 dB) but is heavily masked by stream and river noise by up 49 dB in the hearing range 0.5–1.0 kHz. In contrast, perch auditory capacity is not impacted by the noise of all habitats.

6.13 Acoustic Masking in Marine Ecosystems

The increase of human intrusion in the marine ecosystem has been accompanied by noise intrusion with important effects of masking for marine animal communication. This physical process has important consequences for the life of pelagic mammals, as recently stressed by Clark et al. (2009), who proposed a model to quantify changes of animal communication space as a result of spatial, spectral, and temporal changes in background noise. They provided a sonar equation to calculate time-varying spatial maps of potential communication masking according to the frequency bands considered (Fig. 6.10).

Pile-driving activity during the construction of bridges or offshore windfarms represents an important source of disturbance to water life. In particular, Bailey et al. (2010) have measured the sound produced by pile-driving activity at a distance from 0.1 to 80 km. They found that for the bottlenose dolphins, auditory injury would only have occurred within 100 m from the pile driving, but behavioral modification are expected to occur up to 50 km away.

Acoustic measurements were carried out also by Erbe (2009) during the construction of the duplicate Houhton Highway bridge in western Moreton Bay, Queensland (Australia). This area is a protected marine park and a World Heritage site of the Ramsar Wetland where turtles, dugong, sharks, dolphins, and whales are common. This author has not found evident damage to marine fauna but has observed a low abundance of marine fauna close to the pile-driving activity.

Fig. 6.10 Soundscape as determined by 19 recorded positions in a Massachusetts bay at three different frequency bands: 18–28 Hz fin whale (*Balaenoptera physalus*) band (**A**); 224–708 Hz humpback whale (*Megoptera novaeangliae*) band (**B**); 71–224 Hz right whale (*Eubalaena glacialis*) band (**C**). In (**A**) and (**C**), the peak at the center is the result of construction noise. In (**A**) the sonic activity is from a singing fin whale. In (**B**) and (**C**), there was no acoustic activity during the sampling of humpback or right whales (Reproduced with permission from Clark et al. 2009)

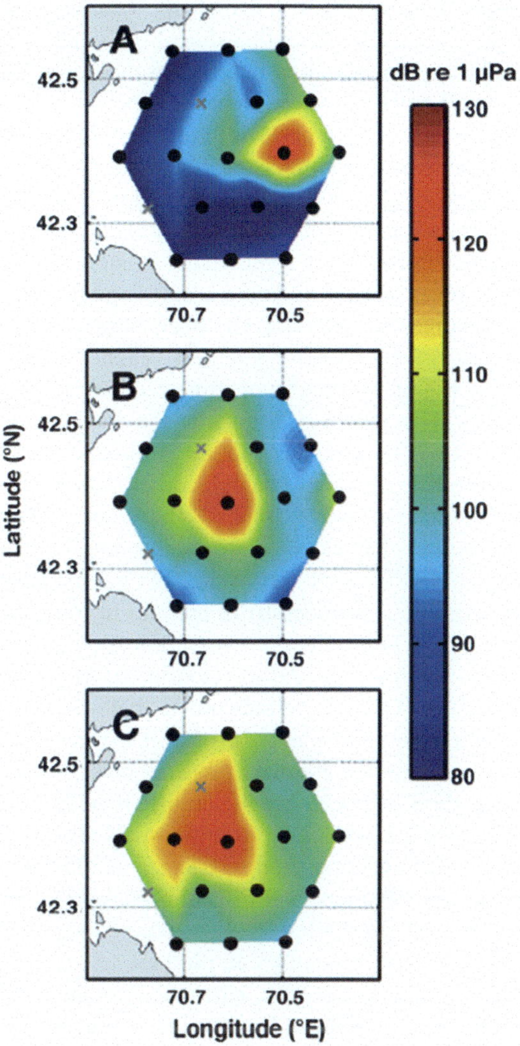

6.14 Ship and Boat Traffic Impacts on Marine Wildlife

In the marine environment, noise produced by outboard and inboard engines creates complete sonic conditions in which many animals can suffer direct hearing injury or masking of communications. Au and Green (2000) have evaluated that the presence of boats close to a group of humpback whales (*Megaptera novaeangliae*) could create problems in these animals during the singing season (February–April) in Hawaiian waters. During the singing season the amount of sound is high, creating continuous choruses, but measurements on visitor boats used to watch whales were lower than the level at which the hearing system of whales could be damaged. For instance, the sound produced by an adult ranges between 170 and 175 dB re 1 μPa at 1 m, which is lower than the noise produced by some investigated boats (120 dB at 91 m). Some changes in behavior have been observed with the arrival of boats close to the whales but is difficult to assess precisely what causes the changes, whether noise or simply the visual cues.

Parks et al. (2007) have discussed the effects of noise on acoustic communication in right whales (*Eubalena glacialis* and *E. australis*) exposed to an increased level of ambient noise. In the right whale, an increase of ambient noise level is accompanied by a lower rate call to avoid a masking effect from low-frequency noise.

Changes in behavior has been observed in the manatee (*Trichechus manatus latirostris*) by Miksis-Olds and Wagner (2011). This species when the ambient noise is high spends more time feeding and less time milling (undirected movement). The increased foraging activity can be explained as an attempt to meet nutritional requirements in the shortest time possible when experiencing noise disturbance that could be considered a menace or a dangerous situation.

In the bottlenose dolphin, Buckstaff (2004) has found a different whistling frequency when boats were approaching, and whistles were more frequent at the beginning of vessel approach. This modification in the frequency of whistles can be caused by the necessity of animals to come close together but also to compensate for the masking effect on communication as the boat approaches.

In *Gobius cruentatus* (Gobiidae), experiments conducted by Sebastianutto et al. (2011) have demonstrated that when males were tested in a round-robin design during the playback of a field-recorded diesel engine noise residents were more submissive and won less encounters than in control (silent) conditions.

In the Caribbean hermit crab (*Coenobita clypeatus*), the noise produced by a boat seems to reduce the attention paid toward potential predators, as confirmed by an experiment conducted by Chan et al. (2010) to verify the hypothesis that ambient noise has an impact on predatory risk. Boat noise playback was able to distract hermit crabs and allowed approaching the crabs more closely.

Anthropogenic noise can produce severe effects not only on behavior of pelagic fishes but also on fish ears, as demonstrated by McCauley et al. (2003) in experiments conducted on the pink snapper (*Pagrus auratus*), in which the hair cells became damaged after exposure to air-gun noise.

6.15 Effect of Traffic Noise on Wildlife

Roads are important infrastructures worldwide, with an increased impact on the environment for habitat fragmentation, chemical and sonic pollution, and mortality from collisions of vehicles with animals. Transportation infrastructures are exponentially increasing around urban areas, affecting many ecosystem processes (Forman et al. 2003).

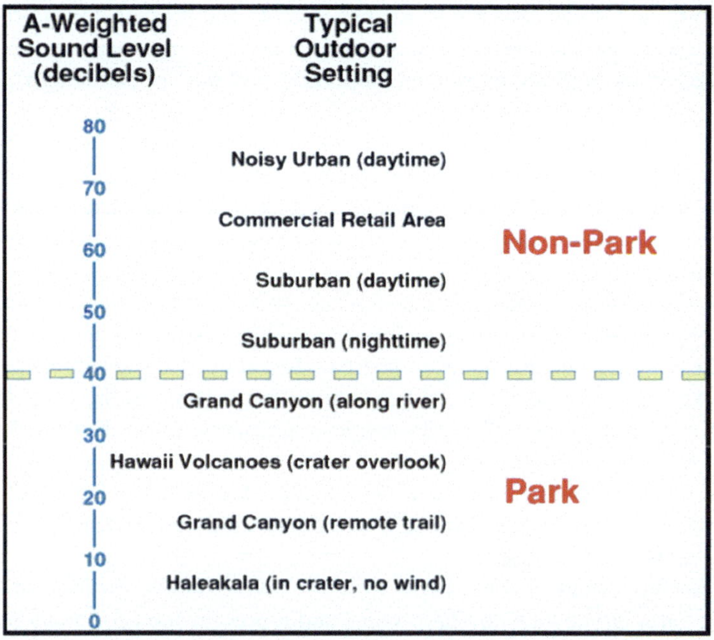

Fig. 6.11 Common outdoor A-weighted levels in different environments (From Miller 2003)

Transportation noise is highly impacting in every developed country. Miller (2003) has presented a detailed report about this problem across the USA (Fig. 6.11). Noise produced by vehicles can travel up to 4 km in distance according to the type and volume of traffic.

The acoustic effect of traffic is well known, and species seem to react to noise masking produced by traffic by modifying acoustic traits and by population shifting.

These effects have been proven by Parris and Schneider (2009) on two species of Australian birds: the grey shrike-thrush (*Colluricincla harmonica*) and the grey fantail (*Rhipidura fuliginosa*). The grey shrike-thrush sings at a higher frequency in the presence of traffic noise. The other species, which normally sings at a higher frequency than the first species, seems to not change its acoustic pattern but to appear less frequently in areas with high traffic noise. These different effects on these two species demonstrate two different strategies to cope with traffic noise: the first can raise the frequency to avoid the masking of the lower frequency of traffic noise, and the second species, which already has a song emitted at high frequency and cannot further accommodate the frequency, reacts by avoiding louder conditions. Birds have been shown to sing at a higher pitch, sing louder (Lombard effect), and change the time of day for singing to reduce the masking effect of traffic.

In other animals, research in this field is not as common. For instance, in frogs it has been found that motorcycle noise changes their chorusing behavior and that noise can reduce the diversity of species and abundance.

Frog songs are innate and not learned as in birds and primates: they are simpler and stereotyped. Females select a male on the basis of song attributes and performance (frequency or pitch, pulse rate, amplitude, call rate, or call length). Females prefer calls that are costly, meaning they prefer calls that are louder, longer, and that have a higher repetition rate. A male that produces a call with these properties is indicating that he has enough energetic reserves and vigorous, healthy, with access to abundant resources; in other words, a "winner." In some species the females prefer lower-frequency calls that are typical of large or older males with more experience.

Masking represents an acoustic interference produced by background noise or by the calls of other individuals. The noise produced by a chorus can reduce the female choice for a male, but this interference can be reduced by selection of a more efficient way of communication. When a noise masks their calls, frogs have two possibilities to reduce this effect: to increase amplitude or to shift the frequency pitch toward higher frequencies.

This second possibility is relative to the fact that noise occurs at the lowest frequencies. Considering that frog calls are really costly energetically, the second solution seems more favorable.

Frogs are important in ecosystems for they have a double position in the food web. Frogs at a larval stage are herbivores and represent food for many predators. At the adult stage, frogs are predators of many insects and pests. Their presence around the world is at risk for reasons such as spontaneous diseases, soil reclamation, climate changes, and increased numbers of roads that not only impact by directly killing individuals during their migration to reproductive ponds but also mask their choruses with increasing noise.

For instance, in birds adaptation to traffic noise can be more rapid than in frogs because a large portion of the song repertoire of birds is learned from their parents, neighbors, and other flock members; this probably does not occur in frogs, which have an innate repertoire.

Parris et al. (2009) have investigated the effect of traffic noise in two species of Australian frogs: the southern brown tree frog (*Litoria ewingii*) and the common eastern froglet (*Crinia signifera*). The effects of traffic noise were different in these two species. In *L. ewingii* calls in presence of traffic noise shifted to 4.1 Hz/dB of traffic noise, with a total size of 123 Hz. In *C. signifera*, there is a call with higher pitch, but the authors recommend more data to confirm this trend.

In evolutionary terms, the strategy to shift toward higher frequencies could produce negative effects in mating strategies. In fact, females normally respond better to lower-frequency calls of large or more experienced males.

Similar to male birds, calling at a higher frequency with background noise from traffic seems a trade-off between communicating with females and reducing the risk of predation for females that could be obliged to move more to localize males whose calls are masked by noise.

In *L. ewingii*, the active distance at which calls can be received increases from 19 to 24 m, but really few occur there when compared with the distance of 74.5 m predicted in absence of traffic. *L. ewingii* changes calls in response to traffic noise but this change is insufficient to modify the signal-to-noise ratio. In conclusion, Parris et al. (2009) underline the risk of local extinction also for populations exposed to traffic noise in suitable habitats.

How to protect frogs from traffic noise is a real challenge. The adoption of solid barriers may seem a first choice but such barriers increase the isolation of managed areas. A ultimate solution could be the adoption of a dense vegetation barrier that is cheaper and more environmentally compatible.

In common marmosets (*Callithrix jacchus*), a New World monkey, the effect of increased noise produced an increase in the sound level of their spontaneous calls, but no noise-induced increase in the number of syllables per call series was observed. This finding suggested that an increased serial redundancy of vocal signals is not utilized by this species when the environment becomes noisy. Marmosets increased the duration of their call syllables when environmental noise increased, demonstrating a vocal plasticity in the communication system (Brumm et al. 2004).

The effect of traffic on grassland birds and the unexpectedly great reactions of birds to the presence of road traffic is well documented in a study conducted by Reijnen et al. (1996) in Dutch agricultural grasslands. Using transect counts these authors have found that of 12 species considered, 7 showed reduced density close to roads. Species showed great variability in the distance from roads, from 20 to 1,700 m for a road with 500 cars a day, and from 65 to 3,530 m from a road with 50,000 cars a day. In the first case the decrease of density ranged from 12 to 56 % within 100 m. For roads with 50,000 cars a day, the decrease of density was 12 to 52 % within 500 m. Figure 6.12 presents the loss of population (%) of the most common open range birds.

Rheindt (2003), investigating the effect of highways on birds, has found a decrease in richness and diversity of birds living close to a German motorway. He found a significant relationship between dominant frequency and the decline in abundance of species toward the motorway. Birds that produce a high-frequency pitch are less susceptible to noise pollution and express a greater abundance in the proximity of the roads.

Noise from anthropogenic origin seems to produce permanent changes in the structure of song in red-winged blackbirds (*Agelaius phoeniceus*). In fact, population living in quiet marshes far from traffic noise, when exposed to traffic noise playback reacted with an increase of signal tonality, demonstrating high acoustic plasticity (Hanna et al. 2011). But in marshes located at the side of traffic-heavy roads, red-winged blackbirds maintain higher tonalities when the area is temporarily not exposed to traffic noise. This last fact demonstrates that anthropogenic noise influences the structure of their song, at least in this species.

In passive noise, for listeners such as the greater mouse-eared bat (*Myotis myotis*) anthropogenic noise could represent a strong limitation to foraging sites. Experiments conducted on this species by Schaub et al. (2008) on caged bats have demonstrated how greater mouse-eared bat avoided entering an area in which noise was produced. Submitted to three types of noise, this species reacted at increasing rate from a noise produced by traffic, noise produced by vegetation movement, and

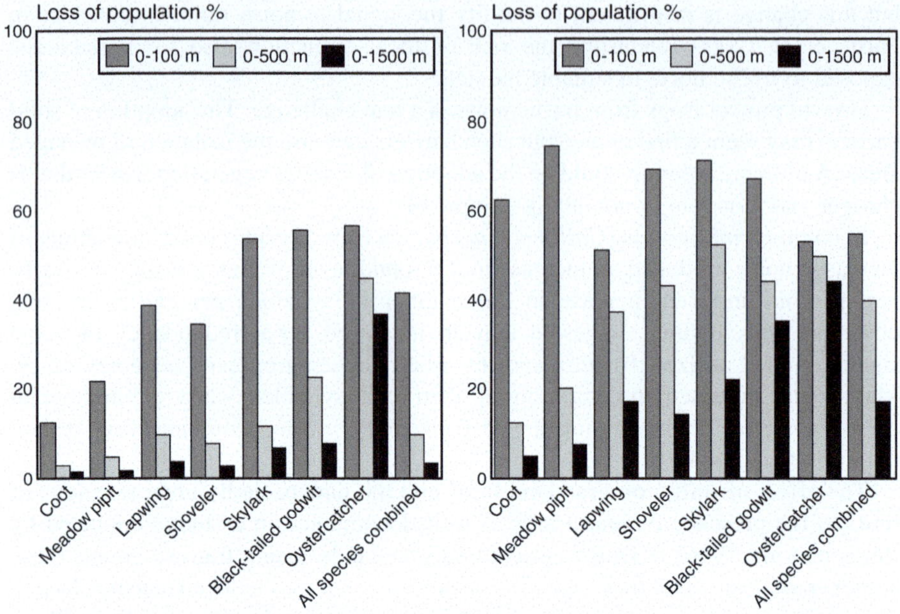

Fig. 6.12 Loss of population of different species of birds according to the number of cars/day at a mean speed of 120 km/h: 5,000 cars/day (**left**); 50,000 cars/day (**right**) (Reproduced with permission from Reijnen et al. 1996)

finally to broadband computer-generated noise (Fig. 6.13). The highly repellent effect of vegetation noise (12 dB below traffic noise) probably depends on the acoustic similarity between prey sounds and vegetation noise with a stronger masking effect.

6.16 Noise as a Signal in Marine Environments

The marine soundscape is modified by anthropogenic intrusion by shipping and small boats, drilling and mining seismic surveys, and by offshore wind farms, and this impact can mask important natural sonic cues. For instance, we expect in future years that the receding of ice will cause further intrusion of human activity in the polar regions, accompanied by increased ambient noise in summer months by seismic air-guns associated with oil and gas exploration, in areas densely populated by blue, fin, and sperm whales (Klinck et al. 2012).

Orientation, habitat selection, and predator avoidance are supported by the use of acoustic information across the animal kingdom. In particular, communication in invertebrates is of great importance for their survival and maintaining the complex food chains.

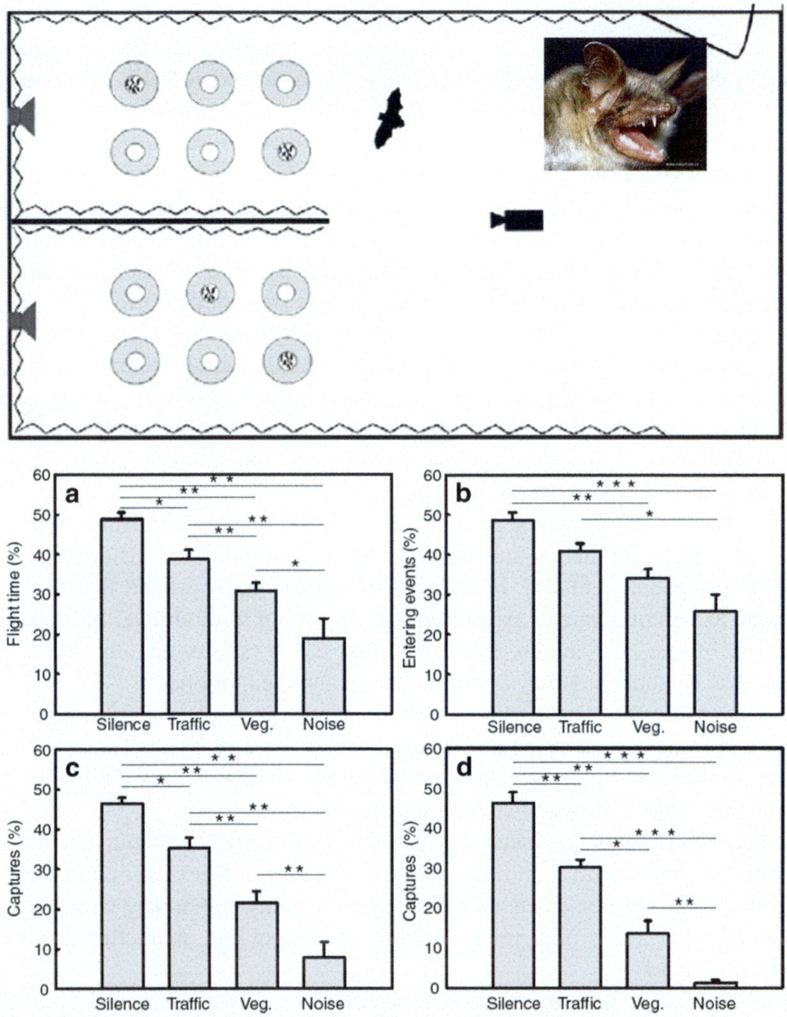

Fig. 6.13 Experimental flight room in which individuals of *Myotis myotis* were offered mealworms in a silent and a noise compartment. Flight time (**a**), entering events (**b**), capture (**c**), and percentage of the first 25 capture events (**d**) are compared with different sonic environment: silence, traffic, vegetation noise, broadband computing noise (Reproduced with permission from Schaub et al. 2008)

In marine coastal habitats, biological and abiotic noise is present at a high level. For instance, the recent diffusion of offshore wind farms have all the potentiality to modify the sonic ambience around the windmills, producing, as observed by Wahlberg and Westerberg (2005), a displacement of fishes that avoid close contact with these devices. This effect largely depends on wind speed, which generates noise on the blades.

In the underwater world, sound has two major components: an acoustic nearfield where particle velocity dominates, and an acoustic farfield where the propagation of pressure dominates. Marine animals perceive these two sonic fields differently: the nearfield is perceived by sensory hair-like receptors, and membranous receptors are used to detect the fairfield particle oscillation. These two systems are well known in fish and marine mammals, but little is known in invertebrates. For instance, larvae of many species of coral reefs can detect and are attracted by the sound produced by breaking waves on the coral reef. For many other species of invertebrates, for instance, many crustaceans, the coral reef is perceived as a hostile or dangerous ambience, because the coral reef is a place at high risk for predation.

The pelagic larvae of reef fishes seem attracted by the noise produced by a coral reef. To prove this hypothesis, Tolimieri et al. (2000) used light traps in a natural condition and with natural noise broadcast by a loudspeaker to capture pelagic larvae. They found more larvae of triplefin (a benthic reef fish) in the noise traps than in silent traps, confirming the hypothesis that acoustic cues are used to some extent by pelagic animals for orientation.

In a series of experiments Simpson et al. (2008) have confirmed that many species of coral reef fishes are attracted by reef noise with different intensity according to species and age. Further investigations have proved that behavioral traits such as recent acoustic experience are important to modulate the response to cliff noise. In fact, settlement-stage larval reef fishes (approximately 20 days old) are attracted by sounds if they had a recent acoustic experience.

Vermeij et al. (2010) have observed that free-swimming larvae of tropical corals respond to acoustic cues of the noise produced by fish and crustaceans of the coral reef and are attracted by this sound, which represents an orientation cue to find a favorable site where they will spend their benthic life.

Further experiments conducted by Simpson et al. (2012) with playback of coral reef noise combined with light traps, in the Great Barrier Reef Lagoon in Australia, have found that reef noise acts as a deterrent for many previously neglected taxa. The animals that avoid reef noise are taxa with a pelagic or nocturnal emergent lifestyle (Fig. 6.14).

This study for the first time demonstrates that a broad set of crustacean taxa with different habitat requirements and life histories are able to detect and respond to acoustic information provided by coral reefs. Furthermore, it has been proved that organisms living close to the coral reefs actively avoid the reef to escape mobile and fixed predators.

Sound produced by wave movements is a fundamental indicator of an unsafe zone for a broad range of taxa. Zoea, the pre-settlement larval stage of Brachyura, are attracted by reef noise, but megalops, the settlement stage of Brachyura, seem not attracted by the reef. The mechanisms used by crustaceans to orient using noise are not fully explained, and much research should be undertaken.

The larvae of a number of crab species have been observed to orientate and swim toward ambient underwater sound produced by coastal settlement habitats (Stanley et al. 2009). Recent laboratory- and field-based experiments conducted by Stanley et al. (2012) have demonstrated that time of metamorphosis in the settlement-stage

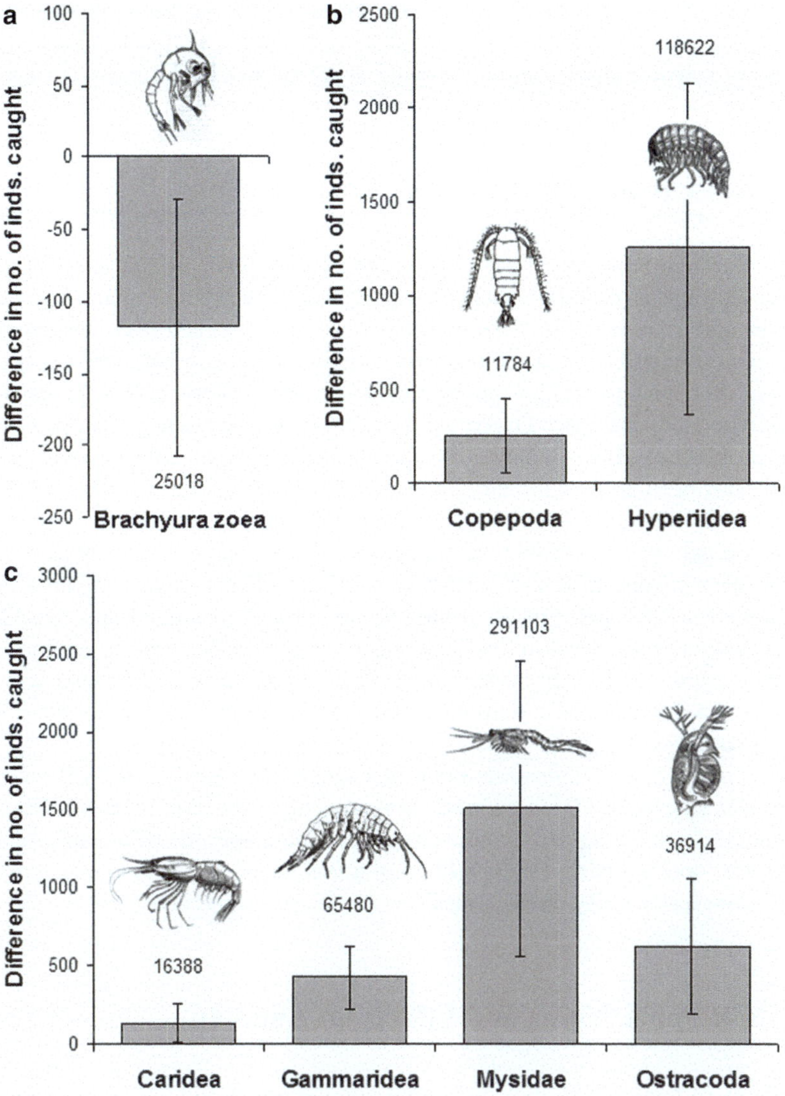

Fig. 6.14 Difference in number of individuals caught in light traps with playback of coral reef noise. Mean number of individuals (± SE) of larval developmental stage of reef-settling Brachyura (**a**), pelagic taxa (**b**), and nocturnally emergent taxa (**c**). Negative values indicate greater numbers in noise traps; positive values indicate greater numbers in control traps. Numbers *above* or *below* bars indicate total number of individuals sampled (Reproduced with permission from Simpson et al. 2012)

larvae of common coastal crab species is related to the acoustic signature of the different habitat types. A decreased metamorphic time was observed in megalopae of five species of temperate and tropical crabs when exposed to the sounds of the optimal settlement habitat type.

The use of sound to orient a broad suite of crustaceans has been demonstrated to be a widespread process, and its degradation by noise pollution should create concern in decision makers and politicians to defend marine biology.

6.17 Noise from a Multi-source Environment

When in the environment there are many sounds, people often are able to select speech of interest from the so-called cocktail party (Cherry 1953 and Pollack and Pickett 1958; see also Litovsky 2012). This ability to segregate speech from other masking sounds is not limited to humans: other nonhuman animals probably experience this capacity, although information on this point is scant. Sounds can be better distinguished when they are spaced (separate in space) than when they are co-localized.

The problem of distinguishing, in a multi-source environment, the correct sound probably has been an evolutionary constraint along the evolutionary path in many animals.

In humans has been verified the capacity to distinguish different speeches by a "spatial release from masking" (SRM) of 6–10 dB when separated by an angle of 90°. To verified this property in nonhuman animals, Nityananda and Bee (2012) have investigated the SRM in the Cope's gray treefrog (*Hyla chrysoscelis*).

In this species, as in many other amphibians, the female must localize a calling male inside a noisy social environment created by several contemporary calls of other competing males.

Experiments have demonstrated that this species has a SRM of 3 dB when the masker was displaced 90° in azimuth from the target and the value of SRM was independent from the spectral composition. In conclusion, the SRM contributes to solving the "cocktail party-like" problem of a female that must recognize a male in a very biologically noisy environment.

6.18 Oil Spills, Noise, and Effects on Animals

The effects of human activities on animal populations of wild areas are evaluated generally in term of abundance and distribution. In reality, however, local effects create modification of behavior and of reproductive activity as well. Short-term changes in behavior can produce on a larger time scale effects on abundance and distribution, and consequently it seems important to have access to this information in time to modify policies and management strategies. The acoustic method allows detecting local subtle changes in behavior that may have effects on long-term and large spatial range.

The extraction of natural resources such as gas, oil, and minerals requires geophysics prospecting by the use of dynamite detonation and the increased presence of people in remote areas. In the marine environment, seismic shooting with air-guns is used to map oil and gas resources.

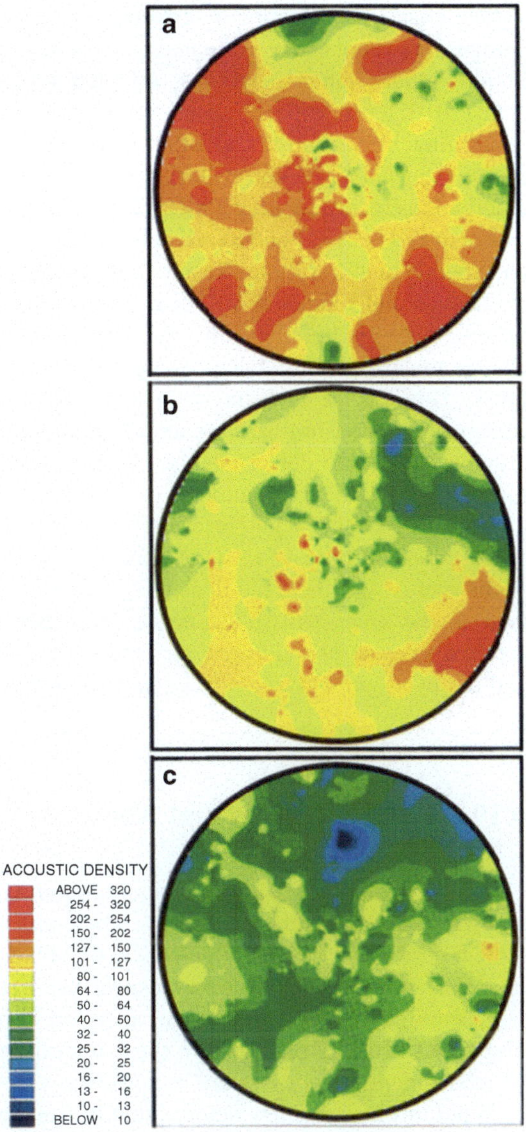

Fig. 6.15 Distribution of cod and haddock expressed in acoustic density ($m^2 \times nm^{-2}$) before (**a**), during (**b**), and after (**c**) seismic shooting. The region represented has a diameter of 40 nautical miles (Reproduced with permission from Engas et al. 1996)

In these systems, evidence has confirmed that oil and gas prospecting by using air-guns has severe consequences on the local abundance of pelagic fishes of strategic importance for the fishery such as cod (*Gadus morhua*) and haddock (*Melanogrammas aeglefinus*). In fact, after acoustic mapping and fishing trials with trawl and longlines in the central Barents Sea, Engas et al. (1996) have found that seismic shooting severely reduced distribution, local abundance, and catch rate in an area of 40 × 40 nautical miles. Close to the shooting area the reduction of catches was 70 % for trawls and 45 % for longline technique (Fig. 6.15).

The seismic prospecting in terrestrial ecosystems produces a disturbance to wild animals that is not easily evaluated by the traditional methods of line transect counts of individuals, dung, nest sites, or other biological signs. For instance, Wrege et al. (2010) have proposed an automatic acoustic survey able to collect information about movements and vocalization of populations of forest elephants (*Loxodonta cyclotis*) in Gabon, Central Africa in an area disturbed by oil prospection. Ten autonomous recording units (ARU) were dispersed in the area in which oil research was active. The ARU had a microphone and a geophone to collect seismic activity and sounds in atmosphere. Results indicate that elephant distribution and seasonal abundance and frequency of vocal emission apparently were not changed during blasting activity, but elephants had shifted their activity during the night, and the change toward nocturnal activity was increased with the increase of human proximity. This change in behavior would have escaped the traditional approach of censusing, demonstrating how acoustical monitoring represents an innovative and efficient system to detect changes in animal behavior that are the forerunner of long-term population dynamics.

Evidence of the negative effect of energy facilities was reported by Baynes et al. (2008) for boreal forests of Alberta, Canada. In the areas in which there were silent-energy facilities, the density of passerine community was 1.5 times greater than in the areas with noise-energy facilities. In particular, the white-throated sparrow (*Zonotrichia albicollis*), yellow-rumped warbler (*Dendroica coronata*), and red-eyed vireo (*Vireo olivaceous*) had lower density in noise areas.

6.19 Wind Turbines and Noise

Electricity-generating wind turbines (windfarms) are a modern system to produce renewable and no-emission energy. Their impact on the landscape is at least of three types: visual (aesthetic), physical (the blades impact with migratory birds and raptors), and acoustic. This last aspect has been for a long time neglected but represents in wild and mountains areas an important source of sonic pollution. The effect of noise produced in these windfarms alters the sonic environment to which residential animals may be sensitive.

Wind turbines are new devices that in some areas become the dominant scenic and sonic elements. Their rotation capture human eyes and their sonic emissions are important for the soundscape and the landscape. It is well known that vision influences environmental noise perception and vice versa, and that the noise annoyance is greater when the source of noise is visible.

In a study conducted on the form of noise barriers, the more pleasant barriers were considered a minor annoyance that was attributed to the noise traffic. After this example it seems reasonable to predict that seeing wind turbines and hearing their noise should produce the highest rate of annoyance, but if the wind turbines are evaluated positively by a social assessment, the annoyance should decrease.

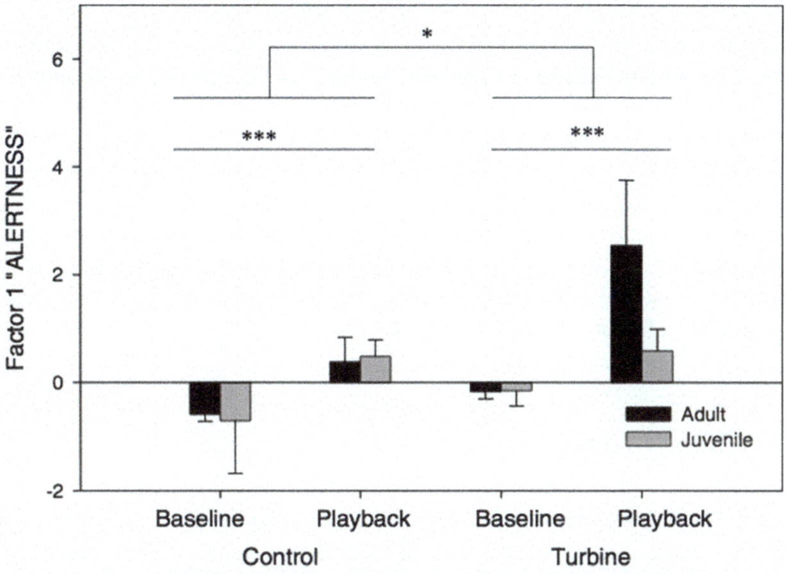

Fig. 6.16 Adults and juveniles of California ground squirrel (*Spermophilus beecheyi*) close to wind turbines have an increased alertness response to conspecific alarm calls. The difference between baseline and playback results was highly significant statistically both in control site and near turbine. Values were higher at the turbine than at the control site (Reproduced with permission from Rabin et al. 2006)

A Danish study has demonstrated that height and closeness, expressed as vertical virtual angle (VVA), were better predictors of the noise than the measurement of the noise level per se. When a turbine is placed close to the home of a respondent, it becomes clear that the feeling of intrusiveness could be very strong (Pedersen and Larsman 2008).

Complex reactions have been observed in wild animal populations. Rabin et al. (2006) have investigated the effect of such devices on a population of California ground squirrels (*Spermophilus beecheyi*) at the Altamont Pass Wind Resource Area of Northern California (Fig. 6.16).

This species alerts the arrival of predators with specific alarm calls and the noise of turbines significantly masks this signal that is important for the entire population. Two populations were selected for this study, one population close to the turbines and another far from the windfarm. The population living close to the turbine was noted to have a new behavior consisting of an elevated rate of vigilance and a return to their burrows when alarm calls were uttered. This difference was proved not to depend on other environmental factors such as population density, predator pressure, or environmental characteristics.

At the beginning of the use of wind turbines, displacement noise annoyance was caused by mechanical noise, but actually with technological improvement of the design, only a residual noise is produced by a broadband aerodynamic noise from the blades close to a turbine. The noise (a distinct "swooshing") is similar to a aircraft background noise.

To limit the annoyance effects of the wind turbines, their noise has been entered into an emission regulation protocol. For instance, in Germany a noise of 45 dB is allowed. This tradeoff could be modified because the limitation of the power by a turbine during strong wind could be masked by the noise of wind itself.

Noise level changes greatly in different countries according to the character of the landscapes. If the wind turbines are functioning in suboptimal regime to respect the noise regulation, this requires a displacement of a major number of turbines per unit of surface. In the UK the noise of turbines is assessed 5 dB higher than the environmental noise level at different wind speeds. This approach seems very interesting because the noise limit is adaptive, but it is expensive to have data for each turbine to calculate the possible noise.

6.20 Military Noise

Few studies are concerned with military noise. The major source of noise in peacetime is represented by fixed-wing aircraft (sonic boom, turbine noise, propeller noise, exploding bombs/missiles), helicopters (rotary-wing aircraft: rotor blade, turbine), tanks (engine and road noise), artillery, guns (muzzle blast, shock wave of projectile, explosion) (airborne and substrate-born), and infantry (small arms).

The concern is that most of the military noise goes on in wild areas far from people and urban settlements, and this can cause great impacts on animal populations.

The effect of military noise covers a broad spectrum of processes and dynamics. Animals are really impacted by military activity at ground level when the movement of tanks and infantry is accompanied by explosions of bombs, artillery, and rifles. The amount of sound level of a small projectile (60 g TNT) exploding generates a peak of 144 dB, and a larger projectile of 20 kg TNT has a peak of 163 dB at 100 m distance.

Blast noise represents an impulse noise with a shock wave that can be >150 dB SPL.

Further research is necessary, but military noise, excluding the blasts, is not different from civilian vehicle-borne noise.

6.21 A Psychoacoustic Approach to Noise

The perception of the everyday world is an exceedingly complicated phenomenon, and especially the perception of sound, which is an omnipresent source of information and a signal per se, cannot be voluntarily excluded by our sensorial input process. In fact, sounds are perceived when our mind is immersed in other fields of cognition; in other words, we cannot switch off the sound sensations. This fact has great consequences on everyday life, and a sound assessment requires not only physical parameters such as frequency and sound pressure level but also psychoacoustics parameters such as loudness, sharpness, roughness, fluctuation strength, the informative content of the sound signal, and a cultural filter.

Noisiness and annoyance are subjective attributes of perceived sound, and definitively the social and cultural background are important more than the energy of sound. Thus, noise evaluation depends on the physical characters of the sound, the psycho-acoustical features of the human ear, and simultaneously the psychological and social aspects of a human being. Human sensation and psychoacoustic parameters are connected by complex interactions, and manipulating sounds is a premise for design and managing sounds indoor and outdoors also.

Noise annoyance that represents a so strategic and relevant issue can actually be mapped using the A-weighted sound pressure level, but this metric does not reflect all the complexity of the soundscape domain (Genuit and Fiebig 2005). The appreciation of annoyance requires characterization of the sound, its time structure, and finally the attitude and expectation of the listener. Annoyance that is considered the overall evaluation of the disturbance and the unpleasantness of environmental noise remain highly subjective, linked to social and cultural backgrounds. Finally, annoyance and all the effects of noise on human life (but also on animal life) cannot be simply described by a few physical parameters. In this direction is addressed the psychoacoustics approach and methodologies that considers other aspects beyond the physical sonic components, such as context, kind of information, individual expectation, and attitude to the sound.

Unfortunately, masking effect, sound impression, spatial distribution, and complex phase relationships are factors that limit the conventional one-channel measurements. Binaural technology seems to solve such limits. Binaural reproduction allows the listener to be in approximately in the same condition where the sound occurred in the living sonic environment. The use of binaural recording for reproducing a sonic environment in psychoacoustic tests is very effective. Spectral band and time integration are two possible parameters that when changed produce a different reply to annoyance.

Conventional noise maps based only on the sound pressure level A (averaged) do not describe completely the environmental noise quality and complexity. The adoption of binaural technique improves the capacity to record more realistic sonic patterns useful for formulating, after a post-processing, the level of annoyance of listeners.

6.22 Noise from Urban Soundscapes

Urban can be defined any area of high-density human population and human-built structures in which the noise produced by human machines creates an unrivaled effect that cannot be found in the natural world (Warren et al. 2006). Sound in urban areas can originate from traffic sounds, mechanical sounds, human sounds, and natural sounds.

The urban soundscape is very heterogeneous in the world but has common features such as elevated noise level, predominance of low frequencies, diurnal variation in noise, abundant large flat impervious surfaces, and modified sound channels, although these characters can be also found in nature (e.g., in streams, canyons, and windy environments). The same acoustic energy of a metropolitan area can be found in the great falls (Niagara Falls, for instance), or before the Southern Hemisphere marine coast (Cape of Good Hope) during a strong wind, or during Caribbean hurricanes or East Asia typhoons.

6.23 A Historical Perspective of Urban Noise

Noise in urban areas is truly ancient. At the time of the Roman emperors a directive banned the chariots from crossing the streets of Rome at night to prevent the noise created by wheels on paving stones. In Medieval Europe, in some cities transit by chariots and horses was banned at night, or the streets were covered with straw to reduce the noise and allow residents to sleep.

The framers of the American Constitution in the eighteenth century, working in Philadelphia, established that covering the nearby cobblestone streets with earth would avoid reducing their concentration.

In conclusion, noise from roads is an old problem that has persisted across the centuries despite the improvement of technology and lifestyles. In New York City, the maximum sound pressure level was measured on subway platforms as 106 dBA and 112 dBA inside subway cars. In the 10 years from 1980 to 1990 the noise increased by 10 %, and in 2000 30 % of Americans complaints regarded noise. The noise was considered so annoying that 40 % of Americans want to change their place of residence. In New York City in 1998 the noise complaints were at first place of the Quality of Life Hotline. In 1971 the World Health Organization (WHO) Working Group announced that noise is a major threat to human well-being. A technological and useful world demands energy and produces entropic noise that in the end is considered necessary and regulated. Some activities such as garbage disposals, dishwashers, clothes washers and dryers, furnaces, refrigerators, air conditioners, yard maintenance equipment (mechanical shears, chainsaw, mowing machines) are considered necessary to our life and their sound tolerated. But we are exposed also to television, radio, and related technologies. The sounds of internal combustion engines, trains, and ship horns are from necessary devices that produce a significant dose of noise and require strategic regulation. Noise from societal purposes, such as emergency vehicles and desired activities, are tolerated by most people. Ultimately during our life we consider every unwanted sound to be noise, and noise remains intrusive, pervasive and ubiquitous, and finally unhealthy.

6.24 The Cognitive Component in Noise Evaluation

The soundscape term that appears in the literature in a inflationary way after the first use by Schafer (1977) is not an objectively existing reality but the result of cognitive processes, resulting in a culturally affected context connected to human perception (Genuit and Fiebig 2006).

The evaluation of a soundscape, defined also as "*perception of the acoustic environment as perceived by people in that place, in context*", is based on the discriminant of wanted or unwanted sound into a specific context and not simply by the metric of sound level. Unpleasant–pleasant, uneventful–eventful, chaotic–quiet, and boring–exciting are some terms used to describe the sonic context according to the personal attitude and sensitivity of the subject.

In fact, the common acoustical and psycho-acoustical metrics are not related to personal evaluation of sound quality, as resulted from experimentation carried out in Hong Kong by Lam et al. (2010).

In research conducted in the urban area of Rotterdam, found no correlation between annoyance from various sound sources and the pleasantness of these sound sources.

In parks, a discrimination was found for music (both annoying and pleasant), airplane (pleasant), water (pleasant), animals (pleasant), city sounds (pleasant), road traffic (annoying), people (annoying), and scooters (annoying).

In natural areas the sound discriminative sources were people, children, and water. Water and birds were discriminative for annoyance. Construction noise in the parks was surprisingly of little interest to the subjects interviewed, which fact remains unexplained.

An example of psychological conditioning has been reported by Schuemer-Kohrs et al. (1998) comparing noises perceived by train traffic and noises perceived by highway traffic. These authors carried out 16,000 interviews after having selected eight areas in which train and traffic noise were equally distributed. The results confirmed that people were more tolerant toward train noise than toward road traffic noise. The authors warn that the investigation was conducted along a railway with fewer than 260 trains in 24 h, with a speed not greater than 200 km/h and with a proportion of freight traffic not exceeding 67 %. Different results could be expected with high-speed trains and modern intense passenger and freight traffic.

In urban areas, the natural remnants or the artificial natural places are very much appreciated and the relationship between soundscape quality and tranquility is very strong.

In fact, natural components receive high appreciation when the context is urban, but these features are neglected in natural areas because they are abundant and diffuse or a higher profile of natural integrity is required for explicit appreciation. In other words, in natural systems the integrity is absolutely necessary to appreciate quality and tranquility.

The soundscape strongly reflects people's expectations according to the individual profile, experience, and cultural background. Thus, the soundscape rating will be different for residents and for visitors. The time of the day that for visitors is generally more restricted, for instance, confined to the business hours (shops, museum, gallery, events), for residents the sonic context especially of the outdoors is longer lasting.

Noise maps and psychoacoustic maps combined together could improve the further interpretation of the complexity of the sonic ambience and the non-sonic context, especially where the sound level is not very high. In particular local experts, people who have lived for a long time in the same place and have a good knowledge of the sites and of the natural and anthropic dynamics, may have a different idea of the sonic ambience.

The urban context represents an impressive laboratory to experiment with new epistemic principles and methodologies in soundscape analysis.

Investigations based on past experience, collection of visual stimuli, consideration of potential sonic context, and the contemporary collection of physical characters of the sonic ambience, spectral analysis, and acoustic complexity metrics represent elements for a new research agenda to be applied not only in urban areas but also in rural and natural environments.

6.25 Effect of Noise on Humans and Society Health

Noise that may be defined also as "a sound that creates problems" has received growing attention during the last decennia, becoming a very popular theme in soundscape and acoustic ecology. The effects of noise pollution on human societies are numerous, pervasive, persistent, and medically and social relevant (Goines and Hagler 2007). Noise produces tangible (economic) and intangible (well-being) losses. Noise interferes with communication, recreation, concentration, and sleeping. Following airborne pollution, noise represents a non-secondary factor that should enter the political agenda of decision makers. Homes, yards, public spaces, cars, theaters, restaurants, parks, arenas, beaches, and mountain ridges: there are no spaces that can be subjected to unwanted noise pollution.

Noise poses health problems because doses that are not harmful for our hearing apparatus influence the subconscious, which transforms sounds as danger signals even during the sleeping hours.

The health problem is becoming important but is underestimated, and some NGO organizations such as Noise Free America (www.noisefree.org) or Noise Pollution Clearinghouse (www.nonoise.org) are taking action to prevent noise and to educate. Also the web is active on this subject; see, for example Noise Off (www.NoiseOFF.org).

Health and well-being are two components necessary to assure quality of human life. Among the factors that can negatively interact with health and human well-being, growing importance is represented by sonic ambient noise. Noise increases the health deficit, especially in urban societies. Noise interacts with health on two levels: a primary health effect and a secondary health effect. In Fig. 6.17 are

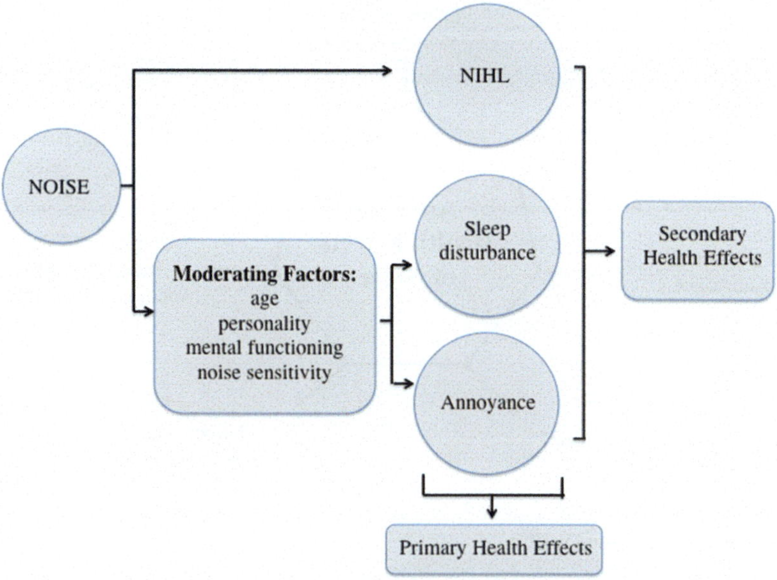

Fig. 6.17 Relationship between sound and primary and secondary health effects

indicated the physical effects of noise (noise-induced hearing loss, NIHL) and the nonphysical effects. The cumulative effects of the different components of noise parameters and life trait such as age, personality, mental functioning, time of day, and noise sensitivity represent the moderating factors. Sleep disturbance and annoyance are the mediators between physical factors and secondary health effects (reduction of health quality, diseases). The noise-induced hearing loss acts directly on the secondary health effects.

It is interesting to note that there is a strong correlation between noise and annoyance, but the correlation is weak between noise sensitivity and noise level.

According to the World Health Organization (WHO), seven categories of adverse health effects of noise have been recognized (Fig. 6.18):

Hearing impairment is the result of many causes such as trauma, ototoxic drugs, infection, and heredity. Below 70 dB also prolonged exposure can produce damage, but when the sound pressure is above 85 dB for more than 8 h the risk of hearing damage increases.

Hearing loss is largely caused by exposure during work activity, and children seem more sensitive to damage to hearing. Hearing impairment in some cases is accompanied by loudness perception (loudness recruitment), distortion (paracusis), and tinnitus. Hearing losses produce in the subjects, according to age, different effects such as sense of isolation, depression, loneliness, impaired speech discrimination, impaired school and job performance, and later limited job opportunities.

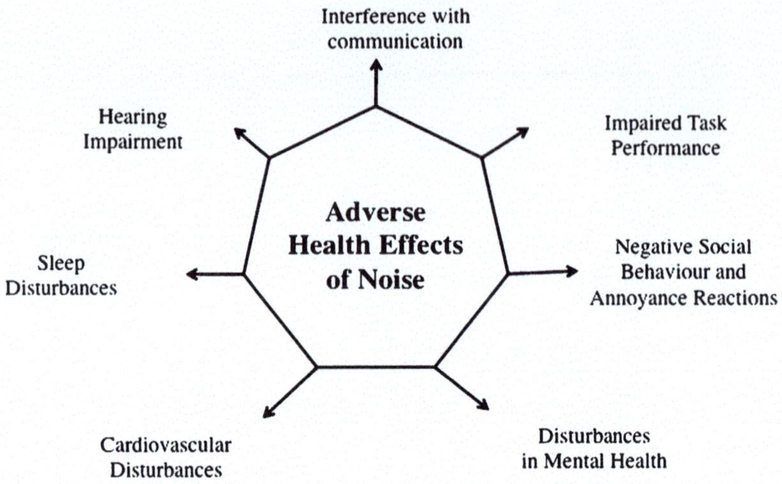

Fig. 6.18 Schematic representation of the major adverse health effects of noise according to the World Health Organization (WHO)

In particular, children who receive wanted or unwanted high doses of sounds can have hearing losses that affect communication, cognition, social-emotional development, social behavior, and academic outcomes.

People working in bars, nightclubs, and other places of entertainment are at risk as are those who work in a marble mine or sawmill.

Special recommendations are devoted to workers who use tools such as a jackhammer or snowmobile (>100 dB) to reduce their working time. The threshold of pain is fixed at 140 dB. Firecrackers and cap pistols can produce permanent hearing loss.

Interference with communication. When noise is high there are difficulties in understanding speech and this behavioral limitation has consequences for concentration, fatigue, uncertainty, lack of self-confidence, irritation, misunderstandings, decreased working capacity, disturbed interpersonal relationships, stress reactions, etc.

Sleep disturbances. A noise dose of 30 dB is sufficient to disturb sleep in most people. Intermittent noise increases the probability of being awakened. Noise has several effects on our physiology during the sleeping period: increased blood pressure, increased heart rate, increased pulse amplitude, vasoconstriction, changes in respiration, cardiac arrhythmias, and increased involuntary body movements. Secondary effects during the day after the exposure are fatigue, depression, and decreased performance. The effects last for the next 24 h overall. These effects are particularly evident in the elderly, shift workers, persons vulnerable to physical or mental disorders, and people with sleep disorders.

Low-frequency sounds are more disturbing than high-frequency and louder sounds.

Cardiovascular disturbances. The endocrine and autonomous nervous systems are particularly exposed to noise effects. This effect can act on the cardiovascular system and may create a risk for cardiovascular diseases. A long-term noise dose above 65 dB or acute exposure to 80–85 dB can initiate cardiovascular disorders.

In children, increase of blood pressure and an elevated level of stress-induced hormones are particularly common.

Disturbances in mental health. Noise exposure accelerates and intensifies the development of mental disorders: stress, anxiety, nervousness, nausea, headache, emotional instability, argumentativeness, sexual impotence, changes in mood, increase in social conflicts, neurosis, hysteria, and psychosis. In a noisy context some interpersonal disputes end in injury or death. The dehumanization complaint in a modern urban context may the result of such a noise-polluted context.

Impaired task performance. Noise reduces the capacity of concentration and more in general in performance at school and at work, with an increase in errors and a decrease of motivation.

Problem-solving capability, reading attention, and memory are strongly reduced by noise. All these facts suggest the necessity to better control the quality of the sonic context to reduce the risk to have handicapped future generations. The longer the exposure to noise, the greater are the effects in children, and this suggests that schools and recreational areas should be located in quiet areas rich in pleasant sonic information.

Negative social behavior and annoyance reactions. Annoyance is defined as a feeling of displeasure created by an agent or condition that is believed to be affecting the subject adversely. Vibration and low frequencies increase annoyance, but other effects are companions of annoyance, such as anger, disappointment, dissatisfaction, withdrawal, helplessness, depression, distraction, anxiety, exhaustion, and agitation.

People exposed to noise change behaviors such as closing all windows and doors to impede the outdoor noise from entering the home, avoiding balconies, yards, and patios, turning up the volume of radio and televisions, change social behavior (e.g., aggressiveness, unfriendliness, nonparticipation, or disengagement), and change in mood.

The degree of annoyance depends by several factors such as the time of day, the unpleasant characteristics of the noise, duration and intensity of the noise, and the meaning associated with the noise. Annoyance increases as the noise increases: such progression is strongly perceived.

The human population is not homogeneous in terms of effect of noise. Patients with various diseases, patients in hospitals, or those subjected to rehabilitation after injury, blind people, those hearing impaired, fetuses, infants and your children, and the elderly are more sensible to noise impact.

Children, which are more vulnerable to noise, require special attention and noise protection. A recent investigation carried out by van Kempen et al. (2010) on 2,844 schoolchildren (aged 9–11 years) attending 89 primary schools around three European airports did not confirm a relationship between transportation noise exposure and perceived health in these children, but no definitive conclusions can be drawn. Children that denied annoyance reported more symptoms than children without annoyance. Annoyance produces a reduction of blood pressure, but in the children investigated in this project there were no differences between blood pressure of annoyed and unannoyed children.

In urban areas especially noise is the result of multiple sources of sound (e.g., aircraft, motor vehicles, trains, boom cars, car horns, car alarms, sirens). The negative effects on health seem more the result of the combination of different sources than the effect of a specific noise. In particular, a low-frequency noise that is pervasive, accompanied by a more efficient propagation, is difficult to be mitigated or reduced. This low-frequency noise that is underestimated by measurements may have severe consequences on health.

The World Health Organization (WHO) has produced important guidelines for dwellings, schools and preschools, hospitals, for industrial, commercial, shopping, and traffic areas, for ceremonies, festivals, and entertainment events, and the use of headphones or music, impulse sounds from toys, fireworks, and firearms.

6.26 Urban Soundscape Design for Biodiversity

As argued by Katti and Warren (2004), the urban environment represents a challenge for integration between humans and nature. Urban noise offers a unique opportunity for great experimentation on how a signal changes under the pressure of such a dynamic sonic environment. Urban means heterogeneous in space and time for the distribution of houses, roads, gardens, parks, etc. This heterogeneity is found also in the soundscape, which creates new conditions in which species can evolve and adapt coping cultural and genetic mechanisms. The adaptation of species to the urban environment can suffer physiological constraints such as the impossibility of changing the frequency or suffering from the shift in acoustic rhythms to escape the masking effect of noise. These adaptations simultaneously could produce desynchronization of mating or increased exposure to predators.

In a recent investigation in 27 urban parks of cities and villages across Spain and Portugal, Paton et al. (2012) have found that of the 91 species of birds found in the urban parks at least 10 (*Regulus regulus*, *Streptopelia turtur*, *Dendrocopos minor*, *Buteo buteo*, *Hirundo daurica*, *Corvus corax*, *Oriolus oriolus*, *Cettia cetti*, *Passer hispaniolensis*, *Sylvia melanocephala*) seem affected by noise level and that a reduced value of noise (below 50 dB) can allow the presence in the urban parks of rare species (Fig. 6.19). The authors conclude that the reduction of urban noise could open a new era of biodiversity conservation in urban green areas. In fact, modern urban ecosystems, because of their intrinsic heterogeneity that reflects the great surface occupied, are attractive to a growing number of synanthropic animals. Animal communication is affected by urban noise only when the intensity is high, but when this value is medium or low, adaptive mechanisms prevail.

For instance, from the different reactions that animals have to incorporate the effect of noise in the communication system, the shift of frequency has been the one

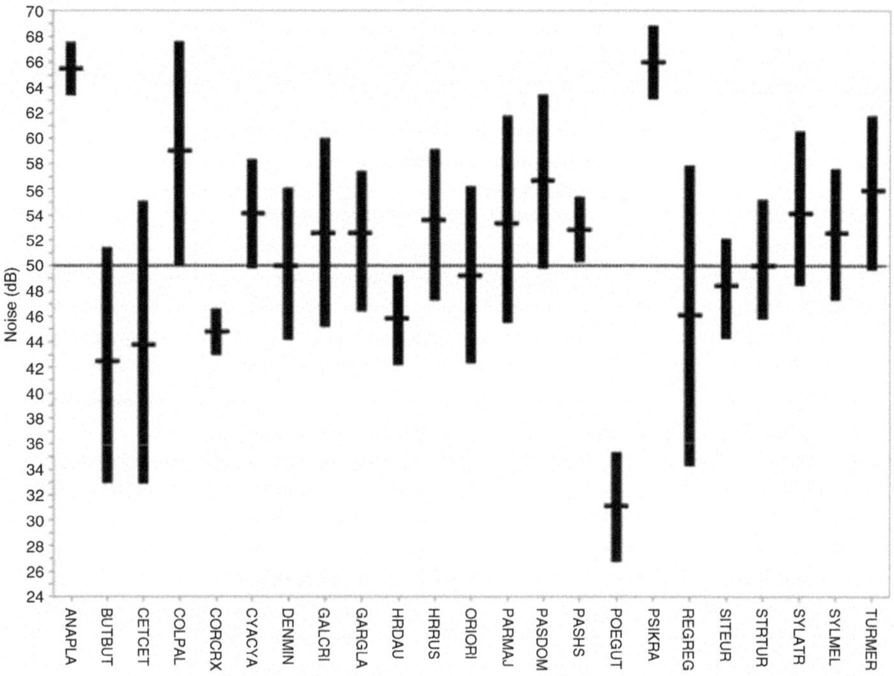

Fig. 6.19 Noise preferences (means) and noise tolerance (standard deviations) of bird species that have a significant sensitivity to noise level. The line at the threshold of 50 dB separates rare from common species. ANAPLA *Anas platyrrhynchos*, BUTBUT *Buteo buteo*, CETCET *Cettia cetti*, COLPAL *Columba palumbus*, CORCRX *Corvus corax*, CYACYA *Cianopica cyana*, DENMIN *Dendrocopos minor*, GALCRI *Galerida cristata*, HRDAU *Hirundo daurica*, HRRUS *Hirundo rustica*, ORIORI *Oriolus oriolus*, PARMAJ *Parus major*, PASDOM *Passer domesticus*, PASHIS *Passer hispanolensis*, POEGUT *Taeniopygia guttata*, PSIKRA *Psittacula krameri*, REGREG *Regulus regulus*, SITEUR *Sitta europaea*, STRTUR *Strepopelia turtur*, SYLATR *Sylvia atricapilla*, SYLMEL *Sylvia melanocephala*, TURMER *Turdus merula* (Reproduced with permission from Paton et al. 2012)

observed by Luther and Baptista (2010) in the three dialects of the white-browed sparrow (*Zonotrichia leucophrys*) living in the San Francisco peninsula (California). Over a 30-year time span urban noise, which is louder at the low frequencies, has increased, and this has affected the three adjacent dialects with an increase of the minimum frequency. This mechanism seems culturally transmitted to counteract the increase of noise at lower frequencies.

At the conclusion of a long review Patricelli and Blickley (2006) argued that birds can react in a quite complex way to urbanization and according to all the degree of changes that happen when humans organize urban areas. Some species disappear, but others receive some advantages. It remains an open question if species that have the capacity to adjust their songs could be less affected by urbanization processes.

A point not yet discussed is whether the species present in a rural area are favored by the transformation in urban areas. When these areas are definitively converted into metropolitan areas, biodiversity probably is more adapted to persist.

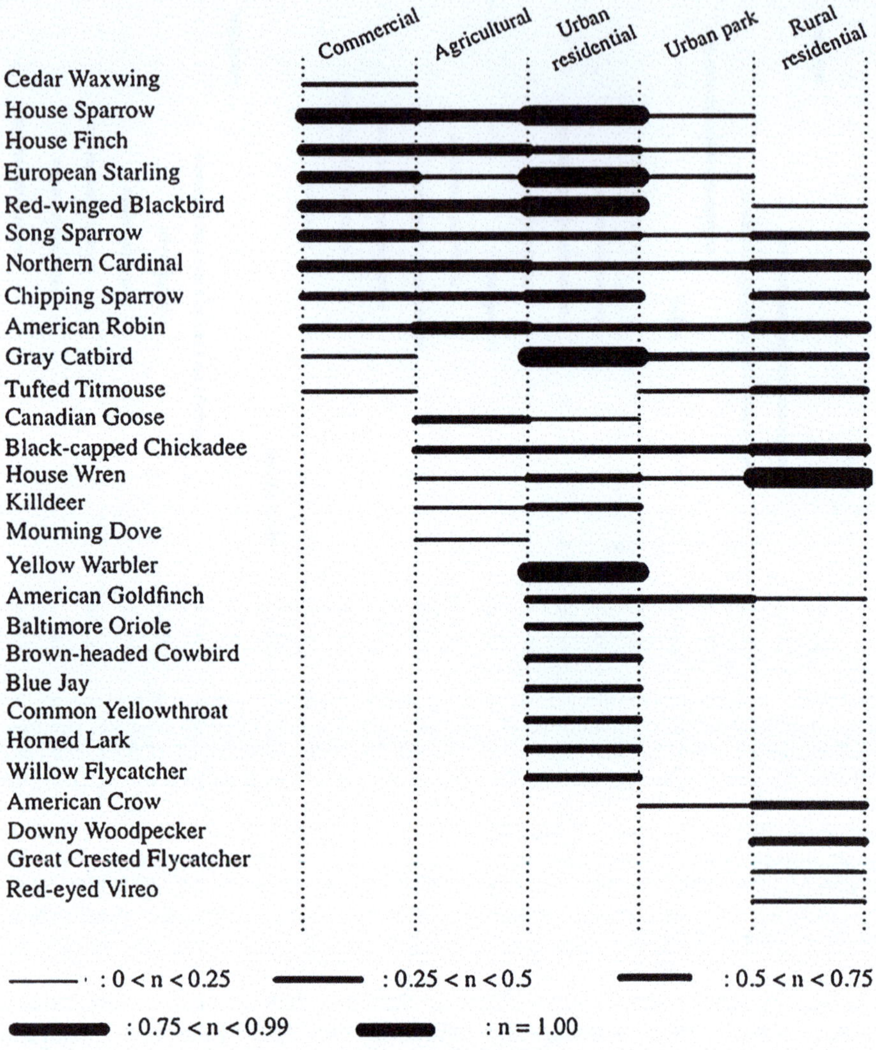

Fig. 6.20 List of birds listening across a landscape gradient ranging from commercial to rural residential in Greater Lansing area, MI (USA), ordered according to the highest occurrence probability from commercial to rural residential (Reproduced with permission from Joo et al. 2011)

For this, it would be of great interest to investigate which are the species that are better adapted to urban areas, and if these species are characteristic of rural areas, where rural areas could represent an intermediate stage of animal adaptation to human-dominated ecosystems.

In a study conducted by Joo et al. (2011) in the Greater Lansing Area (MI, US), house sparrow, house finch, and European starling were found dominant (>75 %) in the high or moderate development areas (Fig. 6.20). Cedar waxwing, a species that usually lives in open rural landscapes, has been found in commercial areas rich with fruit trees planted for ornamental purposes. It seems that areas with a moderate level of urbanization have the maximum level of diversity. Although a direct relationship between bird diversity and soundscape diversity is not clearly demonstrated, nevertheless there is good evidence that biophonies increase with the increase of landscape complexity where such complexity is strictly related to human land use. Decrease of the level of noise probably would have a beneficial effect on a permanent urban settlement of wild populations of birds.

6.27 Perspectives in Noise Research

Anthropogenic activity represents one of the most influential sources of noise with wind, water, and animals. This process is pervasive and is in continuous expansion in terms of diversity and intensity, and the resulting noise masks many other natural sounds during daytimes and with the seasons. In fact, anthropogenic noise is louder, more frequent, and locally more common than the other natural sources of noise. Although the majority of research on noise has been conducted in terrestrial systems, noise intrusion is important also in aquatic systems and in particular along coasts and also in the open oceans.

Noise, extensively discussed in this chapter, is a complex subject whose effects extend from gene to landscapes (Fig. 6.21). In a recent review Kight and Swaddle (2011) have focused on eight categories of human systems that are potentially exposed to noise effects: the neuroendocrine system, the reproduction and development systems, metabolism, cardiovascular health, cognition and sleep, audition and cochlear morphology, the immune system, and DNA integrity and genes.

Anthropogenic noise creates a diverse and complex impact on wildlife, influencing multiple biological processes in either direct or indirect ways.

For instance, animal behavior and reproduction are important processes in which noise can determine significant changes and shifts in several living traits.

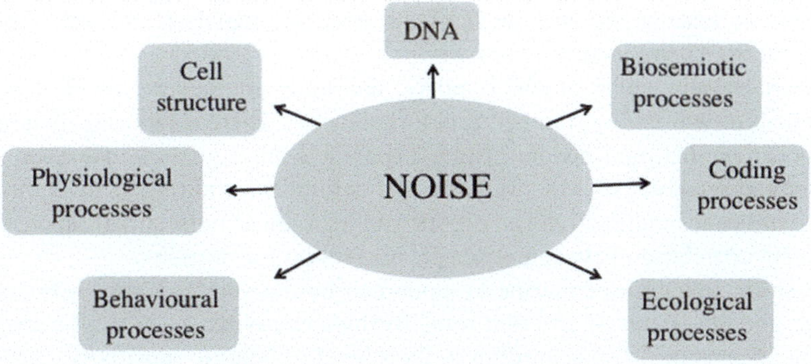

Fig. 6.21 Noise interacts at different scales with all life systems

Future research should better represent groups of animals such as amphibians, invertebrates, and reptiles that are not sufficiently studied. A comparative analysis of the behavior of species exposed to equivalent noise pressure could allow us to understand the different effects of noise and to suggest strategic procedures of mitigation and of conservation management. Extension of the investigations to infrasonic and ultrasonic frequencies can be extremely useful for better understanding the adaptations of species that are really specialized in acoustic communication not audible by humans. Finally, it remains of central importance to understand the effects of accumulated noise in the short- and long term in populations and communities.

Summary

Acoustic noise is a diffuse phenomenon in nature and especially in human-modified ecosystems. Noise can be defined a sound with poor information or simply an unwanted sound that can mask other sounds. The source noise may be of natural (geophonies, biophonies) or manmade (anthrophonies) origin. The noise produced by human activities has grown quite rapidly in the past century, covering the larger part of the Earth. In particular, transportation facilities and logistics are the major sources of noise (pollution). Noise reduces the distance from the sender at which a signal can be detected by a receiver: this distance is called active space and represents an important parameter in communication. Noise is a common word in modern life and has been substituted for the generic term sound. Such substitution is largely because of the scant interest that human society has reserved for the sonic context. The patchy distribution of the sonic environment creates sonotopes with variable noise conditions, which determines the final distribution of vocal species according to their level of tolerance and adaptive plasticity.

Noise pollution may have long-term effects on populations and communities, reshaping communities according to the level of noise tolerance between prey and predators. Several effects of noise on vocal animals include direct stress, masking predatory menace, and interference with the communication mechanisms in general.

There is evidence that vocal animals try to reduce the effect of background noise of either natural origin or anthropogenic origin, adopting at least two distinct strategies: increasing the amplitude of the acoustic signals (Lombard effect) or shifting the signal frequency.

The underwater environment is noisy: the sonic sources are in large part of natural origin (wind, tides, rain, current turbulences). There is evidence that marine systems can be distinguished according a specific sonic signature that results from the combination of disparate sonic sources of natural and anthropogenic nature. The increase of ocean acidification can produce an increase of noise diffusion caused by the inverse correlation between water pH and sound transmission.

The sonic signature of marine systems can be documented also in freshwater systems, and the effects of different sonic freshwater environments can be proven in different species of fishes according to their level of hearing sensitivity.

Acoustic masking in marine ecosystem from human building activity (e.g., pile-driving) may have severe consequences for pelagic fauna for a range up to 80 km away, impacting, for instance, on the behavior of social marine mammals. Ships and boats seem have a minor impact on pelagic fauna, but few studies have been devoted to the potential effects on their behavior. For instance, the presence of boats close to whale pods produces an increase of their acoustic communication.

The noise produced by road traffic is of great concern in conservation biology. Species can react differently to traffic noise, by avoiding noisy areas or by increasing the frequency of their acoustic signals. Accordingly, the groups of animals considered for the effects of traffic noise may have dramatic differences. In frogs, acoustic activity is innate and costly in terms of energy invested in reproductive choruses. When choruses are masked by ambient noise, males shift their repertoire toward higher frequencies, but this contrasts with the fact that females select the largest males, which usually have a low-frequency repertoire. Thus, the shift toward higher frequencies alters the relationship between male selection and acoustic frequency.

In marine systems, noise can be considered a sonic context important to orientation. Well documented are the pelagic larvae of coral reef dwellers, which are clearly attracted by the noise produced by wind on coral reef barriers. In this way, many organisms can orient during the delicate trip toward coral reef settlement locations. On the other hand, permanent pelagic organisms can use the sonic ambience perceived close to a coral reef as a repellent. Changes in physiology can be observed in some species of crabs with a decreased metamorphic rate in presence of coastal noise.

In the presence of multiple sources of sounds, species must make a careful selection of the important frequencies, excluding in their evaluation those less informative. This capacity, which has been recognized in people, has been found also in some vocal animals such as the Cope's gray treefrog (*Hyla chrysoscelis*).

Oil spills and other seismic geological prospecting, usually carried out in remote land areas and in the deep seas and oceans, are important sources of noise that reduce wildlife populations, change their behavior, and reduce overall productivity and biodiversity.

Wind turbines are common devices displaced in terrestrial and aquatic systems. This no-emission energy producer has at least three important impacts on the environment: visual, physical, and acoustic. The acoustic impact has complex implications resulting from the wind condition at which the turbines operate. Strong winds increase the noise produced by rotating blades, and regulations are imposed at country scale to limit the negative effects.

Little attention is paid to military noise, but the noise produced by land vehicles, airplanes, vessels, and submarines and by the explosion of bombs, artillery, and rifle shots is an important source of sonic pollution, especially in remote areas.

Health and well-being are two correlated aspects of the everyday metabolism of humanity. Loud sounds and noise contribute to the health deficit and feelings of annoyance. The great concern that noise pollution creates in the technological societies is forcing local and central governments to reduce the effect of noise exposure on health and also in general on the well-being of society.

Noise sensitivity, resembling a level of allergy to sounds, seems the primary mechanism that produces the interindividual different levels of annoyance and other effects such as those on communication, recreation, concentration, and sleeping disturbance.

Noise is generally associated with the urban context where is found the highest concentration of technological sonic devices. Urban noise is on the agenda for mitigation measures and is in first place in urban regulation.

According to the World Health Organization (WHO), at least seven major consequences on human health are expected after prolonged exposure of the human body to a severe sonic environment: hearing impairment, interference in communication, sleep disturbance, cardiovascular disturbance, disturbances in mental health, impaired task performance, negative social behavior, and annoyance reactions.

Noise is not only a physical phenomenon but also a psychoacoustic object that can be analyzed in terms of loudness, sharpness, roughness, fluctuation strength, informative content of the sound signal, and cultural background. The cultural background becomes of primary importance in sound perception more than the energy of the sound per se.

The urban landscape is one of the major contributors to the dose of noise that humans absorb every day. Soundscape rating in an urban context largely depends on the status of the subjects, whether residents or visitors. Just these two categories have a different amount of sonic ambience.

Urban areas because of their increasing extension and intrinsic environmental heterogeneity have a growing attractiveness for biodiversity, which opens new unexpected scenarios to biodiversity conservation in urban areas where sensitivity to natural values meets a broad consensus in urban societies.

The expansion of urban areas and the associated increase of anthropogenic sound intrusion require urgently an intensification of the research agenda on the effects of noise for terrestrial and aquatic systems.

References

Aiken RB (1982) Shallow-water propagation of frequencies in aquatic insect sounds. Can J Zool 60:3459–3461

Amoser S, Ladich F (2005) Are hearing sensitivities of freshwater fish adapted to the ambient noise in their habitats. J Exp Biol 208:3533–3542

Au WWL, Green M (2000) Acoustic interaction of humpback whales and whale-watching boats. Mar Environ Res 49:469–481

Bailey H, Senior B, Simmons D, Rusin J, Picken G, Thompson PM (2010) Assessing underwater noise levels during pile-driving at an offshore windfarm and its potential effects on marine mammals. Mar Pollut Bull 60(6):888–897. doi:10.1016/J.marpolbul.2010.01.003

Barber JR, Crooks KR, Fristrup KM (2009) The costs of chronic noise exposure for terrestrial organisms. Trends Ecol Evol 25(3):180–189

Baynes EM, Habib L, Boutin S (2008) Impacts of chronic anthropogenic noise from energy-sector activity on abundance of songbirds in the boreal forest. Conserv Biol 22(5):1186–1193

Blickley JL, Blackwood D, Patricelli GL (2012) Experimental evidence for the effects of chronic anthropogenic noise on abundance of greater sage-grouse at leks. Conserv Biol 26(3):461–471

Brumm H (2004) The impact of environmental noise on song amplitude in a territorial bird. J Anim Ecol 73:434–440

Brumm H, Slabbekoorn H (2005) Acoustic communication in noise. In: Slater PJB, Snowdon CT, Roper TJ, Brockmann HJ, Naguib M (eds) Advance in the study of behavior, vol 35. Elsevier, San Diego, pp 151–209

Brumm H, Slater PJB (2006) Ambient noise, motor fatigue, and seral redundancy in chaffinch song. Behav Ecol Sociobiol 60:475–481

Brumm H, Todt D (2003) Facing the rival: directional singing behaviour in nightingales. Behaviour 140:43–53

Brumm H, Voss K, Köllmer I, Todt D (2004) Acoustic communication in noise: regulation of call characteristics in a New World monkey. J Exp Biol 207:443–448

Buckstaff KC (2004) Effects of watercraft noise on the acoustic behavior of bottlenose dolphins, *Tursiop truncatus*, in Sarasota Bay, Florida. Mar Mamm Sci 20(4):709–725

Cato DH (1976) Ambient sea noise in waters near Australia. J Acoust Soc Am 60(2):320–328

Chan AAY, Giraldo-Perez P, Smith S, Blumstein DT (2010) Anthropogenic noise effects risk assessment an attention: the distracted prey hypothesis. Biol Lett 6:458–461

Cherry EC (1953) Some experiments on the recognition of speech, with one and with two ears. J Acoust Soc Am 25:975–979

Clark CW, Ellison WT, Southall BL, Hatch L, Van Parijs SM, Frankel A, Ponikaris D (2009) Acoustic masking in marine ecosystems: intuitions, analysis, an implication. Mar Ecol Prog Ser 395:201–222

Curtis KR, Howe BM, Mercer JA (1999) Low-frequency ambient sound in the North Pacific: long time series observations. J Acoust Soc Am 106(6):3189–3200

Cynx J, Lewis R, Tavel B, Tse H (1998) Amplitude regulation of vocalizations in noise by a songbird, *Taeniopigya guttata*. Anim Behav 56:107–113

Dodd G (2001) Noise, noise sensitivity and attention. Internoise 2001. The Hague, The Netherlands

Engas A, Lokkeborg S, Ona E, Soldal AV (1996) Effects of seismic shooting on local abundance and catch rates of cod (*Gadus morhua*) and haddock (*Melanogrammus aeglefinus*). Can J Fish Aquat Sci 53:2238–2249

Erbe C (2009) Underwater noise from pile driving in Moreton Bay, QLD. Acoust Aust 37 (3):87–92

Etter PC (2012) Advanced applications for underwater acoustic modeling. Adv Acoust Vibration Volume 2012, article ID 214839. doi:10.1155/2012/214839, 28 pages

Forman RTT, Sperling D, Bissonette JA, Clevenger AP, Dale VH, Fahrig L, France R, Goldman CR, Heanue K, Jones JA, Swanson FJ, Turrentine T, Winter TC (2003) Road ecology science and solutions. Island Press, Washington, DC

Francis CD, Ortega CP, Cruz A (2009) Noise pollution changes avian communities and species interactions. Curr Biol 19:1415–1419

Francis CD, Ortega CP, Cruz A (2011) Vocal frequency change reflects different responses to anthropogenic noise in two suboscine tyrant flycatchers. Proc R Soc B 278:2025–2031

Frisk GV (2012) Noiseonomics: the relationship between ambient noise levels in the sea and global economic trends. Nature 2:437. doi:10.1038/srep00437

Fuller RA, Warren PH, Gaston KJ (2007) Daytime noise predicts nocturnal singing in urban robins. Biol Lett 3:368–370

Gall MD, Ronald KL, Bestrom ES, Lucas JR (2012) Effects of habitat and urbanization on the active space of brown-headed cowbird song. J Acoust Soc Am 132(6):4053–4062

Genuit K, Fiebig A (2005) Prediction of psychoacoustic parameters. Noise-con 2005, Minneapolis

Genuit K, Fiebig A (2006) Psychoacoustics and its benefit for the soundscape approach. Acta Acust United Acust 92:1–7

Goines L, Hagler L (2007) South Med J 100:287–294

Hanna D, Blouin-Demers G, Wilson DR, Mennill DJ (2011) Anthropogenic noise affects song structure in red-winged blackbirds (*Agelaius phoeniceus*). J Exp Biol 24:3549–3556

Hanson CE (2008) High speed train noise effects on wildlife and domestic livestock. In: Schulte-Werning B et al (eds) Noise and vibration mitigation, NMFM 99. Springer, Berlin, pp 26–32

Hartbauer M, Siegert ME, Fertschai I, Römer H (2012) Acoustic signal perception in a noisy habitat: lessons from synchronizing insects. J Comp Physiol A 198:397–409

Holt MN, Noren DP, Veirs V, Emmons CK, Veirs S (2008) Speaking up: killer whales (*Orcinus orca*) increase heir call amplitude in response to vessel noise. J Acoust Soc Am 125(1): EL27–EL32

Joo W, Gage SH, Kasten EP (2011) Analysis and interpretation of variability in soundscapes along an urban–rural gradient. Landsc Urban Plan 103:259–276

Katti M, Warren PS (2004) Tit, noise and urban bioacoustics. Trends Ecol Evol 19(3):109–110

Kight CR, Swaddle JP (2011) How and why environmental noise impacts animals: an integrative, mechanistic review. Ecol Lett 14(10):1052–1061. doi:10.1111/j.1461-0248.2011.01664.x

Klinck H, Nieukirk SL, Mellinger DK, Klinck K, Matsumoto H, Dziak RP (2012) Seasonal presence of cetaceans and ambient noise levels in polar waters of the North Atlantic. J Acoust Soc Am 132(3):EL176–EL181

Krause B (2012) The great animal orchestra. Little, Brown and Company, New York

Lam KC, Brown AL, Marafa L, Chau KC (2010) Human preference for countryside soundscape. Acta Acust United Acust 96:463–471

Langemann U, Gauger B, Klump GM (1998) Auditory in the great tit: perception of signals in the presence and absence of noise. Anim Behav 56:763–769

Litovsky RY (2012) Spatial release from masking. Acoustics Today 8(2):18–25

Lombard E (1911) Le signe de l'elevation e la voix. Annales des Maladies de l'Oreille e du Larynx 37:101–119

Luther D, Baptista L (2010) Urban noise and the cultural evolution of bird songs. Proc R Soc B 277:469–473

McCauley RD, Fewtrell J, Popper AN (2003) High intensity anthropogenic sound damages fish ears. J Acoust Soc Am 113(1):638–642

McDonald MA, Hildebrand JA, Wiggins SM (2006) Increases in deep ocean ambient noise in the Northeast Pacific west of San Nicolas Island, California. J Acoust Soc Am 120(2):711–718

McNett GD, Luan LH, Cocroft RB (2010) Wind-induced noise alters signaler and receiver behavior in vibrational communication. Behav Ecol Sociobiol 64:2043–2051

Miksis-Olds JL, Wagner T (2011) Behavioral response of manatees to variations in environmental sound levels. Mar Mamm Sci 27(1):130–148

Miller NP (2003) Transportation noise and recreational lands. Noise/News Int 11(1):1–20, International Institute of Noise Control Engineering. http://www.hmmh.com/presentations-papers.html

Mockford EJ, Marshall RC (2009) Effects of urban noise on song and response behaviour in great tits. Proc R Soc B 276:2979–2985

Mokford EJ, Marshall RC, Dabelsteen T (2011) Degradation of rural and urban great tit song: testing transmission efficiency. PLoS One 6:e28242

Narins PM, Feng AS, Lin WY, Schnitzler HU, Denzinger A, Suthers RA, Xu CH (2004) Old world frog and bird vocalizations contain prominent ultrasonic harmonics. J Acoust Soc Am 115:910–913

Nityananda V, Bee MA (2012) Spatial release from masking in a free-field source identification task by gray treefrogs. Hear Res 285:86–97

O'Connor P (2008) The sound of silence: valuing acoustics in heritage conservation. Geogr Res 46 (3):361–373

Parks SE, Clark CW, Tyack PL (2007) Short. and long-term changes in right whale calling behavior: the potential effects of noise on acoustic communication. J Acoust Soc Am 122 (6):3725–3731

Parris KM, Schneider A (2009) Impacts of traffic noise and traffic volume on birds of roadside habitats. Ecol Soc 14(1):29, online. http:www.ecologyandsociety.org/vol114/iss1/art29/

Parris KM, Velik-Lord M, North JMA (2009) Frog call at a higher pitch in traffic noise. Ecol Soc 14(1):25, online. http//www.ecologyandsociety.org/vol114/iss1/art21/

Paton L, Romero F, Cuenca J, Escudero JC (2012) Tolerance to noise in 91 bird species from 27 urban gardens of Iberian peninsula. Landsc Urban Plan 104:1–8

Patricelli GL, Blickley JL (2006) Avian communication in urban noise: causes and consequences of vocal adjustment. Auk 123(3):131–133

Pedersen E, Larsman P (2008) The impact of visual factors on noise annoyance among people living in the vicinity of wind turbine. J Environ Psychol 28:379–389

Pohl NU, Leadbeater E, Slabbekoorn H, Klump GM, Langemann U (2012) Great tits in urban noise benefit from high frequencies in song detection and discrimination. Anim Behav 83:711–721

Pollack I, Pickett JM (1958) Stereophonic listening and speech intelligibility against voice babble. J Acoust Soc Am 30(2):910–913

Potash LM (1972) Noise-induced changes in calls of the Japanese quail. Psychon Sci 26:252–254

Pytte CL, Rusch KM, Ficken MS (2003) Regulation of vocal amplitude by the blue-throated hummingbird, *Lampornis clemenciae*. Anim Behav 66:703–710

Rabin LA, McCowan B, Hooper SL, Owings DH (2003) Anthropogenic noise and its effect on animal communication: an interface between comparative psychology and conservation biology. Int J Comp Psychol 16:172–192

Rabin LA, Goss RG, Owings DH (2006) The effects of wind turbines on antipredator behavior in California ground squirrels (*Spermophlus beecheyi*). Biol Conserv 131:410–420

Radford CA, Stanley JA, Tindle CT, Montgomery JC, Jeffs AG (2010) Localised coastal habitats have distinct underwater sound signatures. Mar Ecol Prog Ser 401:21–29

Reijnen R, Foppen R, Meeuwsen H (1996) The effects of traffic on the density of breeding birds in Dutch agricultural grasslands. Biol Conserv 75:255–260

Rheindt FE (2003) The impact of roads on birds: does song frequency play a role in determining susceptibility to noise pollution? J Ornithol 144:295–306

Schafer RM (1977) Tuning of the world. Knopf, New York

Schaub A, Ostwald J, Siemers BM (2008) Foraging bats avoid noise. J Exp Biol 211:3174–3180

Schuemer-Kohrs A, Schuemer R, Schreckenberg D, Griefahn B, Moehler U (1998) Annoyance due to railway and road traffic noise: first results of an interdisciplinary study. In: Proceedings of the Vortrag 7th international congress on noise as a public health problem – noise effects '98, pp 1–4

Sebastianutto L, Picciulin M, Costantini M, Ferrero EA (2011) How boat noise affects an ecologically crucial behaviour: the case of territoriality in *Gobius cruentatus* (Gobididae). Environ Biol Fish 92:207–215

Seppänen J, Nieminen M (2004) Measurements and descriptions of underwater noise in Finland. Geophysica 40(1–2):23–38

Simpson SD, Jeffs A, Montgomery JC, McCauley RD, Meekan MG (2008) Nocturnal relocation of adult and juvenile coral reef fishes in response to reef noise. Coral Reef 27:97–104

Simpson SD, Radford AN, Tickle EJ, Meekan MG, Jeffs AG (2012) Adaptive avoidance of reef noise. PLoS One 6(2):e16625. doi:10.1371/journal.pone.0016625

Slabbekoorn H (2004) Habitat-dependent ambient noise: consistent spectral profiles in two African forest types. J Acoust Soc Am 116(6):3727–3733

Slabbekoorn H, Peet M (2003) Ecology – birds sing at a higher pitch in urban noise. Nature 424:267. doi:10.1038/424267a

Slabbekoorn H, Ripmeester EAP (2008) Birdsong and anthropogenic noise: implication and application for conservation. Mol Ecol 17:72–83

Staatermasn E, Rice AN, Mann DA (2013) Soundscape from tropical eastern pacific reef and a Caribbean Sea reef. Coral Reef 32(2):553–557. doi:10.1007/s00338-012-1007-8
Stanley JA, Radford CA, Jeffs AG (2009) Induction of settlement in crab megalopae by ambient underwater reef sound. Behav Ecol 21(1):113–120. doi:10.1093/beheco/arp159
Stanley JA, Radford CA, Jeffs AG (2012) Location, location, location: finding a suitable home among the noise. Proc R Soc B 279:3622–3631
Sun JWC, Narins PM (2005) Anthropogenic sounds differentially affect amphibian call rate. Biol Conserv 121:419–427
Tilman D, May RM, Lehman CL, Nowak MA (1994) Habitat destruction and the extinction debt. Nature 371:65–66
Tolimieri N, Jeffs A, Montgomery JC (2000) Ambient sound as a cue for navigation by the pelagic larvae of reef fishes. Mar Ecol Prog Ser 207:219–224
Tonolla D, Lorang MS, Heutschi K, Tockner K (2009) A flume experiment to examine underwater sound generation by flowing water. Aquat Sci 71:449–462
Tonolla D, Acuna V, Lorng MS, Heutschi K, Tockner K (2010) A field-based investigation to examine underwater soundscapes of five common river habitats. Hydrol Process 24:3146–3156
Truax B (ed) (1999) Handbook for acoustic ecology. Cambridge Street Publishing, Burnaby
Van Kempen E, van Kamp I, Nilsson M, Lammers J, Emmen H, Clark C, Stansfeld S (2010) The role of annoyance in the relation between transportation noise and children's health and cognition. J Acoust Soc Am 128(5):2817–2828
Vermeij MJA, Marhaver KL, Huijbers CM, Nagelkerken I, Simpson SD (2010) PLoS One 5(5): e10660
Wahlberg M, Westerberg H (2005) Hearing in fish and their reactions to sounds from offshore wind farms. Mar Ecol Prog Ser 288:295–309
Warren PS, Katti M, Ermann M, Brazel A (2006) Urban bioacoustics: it's not just noise. Anim Behav 71:491–502
Wood WE, Yezerinac SM (2006) Song sparrow (*Melospiza melodia*) song varies with urban noise. Auk 123:650–659
Wrege PH, Rowland ED, Thompson BG (2010) Use of acoustic tools to reveal otherwise cryptic responses of forest elephants to oil exploration. Conserv Biol 24(6):1578–1585
Wysochi LE, Ladich F (2005) Hearing in fishes under noise condition. J Assoc Res Otolaryngol 6:28–36
Wysocki LE, Amoser S, Ladich F (2007) Diversity in ambient noise in European freshwater habitats: noise levels, spectral profiles, and impact on fishes. J Acoust Soc Am 121 (5):2559–2566
Yan HY, Anraku K, Babaran RP (2010) Chapter 3 – Hearing in marine fish and its application in fisheries. In: He P (ed) Behavior of marine fishes: capture processes and conservation challenges. Wiley-Blackwell, Ames, pp 45–64
Zakarauskas P, Chaman DMF, Staal PR (1990) Underwater acoustic ambient noise levels on the eastern Canadian continental shelf. J Acoust Soc Am 87(5):2064–2071

Chapter 7
Sonic Patterns II: The Animal Choruses

Choruses are one of the most mysterious and spectacular acoustic phenomena common to terrestrial and aquatic biomes. Choruses are the result of contemporary vocalization of several individuals and species at a precise time of the day (dawn or dusk).

Several hypotheses have been presented during recent years. First, the beginning of the chorus activity is species specific and can be explained by body size and eye size.

Intrinsic factors, environmental factors, and social factors can concur in the chorus performance.

Among the several hypotheses to explain this phenomenon, the acoustic transmission hypothesis has received much credit. This hypothesis claims a more favorable condition of acoustic transmission at dawn and at dusk for a quieter atmosphere and the lowest background noise. Reduced foraging efficiency at dawn and at dusk and an excess of food storage the day before are further good arguments.

In amphibians choruses are very common, energetically costly, and occur especially during the short reproductive period. In frogs, a cooperative hypothesis according to which a chorus produces benefits in the entire population is opposed to a competitive hypothesis in which a leading caller is preferred in sexual selection. In fact, it has been demonstrated that females are attracted by leading and loud signals.

In birds, there is considerable evidence that choruses can be explained by at least two major hypotheses: an energetic hypothesis and a behavioral hypothesis. According to the first hypothesis, the choruses are performed in coincidence with a food surplus or fat reserves. The second hypothesis considers the choruses as an interactive communication between concurrent males that seem to be a reaction to the past day's vocal interactions between males.

A. Farina, *Soundscape Ecology: Principles, Patterns, Methods and Applications*,
DOI 10.1007/978-94-007-7374-5_7,

7.1 Introduction

Choruses, which represent the most relevant and discussed biophonic phenomena that can be heard in nature, are the result of the contemporary vocalizations of animals prevailing at dawn and at dusk. Choruses have attracted the attention of bioacousticians for a long time, and ethologists and even ecologists often use choruses to assess the abundance of populations (Pellet et al. 2007). The contemporary vocalizations of different species according to a daily dynamic produce a noisy effect that is common to all the biomes.

Choruses occur in frogs, birds, mammals, and insects, with different and distinct strategies and effects on populations and communities. In the marine environment, temporary choruses may be produced by the intense clicking sounds of sperm whales.

In tropical areas, frogs and insects are the major agents, but there are birds that dominate the sonic environment in temperate and boreal biomes.

Choruses have different temporal dynamics at a daily scale according to the species. For instance, choruses in birds occur at dawn and at dusk, and their length and intensity largely depend on the season and weather conditions. In temperate regions, bird choruses reach a maximum in the middle of the reproductive season (May–June) (Fig. 7.1).

Choruses have also been observed as a common phenomenon in aquatic systems (D'Spain and Batchelor 2006). For instance, in coastal waters choruses appear regularly after sunset and before sunrise. In the Salton Sea, California (USA), Fish and Cummings (1972) have recorded nocturnal choruses around 1,000 Hz with an increase of 50 dB resulting from the acoustic activity of the orange-mouth corvina fish (*Cynoscion xanthulus*). In New Zealand Radford et al. (2008) found that most of the acoustic energy emerging from the seabed was the result of feeding noises of sea urchins. These animals produce initially a noise of 800–2,400 Hz that is amplified by the ovoid calcareous skeleton, which acts as a Helmholtz resonator. This phenomenon is important to maintain an ambient noise that can be recognized by pelagic larvae of key reef species, such as fishes, crabs, and lobster, to locate future territories in which to settle.

Cato (1978) reported choruses in tropical waters of the East Indian Ocean, in the West Pacific Ocean, and in the Timor Sea near Australia. These choruses were present for a few hours at a time with an energy distributed between 400 Hz and 400 kHz with spectral peaks at 800 Hz, 1.2–2 kHz, and 3 kHz, respectively. The sonic intensity was on average 30 dB higher than the background noise. Most of the choruses occurred at evening but some choruses were observed at various periods of the day. The principal sources of such choruses were fish and sea urchins.

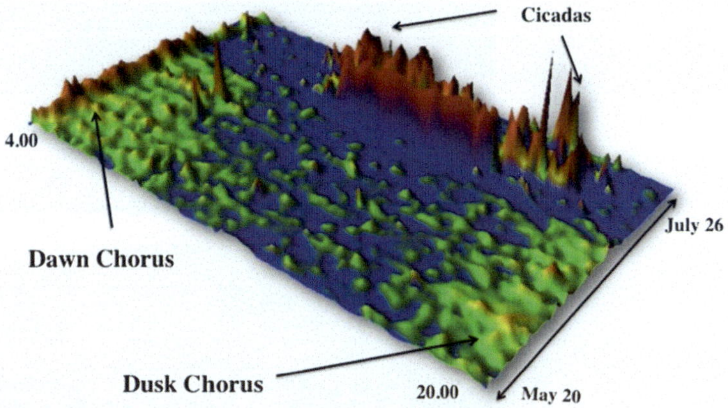

Fig. 7.1 Distribution of choruses at dawn and dusk during 1 year of sound recording in a Mediterranean maqui. The three-dimensional (3D) model was obtained by interpolating the values obtained by the acoustic complexity index (ACI) (Reproduced from Farina et al. 2011)

7.2 The Chorus Hypotheses

Several hypotheses have been proposed during these past years to explain the origins and causes of choruses, which seems an acoustic paradox in terms of energy cost against an apparently modest individual benefit. In fact, choruses create a considerable background and foreground noise from the multiplicity of individuals or species that are vocalizing at the same time.

In birds, the beginning of singing activity varies according to the species. In general larger species with large eyes (and greater pupil diameter) are the first species to initiate the chorusing: they start to sing when the light level is low, as confirmed by the work of Thomas et al. (2002). For instance, in a Mediterranean maqui the common blackbird (*Turdus merula*) is the first songbird to sing, followed immediately by the European robin (*Erithacus rubecula*) (Fig. 7.2).

Several hypotheses have been reviewed by Staicer et al. (1996) to explain the choruses, and these hypotheses can be grouped in three broad categories of proxies: intrinsic, environmental, and social factors.

The intrinsic factors are driven by the circadian cycles of testosterone.

Environmental factors invoke a low predation risk, good acoustic transmission in quiet air, inefficient foraging at low ambient light, and unpredictable nighttime conditions that produce a probable excess of resources in the early morning.

The social factors are based on mate attraction, territory defense, and resolution of social dynamics. Females can receive important information on the quality of the male considering the song as an honest signal.

Among the environmental factors, the acoustic transmission hypothesis (ATH) claims that birds sing most intensively in the quieter time of the day when song is not degraded by air turbulence or anthropogenic noise. During the day the local

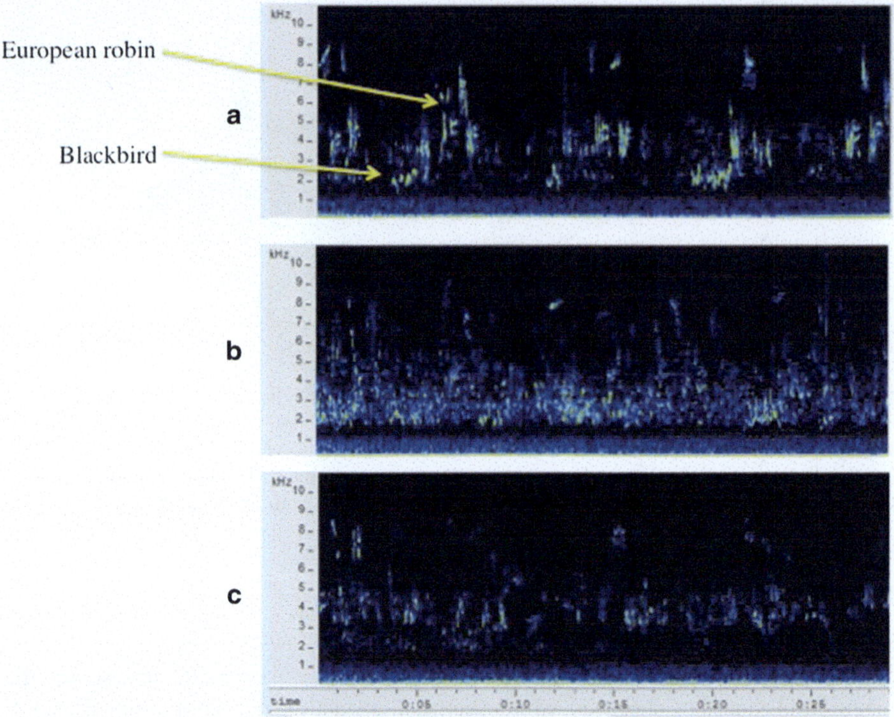

Fig. 7.2 A dawn chorus sequence of spectrograms at 4:00 a.m. (**a**), 4:22 a.m. (**b**), and 5:06 a.m. (**c**) on 6 June 2011 in a Mediterranean maqui. The common blackbird (*Turdus merula*) is the first species to sing at dawn, immediately followed by the European robin (*Erithacus rubecula*)

winds increase in intensity, and the air temperature increases as well, and other biophonies from insects become dominant. More in general, the background noise increases consistently, reducing the signal-to-noise ratio (SNR), producing attenuation of spherical spreading (6 dB for doubling distance) and excess attenuation (EA) from absorption and multiple scattering. Other effects are produced by filtering, atmospheric turbulence, and reverberation. Definitively, during the day the sonic ambience becomes acoustically more complex with a higher risk of sound attenuation by wind turbulence, vegetation rustling, air temperature, and humidity increase. The background noise increases, and this contributes to reducing the signal-to-noise ratio (SNR).

The acoustic transmission hypothesis has been experimentally tested by Brown and Handford (2003) by transmitting the song of the white-throated sparrow (*Zonotrichia albicollis*) and swamp sparrow (*Melospiza georgiana*) both at dawn and at midday and re-recording at distance of 25–100 m in open and closed habitats. No significant differences of mean overall absolute transmission were found at dawn or at midday for both habitats, but the signal transmission quality was higher at dawn, which represents an advantage. The variability of the transmission quality increased with the distance at which the signals were re-recorded.

As every hypothesis, the acoustic transmission hypothesis remains quite controversial, as argued by Dabelsteen and Mathevon (2002) after conducting experiments on the propagation of two types of songs of the blackcap (*Sylvia atricapilla*), a small warbler very common across the shrubland and forest habitats of Europe. They carried out experiments in a broadleaf forest, recording the sounds at dawn, in midmorning, and early in the afternoon, without finding environmental conditions more favorable to acoustic propagation than those at dawn, rejecting this hypothesis.

It has been observed in a Mediterranean scrubland that background noise increases outside the breeding season (Farina et al. 2013) and that the period from March to September has a significant higher SNR, although some temporal peaks of noisy sources have been observed, especially during the second part of the day.

Among other possible explanations of the chorus phenomenon, at least two have received credit:

The first stated that foraging is inefficient in poor light at dawn (Kacelnik's explanation; Kacelnik 1979) and that the dawn time is reserved for social communication.

The second explanation argues that stochasticity in foraging success and overnight energy requirement allow an energetic surplus (Hutchinson 2002).

Considering the great number of hypotheses advanced by many authors (e.g., Burt and Vehrencamp 2005), it is reasonable to admit that social hypotheses are not mutually exclusive and that males can use this strategy to increase social interactions. The dawn chorus seems to create a communication network where there are at least three individuals in continuous contact with each other. Three types of communication network can be described:

A broadcasting network in which one individual sends the vocal communication to two or more receivers.

An eavesdropping network where two senders interact and two or more receivers eavesdrop to obtain information from the two interacting subjects.

Finally, an interactive network where all the individuals are sending information and in turn they eavesdrop the others (Fig. 7.3). This system in the real world seems a little more complex, where probably there are non-interacting females and floater males that are eavesdropping.

However, to find male-to-male interactions is not an easy task at dawn when many individuals are singing simultaneously and where is extremely difficult to distinguish foreground from background threshold.

Henwood and Fabrick (1979) have processed a quantitative analysis of the dawn chorus according to the time of the day. They arranged a mathematical model to test the various attenuation mechanisms in action under different micrometereological conditions in different habitats.

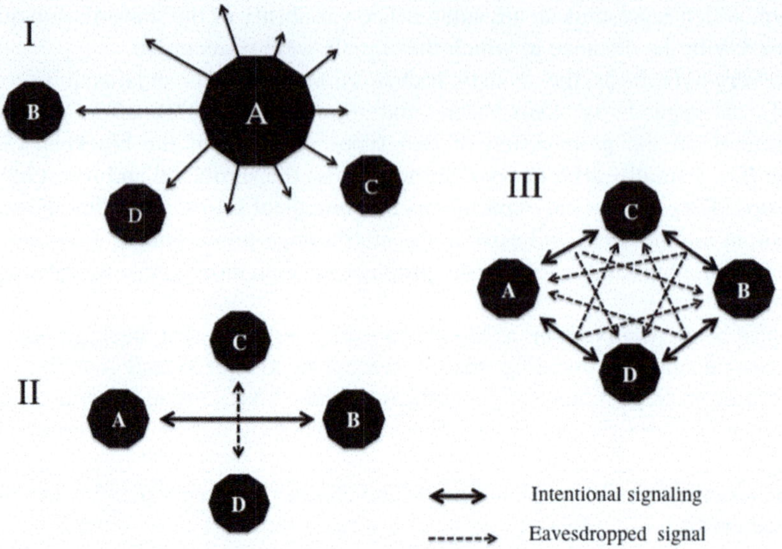

Fig. 7.3 Three network models to explain the function of dawn and dusk choruses in birds. *I* Broadcast network: an individual sends intended signals to two or more neighbors. *II* Eavesdropping network: two individuals are in active reciprocal connection and two or more other individuals eavesdrop the circulating information. *III* Interactive network: all the individuals are contemporary senders and receivers of direct intended signals and in turn are eavesdropping (Adapted with permission from Burt and Vehrencamp 2005)

The variation in sound intensity with distance can be expressed by the equation

$$I = \frac{I_0}{r^2} e - (\alpha_v + \alpha_c + \alpha_d + \alpha_r + \alpha_m) r$$

This expression can be converted to decibel attenuation in sound level (assuming a reference level of 1 m from the source), and the foregoing expression can be converted in the following equation:

$$N = 20 \log r + (\beta_v + \beta_c + \beta_d + \beta_r) r + \beta_t r + \beta_m r + \beta_D (r - r_s)$$

where

r = distance between the source and the receiver, and α and β (where $\beta = 4.343\alpha$) are the absorption coefficients corresponding to the different mechanisms of atmospheric attenuation

β_v = viscosity, β_c = heat conduction, β_d = molecular diffusion, β_r = thermal radiation

$\beta_t r$ = represents the attenuation caused by scattering of sound from atmospheric turbulence

$\beta_m r$ = the attenuation caused by atmospheric moisture content

$\beta_D (r - r_s)$ = represents attenuation from the curvature of the soundpath by wind velocity or temperature gradient

After the application of this model these two authors have found that the early morning is the most favorable period for sound propagation when compared with other times of day and night. The biophonies produced at that time are 20 fold more effective than the same signal with the same amplitude produced at midday. The height above the substrate at which a sound is produced increases the area of broadcast coverage. The frequency of the signal is another important element. Lower frequencies increase the area of broadcast coverage, although in open areas where wind is frequent and also there is higher turbulence the signals are transmitted better at high frequency. This last fact is discussed in terms of communication theory by which the transmission in a turbulent medium may be enhanced by the use of higher frequencies.

7.3 Choruses in Amphibians

In amphibians, choruses are very common phenomena. The phonotaxy of some species of amphibians has been demonstrated, but the chorus seems to represent a complex system of communication and not just a stochastic system of interindividual acoustic competition, overlap, and masking. Especially in species that breed in ephemeral ponds the necessity to breed in a short time before the water disappears obliges males and females to concentrate, which produces a high level of intra- and interspecific acoustic interference: at first sight, the amphibian chorus seems a paradox. Many males singing simultaneously create a background noise that masks the individuals, which are obliged to increase their acoustic performances.

Anuran males produce at least two different calls: an aggressive call used to interact with males and calls directed to females. The aggressive calls generally are less attractive for females. In some species there is a gradual shift from one type to another accordingly. In a few species it has been observed that the same call is composed of two parts: the first is directed to males and the second to the female. This two-note system appears, from an energetic point of view, extremely wasteful.

When a chorus is analyzed at fine temporal resolution, different patterns can be described (Grafe 2005). In general the degree of overlap between calling males increases with the number of contemporarily calling males, but the level of overlap remains lower when compared with a random comparison. This observation means that males have the capacity to place their calls in silent gaps, reducing the masking effect of other males. In some species the capacity to adjust the call timing has been observed. For example, in the neotropical treefrog (*Hyla microcephala*), Schwatz (1993) observed that a male when interrupted by the call of a neighboring male increases the spacing between calls to reduce overlap of the subsequent note.

Different strategies can be observed to reduce the competition: males can repeat their calls several times, may increase the complexity of their calls, and can protect the calling site and the sonic space around it, preventing the intrusion of other males. Increase of the complexity of the acoustic repertoire has been observed in

Boophis madascariensis, a frog endemic to Madagascar that has 28 distinct calls, as reported by Narins et al. (2000).

7.4 Cooperative Versus Competitive Hypothesis in Amphibian Choruses

It emerges from several studies that a cooperative hypothesis to explain the chorus pattern could be presented and argued, at least in insects and anurans. Cooperation means to produce a synchronized chorus with effects such as confusion of predators, degrading the locability of signals, and the enhancement of female detection because synchronization increases the acoustic peaks. This hypothesis can be easily dismantled in favor of the competition hypothesis. In a competitive scenario, leading calls have been observed to be more attractive for females. The mechanism for a male to reduce the overlap is to wait when there is a gap in the chorus, but this cannot always happen and a male could be silent for a long time. In effect, a male interrupts calling just waiting for a temporal subset of a nearby male. Brush and Narins (1989) have found in the Puerto Rican treefrog (*Eleutherodactylus coqui*) that males were responding, avoiding the overlap, to two or three neighboring males. This attention to avoid overlap can be found also in species that do not adopt this strategy.

Calling is expensive in ectothermic species, and a trade-off must exist between rate of calling and extension of the period of calling to reduce the risk for a male to stop calling before the other males do because of exhaustion of the energy pool. In many anurans, the calling effort increases with the increase of individual density, but to compensate for a major requirement of energy the rate of calling is reduced, as observed in *Hyla versicolor* by Grafe (1997).

Mate choice is of first importance in anurans, where females must be able to distinguish the more energetic male in a noisy background. Several studies have demonstrated that females select males that utter loud and conspicuous signals. The leading signal seems the preferred and is considered a "precedence effect" (for review, see Litovsky et al. 1999). This effect states that when two separate sounds are presented with a brief delay in onset, the leading sound is the first to be localized.

The habitat conditions are important to establish a profitable communication network in anuran choruses. The distance between the sender and the receiver and the height at which the call is emitted are factors that affect the efficiency of acoustic transmission, although, as argued by Grafe (2005), "*The evidence for the influence of transmission properties of the environment and network structure on signal design and signaling behaviour in anurans is equivocal at best.*"

It is evident that females are attracted by leading and loud signals, and this has been verified in the laboratory under conditions of few males and low background noise: larger males with lower frequency capacity were preferred. In nature this can happen, and females must select not just the more energetic male but they must decide between calls of lower frequency and call rate. In natural conditions, females must decide in acoustically unfavorable conditions, and they must make the best choice under these real conditions.

Females seem able to eavesdrop on interacting males and to extract the best information on the quality of males. Most of the selections are for leading calls, although in some cases the follower call is preferred.

A simplified model in which a female approaches the first male that it encounters is counterbalanced by evidence that females monitor the different calls of more males.

The eavesdropping strategy seems to be utilized also by silent males that intercept females after other males have intensified their calling for the presence of a female.

So, in conclusion silent males monitor the situation of other males and then decide to enter into the mating arena, privatizing with an ad hoc call, as in *Eleutherodactylus coqui*.

Choruses in frogs are the result of contemporary behavior of a signaler and a receiver, and in addition to the function of sexual selection choruses in frogs also act as a beacon that allows frogs to locate the breeding pond precisely.

The wood frog (*Rana sylvatica*) distributed across North America is an explosive breeding amphibian. Timing and location vary according to the local climatic factors (rainfall and change in temperature). Bee (2007) has conducted experiments on males of this frog demonstrating that the phonotaxis expressed by wood frog males is not only connected with the sexual mechanism to attract females but is also a useful tool (sonic eco-field) or beacon to intercept accurately and in a short time the breeding assemblage, which occurs only in favorable conditions and for a very restricted time period.

7.5 Choruses in Birds

In birds, physiological requirements and behavioral traits are the two major proxies at the basis of the chorus pattern. We call the first point simply an energetic hypothesis, and the second a behavioral hypothesis that must be added to and integrated with the hypotheses discussed in Sect. 7.2.

7.5.1 The Energetic Hypothesis

The energetic hypothesis states that the amount of fat reserves accumulated in the previous day is responsible for the quantity and loudness of the song broadcast by an individual at dawn or at dusk.

Thomas and Cuthill (2002) have adopted a stochastic dynamic programming (SDP) model that in the European robin (*Erithachus rubecula*) the daily singing and foraging routines largely depend on an individual's fat reserves, advancing two further hypotheses that overnight loss of fat reserves is higher on colder nights and that birds sing more at dawn when their fat reserves are high.

In silveryes (*Zosterops lateralis*), a small passerine of 11–14 g distributed throughout Australia, New Zealand, and many South Pacific islands, Barnett and Briskie (2007) have observed that the dawn chorus largely depends on the amount of fat accumulated in a singer's body. They proved that artificially fed males were singing longer and with a higher-quality repertoire.

Dawn and dusk choruses are largely considered phenomena of diurnal animals (birds, frogs), but this process has been observed also in nocturnal animals such as owls.

In the little owl (*Athene noctua*), the dusk chorus has been interpreted by Hardouin et al. (2008) as a distinct phenomenon independent of air temperature and more intense than at dawn. An increased variability in vocal activity has been observed at dusk compared to dawn. From playback experiments conducted in February and March, these authors have hypothesized that food rather than air temperature is a dusk constraint. Nocturnal owls are inactive in diurnal time and at dusk they have the minimum energetic budget, but these authors are also inclined to consider the dusk chorus in the context of social functions.

7.5.2 The Behavioral Hypothesis

According to this hypothesis the choruses are the result of interindividual acoustic relationships, a semetic network that is established between close individuals.

The behavioral hypothesis has been proven in black-capped chickadee (*Poecile atricapillus*) dawn choruses. These choruses represent an interactive communication network, as shown by the study carried out by Foote et al. (2010), who when working on 19 focal males have observed that the levels of frequency matching was higher than expected. All the males were observed to match two or three males simultaneously and sequentially.

In playback experiments conducted by Amrhein and Erne (2006) on the winter wren (*Troglodytes troglodytes*), in which the presence of an intruder was simulated inside the territory of a singing male before sunrise (dawn chorus time), it was found that the day after the supplementary song the resident male was singing at dawn more vigorously but not after sunrise. A preventive territorial proclamation seems the mechanism triggered by the administration of a simulated intruder and demonstrates, at least in this species, that the dawn period is used extensively for territory defense. Other playback experiments conducted on this focal species in spring and autumn by Erne and Amrhein (2008) have confirmed that intruder simulation before sunrise and after sunset produces a significant increase of song output the next day, although in autumn the effect is less important (Fig. 7.4). This

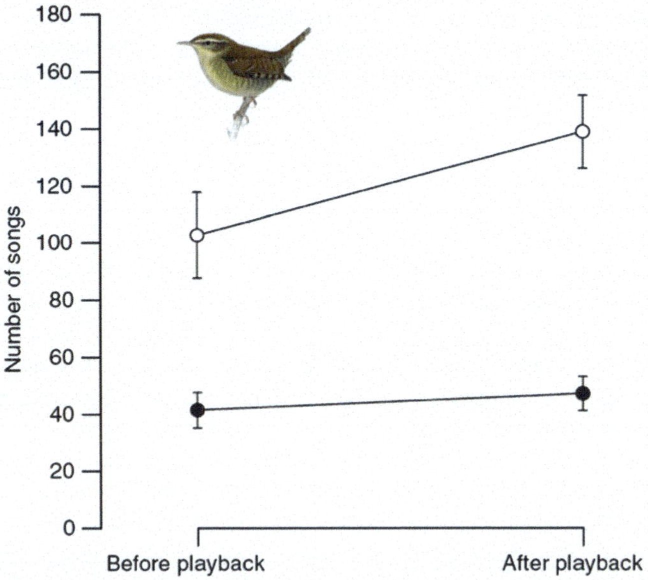

Fig. 7.4 Spring (*open circles*) and autumn (*black circles*): number of songs broadcast by males of winter wren (*Troglodytes troglodytes*) before and after playback of mating male intrusion at dawn (Reproduced with permission from Erne and Amrhein 2008)

work demonstrates the importance in territory defense of the acoustic communication in those two periods. Playback experiments conducted after sunrise and before sunset have not found a significant difference in song output.

Morse (1989) discussed the role of song patterns of northern parula (*Parula americana*), magnolia warbler (*Dendroica magnolia*), yellow-rumped warbler (*D. coronata*), black-throated green warbler (*D. virens*), and the blackburnian (*D. fusca*) warbler at dawn and dusk, represented by unaccented ending song as an advertisement display utilized by males that are singing well spaced, and with females not in evidence, as a communication tool to advertise of presence of singers to eventual intruders.

7.6 Reproduction Mechanisms and Choruses

In some species song activity at dawn is coincident with the fecundity period of females. Song is a phenotypic character that is connected with sexual selection, and we can expect that a complex song, louder and repeated for a long time, represents a honest signal in mate selection.

In the great tit (*Parus major*) and in the blue tit (*Cyanistes caeruleus*), males repeat a particular song type for some time, then they adopt another song type. This phenomenon has been called "drift" and could be dependent on neuromuscular exhaustion or reduced motivation. The males that showed less drift, sang longer strophes, and had a longer song repertoire are considered the "better" singers.

Poesel et al. (2001) have studied the song patterns at dawn in 20 blue tit males, demonstrating that its song output, more than song complexity, is a trait connected with sexual selection. In fact, males with a higher song output without drift were found to participate in pairs in which females started laying earlier and laid larger clutches.

Experimental manipulation of the dawn and dusk chorus in the common blackbird (*Turdus merula*) conducted by Cuthill and Macdonald (1990) has demonstrated that both the nutrition status of the male and his fertility affect the timing and intensity of the dawn and dusk chorus. Providing supplementary food has produced in paired males an earlier song activity: they sang longer and at higher peaks at dawn and dusk. Close to their fertility peak, males start to sing early, and the song has longer duration at dusk and dawn.

Dawn is a period in which birds utilizing the song as an honest signal of male quality transmit such information to females. This point has been confirmed by Murphy et al. (2008) studying the eastern kingbird (*Tyrannus tyrannus*) (Fig. 7.5). In this species a distinctive song is sung at dawn. Morphological character and song complexity were correlated. Earlier singing males had larger and longer flight feathers, and they sang at high rates. Early-singing males were mated with earlier-breeding females.

The dawn chorus in the nightingale (*Luscinia megarhynchos*) remains largely constant during the breeding season, confirming that at least in this species the dawn song has the function of defending territory and is not used by males to attract females. In fact if the song of the male was utilized to maximize pair formation, we would expect variation in the singing performance before and after the mating period, but this has not observed, as reported by Kunc et al. (2005).

The song rate and the proportion of whistle songs were found to be low at the beginning of the reproductive season and cannot be used to predict the mating status. The male performance remained stable also after the arrival of females.

Synchronous and alternating choruses are common in insects and anurans. The interindividual interactions between contemporary calling males generate epiphenomena of synchronization or alternation (Greenfield 1994). Ignoring the farthest individuals may represent a trade-off strategy to reduce a too demanding alternation.

In many birds song is a communicative signal used for multiple purposes such as like territorial defense or female attraction. The song activity of the chiffchaff (*Phylloscopus collybita*), a small songbird of the Palearctic region, has been investigated by Rodrigues (1996) in southeastern England (Wytham Woods, Oxford). Despite evidence from other species, the chiffchaff did not sing more during the fertile period of the female, and the song rate maintained constant during egg deposition and nestling feeding. Probably the male continues to sing to attract a secondary female or neighboring female searching for an extra-pair copulation. This species sings after 30 min from dawn but it sings more during the post-fertile period than at other stages, and this contradicts the "announcement hypothesis" of

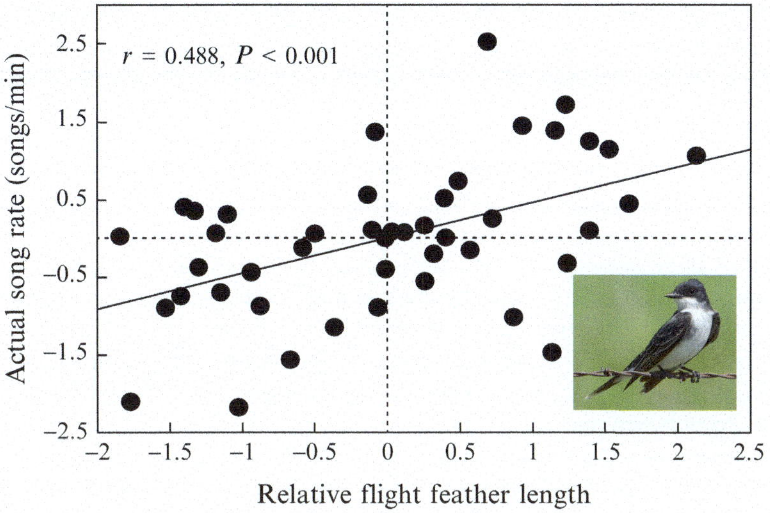

Fig. 7.5 Song rate compared against flight feather length in the eastern kingbird (*Tyrannus tyrannus*) (Reproduced with permission from Murphy et al. 2008)

Moller (1991) in which this behavior is interpreted as a paternity guard to reduce the risk of cuckoldry. During the fertile period the volume of song at dawn remained low.

The chiffchaff seems to not use the song as a paternity-guard mechanism, but singing activity has an intersexual function during the fertile period. In fact, the chiffchaff has a higher song rate at dawn; this song rate decreases after establishment of the pair, and song volume is low during the fertile period. The male uses several song spots after the fertile period to attract more females. The function of song activity after the fertile period could be as an intrasexual signal to inform other males about the territory. This song activity can also be used to have easier access to re-mating after nest predation or female desertion, reducing simultaneously the effort of territory patrolling.

Summary

Choruses are one of the most mysterious and spectacular acoustic phenomena common to terrestrial and aquatic biomes. Choruses are the result of contemporary vocalization of several individuals and species at a precise time of the day (dawn or dusk). Most of the vocal animals have choruses. Choruses are an involuntary source of orientation for pelagic larvae during their definitive settlement along coasts and coral reefs.

Several hypotheses have been presented in recent years. First, the beginning of chorus activity is species specific and can be explained with body size and eye size. The species able to capture more light at dawn (such as the blackbird) are the first singers.

Intrinsic factors, environmental factors, and social factors can concur in the chorus performance.

Among the several hypotheses to explain this phenomenon, the acoustic transmission hypothesis has received much credit. This hypothesis claims a more favorable condition of acoustic transmission at dawn and at dusk for a quieter atmosphere and the lowest background noise. Reduced foraging efficiency at dawn and at dusk and an excess of food storage the day before are further good arguments.

In birds, there is evidence that choruses can be explained by at least two major hypotheses: an energetic hypothesis and a behavioral hypothesis. According to the first hypothesis, the choruses are performed in coincidence with a food surplus or fat reserves. The second hypothesis considers the choruses as an interactive communication between concurrent males and seems a reaction of the past day's vocal interactions between males.

In amphibians choruses are very common, energetically costly, and occur especially during the short reproductive period. The acoustic interference seems quite high, but analysis at a fine temporal scale demonstrated in many cases how males have the capacity to collate the calls in acoustic gaps, reducing the apparent overlap.

A cooperative hypothesis according to which a chorus produce benefits in the entire population is opposed to a competitive hypothesis in which a leading caller is preferred in sexual selection. In fact, it has been demonstrated that females are attracted by leading and loud signals.

A further function of amphibian choruses is discussed in terms of an acoustic beacon for individuals to find the adopted pond in a short time, considering the ephemeral condition of many vernal ponds.

Songs during choruses have in many cases a reproductive function and sexual selection. Individual singing early, louder, and longer receives the advantage of the best mating process with more healthy females. The high singing activity at chorus is for some species a strategy to defend their territory, and in others to assure a extra-copulatory opportunity or nesting replacement after failure caused by nest predation or female desertion.

References

Amrhein V, Erne N (2006) Dawn singing reflects past territorial challenges in the winter wren. Anim Behav 71:1075–1080

Barnett CA, Briskie JW (2007) Energetic state and performance of dawn chorus in silvereyes (*Zosterops lateralis*). Behav Ecol Sociobiol 61:579–587

Bee MA (2007) Selective phonotaxis by male wood frogs (*Rana sylvatica*) to the sound of a chorus. Behav Ecol Sociobiol 61:955–966

Brown TJ, Handford P (2003) Why birds sing at dawn: the role of consistent song transmission. Ibis 145:120–129

Brush JS, Narins PM (1989) Chorus dynamics of a neotropical amphibian assemblage: comparison of computer simulation and natural behaviour. Anim Behav 37:33–44
Burt JM, Vehrencamp SL (2005) Dawn chorus as an interactive communication network. In: McGregor PK (ed) Animal communication networks. Cambridge University Press, Cambridge
Cato DH (1978) Marine biological choruses observed in tropical waters near Australia. J Acoust Soc Am 64(3):736–743
Cuthill IC, Macdonald WA (1990) Experimental manipulation of the dawn and dusk chorus in the blackbird *Turdus merula*. Behav Ecol Sociobiol 26:209–216
D'Spain GL, Batchelor HH (2006) Observations of biological choruses in the southern California Bight: a chorus at mid frequencies. J Acoust Soc Am 10(4):1942–1956
Dabelsteen T, Mathevon N (2002) Why do songbirds sing intensively at dawn? A test of the acoustic transmission hypothesis. Acta Ethol 4:65–72
Erne N, Amrhein V (2008) Long-term influence of simulated territorial intrusions on dawn and dusk singing in the winter wren: spring versus autumn. J Ornithol 149:479–486
Farina A, Pieretti N, Piccioli L (2011) The soundscape methodology for long-term bird monitoring: a Mediterranean Europe case-study. Ecol Inform 6(6):354–363
Farina A, Pieretti N, Morganti N (2013) Acoustic patterns of an invasive species: the red-billed Leiothrix (*Leiothrix lutea* Scopoli 1786) in a Mediterranean shrubland. Bioacoustics 22(3):175–194
Fish JF, Cummings WC (1972) A 50-dB increase in sustained ambient noise from fish (*Cynoscion xanthulus*). J Acoust Soc Am 52(4):1266–1270
Foote JR, Fitzsimmons LP, Mennill DJ, Ratcliffe LM (2010) Black-capped chickadee dawn choruses are interactive communication networks. Behaviour 147:1219–1248
Grafe TU (1997) Use of metabolic substrates in the gray treefrog, *Hyla versicolor*: implications for calling behavior. Copeia 1997:356–362
Grafe TU (2005) Anuran choruses as communication networks. In: McGregor P (ed) Animal communication networks. Cambridge University Press, Cambridge
Greenfield MD (1994) Synchronous and alternating choruses in insects and anurans: common mechanisms and diverse functions. Am Zool 34:605–615
Hardouin LA, Robert D, Bretagnolle V (2008) A dusk chorus effect in a nocturnal bird: support for mate and rival assessment functions. Behav Ecol Sociobiol 62:1909–1918
Henwood K, Fabrick A (1979) A quantitative analysis of the dawn chorus: temporal selection for communicatory optimization. Am Nat 114:260–274
Hutchinson JMC (2002) Two explanations of the dawn chorus compared: how monotonically changing light level favour a short break from singing. Anim Behav 64:527–539
Kacelnik A (1979) The foraging efficiency of great tits (*Parus major*) in relation to light intensity. Anim Behav 27:237–242
Kunc HP, Amrhein V, Naguib M (2005) Seasonal variation in dawn song characteristics in the common nightingale. Anim Behav 70:1265–1271
Litovsky R, Colburn HS, Yost WA, Guzman SJ (1999) The precedence effect. J Acoust Soc Am 106(4):1633–1654
Moller AP (1991) Why mated songbirds sing so much: mate guarding and male announcement of mate fertility status. Am Nat 138:994–1014
Morse DH (1989) Song patterns of warblers at dawn and dusk. Wilson Bull 101(1):26–35
Murphy MT, Sexton K, Dolan AC, Redmond LJ (2008) Dawn song of the eastern kingbird: an honest signal of male quality? Anim Behav 75:1075–1084
Narins PM, Lewis ER, McClelland BE (2000) Hyperextended call note repertoire of the endemic Madagascar treefrog *Boophis madascariensis* (Rhacophoridae). J Zool 250:28–298
Pellet J, Helfer V, Yannic G (2007) Estimating population size in the European tree frog (*Hyla arborea*) using individual recognition and chorus counts. Amphibia-Reptilia 28:287–294
Poesel A, Foerster K, Kempenaers B (2001) The dawn song of the blue tit *Parus caeruleus* and its role in sexual selection. Ethology 107:521–531

Radford C, Jeffs A, Tindle C, Montgomery JC (2008) Resonating sea urchin skeletons create coastal choruses. Mar Ecol Prog Ser 362:37–43

Rodrigues M (1996) Song activity in the chiffchaff: territorial defence or mate guarding? Anim Behav 51:709–716

Schwartz JJ (1993) Male calling behavior, female discrimination and acoustic interference in the Neotropical treefrog *Hyla microcephala* under realistic acoustic conditions. Behav Ecol Sociobiol 32:401–414

Staicer CA, Spector DA, Horn AG (1996) The dawn chorus and other diel patterns in acoustic signaling. In: Kroodsma DE, Miller EH (eds) Ecology and evolution of acoustic communication in birds. Cornell University Press, Ithaca

Thomas RJ, Cuthill IC (2002) Body mass regulation and the daily singing routines of European robins. Anim Behav 63:285–295

Thomas RJ, Szekely T, Cuthill IC, Harper DGC, Newson SE, Frayling TD, Wallis PD (2002) Eye size in birds and the timing of song at dawn. Proc R Soc Lond B 269:831–837

Chapter 8
Sonic Patterns III: Sounds and Vibrations from Soils

Soil is a medium rich in acoustic information. Most of the sounds present in the soil are originated by stridulating animals or are the result of the digging activity of fossorial organisms. The acoustic information from soil is an important tool to monitor soil biodiversity in a nondestructive way.

In some reptiles vibrations produced by prey are used for hunting. Vegetation vibration has been proved to orient fossorial moles toward vegetation clumps rich in insect food. Soil percussion is also a common mechanism used by elephants to communicate at great distances.

In insects and arthropods, vibrations are used extensively to communicate or to escape predators. Well documented is the case of *Diplocardia mississippiensis*, an earthworm that emerges from the soil after a produced vibration in the soil using a "grunting device." This behavior is explained as a strategy utilized by the worms to escape hunting by moles.

It is disputed if ants have hearing systems or body vibrational sensing, but their capacity to react to soil vibrations is surprisingly rapid. Substrate-borne vibrations are used extensively by animals to locate mates, prey, and predators, as in the well-documented case of antlion larvae.

Despite the scarcity of studies, soil monitoring appears to be a promising field of research. Sound prospecting of the soil requires that standard microphones be integrated with piezoelectric microphones, geophones, and accelerometers to capture airborne and soilborne vibrations.

8.1 Introduction

Soil is the main receptor of chemical pollutants and is a medium exposed to continuous dramatic anthropogenic-borne modifications of its internal structure such as logging, plowing, reclamation, and covering with impervious surfaces of cement and asphalt. Urban sprawl and the associated logistics and transportation

A. Farina, *Soundscape Ecology: Principles, Patterns, Methods and Applications*, DOI 10.1007/978-94-007-7374-5_8, © Springer Science+Business Media Dordrecht 2014

infrastructures reduce the free surface of soils in forests and in rural and urban areas.

The organisms living in the soils produce numerous sounds, vibrations, and noises. In fact, most of the sounds produced in the soil are represented by vibrations of the stridulating apparatus of insects and other arthropods (Cocroft and Rodriguez 2005) or by the mechanical vibrations produced by fossorial animals (earthworms, mole crickets, moles) during their digging activity or root scraping. Soil is a granular and not a continuous medium, so acoustic transmission across interstices (intergranular passages) is reasonable. Sound attenuation is higher in the soil (600 dB/m) than in the air (0.008 dB/m), but sounds with a frequency <5 kHz can be detected, at least in sandy soil, over a range of 5–50 cm.

Every organism that digs in the soil for creating refuges or to search for prey produces unwanted vibrations that can be eavesdropped by other organisms. Indeed, the soil strata are not a silent environment but an ambient rich in sonic information for its dwellers.

The majority of the sounds from soil are used by soil fauna to communicate, although these sounds are not easily captured and interpreted by the human ear. For this reason, the study of arthropod vibrational communication represents an interesting innovative field of interest for the future. Acoustic techniques are very useful to detect, to investigate, and to assess the amount of life present in the soils, avoiding laborious and destructive techniques.

Evaluating underground biodiversity and monitoring the changes that occur in the soil, especially after human disturbances, by using acoustic cues, represents a new chapter of the soundscape ecology agenda, although "sounds from soil" from the literature still represent a marginal argument in soundscape ecology.

8.2 Seismic Signals from Reptiles

In snakes there are two sensory systems reacting to airborne sound and to substrate vibration. In Colubridae, Crotalidae, and Boidae, the auditory system has a U-shaped frequency–threshold curve that ranges from 150 to 600 Hz. The auditory system of snakes is 20 dB less sensitive than the human auditory system to stimuli produced on the head and along the body. The lung system seems an important part in the auditory system of snakes, representing a novelty in land vertebrates, as experimentally proved by Hartline (1971).

In Saharan sand vipers (*Cerastes cerastes*), Young and Moran (2002) have experimentally demonstrated that mice are captured using the vibrations produced by the prey. In fact, this species has been observed to also launch its strike when its mouth and nostrils are under the sand, and without a tongue-flick. Experimental animals deprived of chemosensory information and temporarily blinded were able to freely locate mobile prey in every experimental trial, proving that vibrational stimuli are used to locate prey (Fig. 8.1).

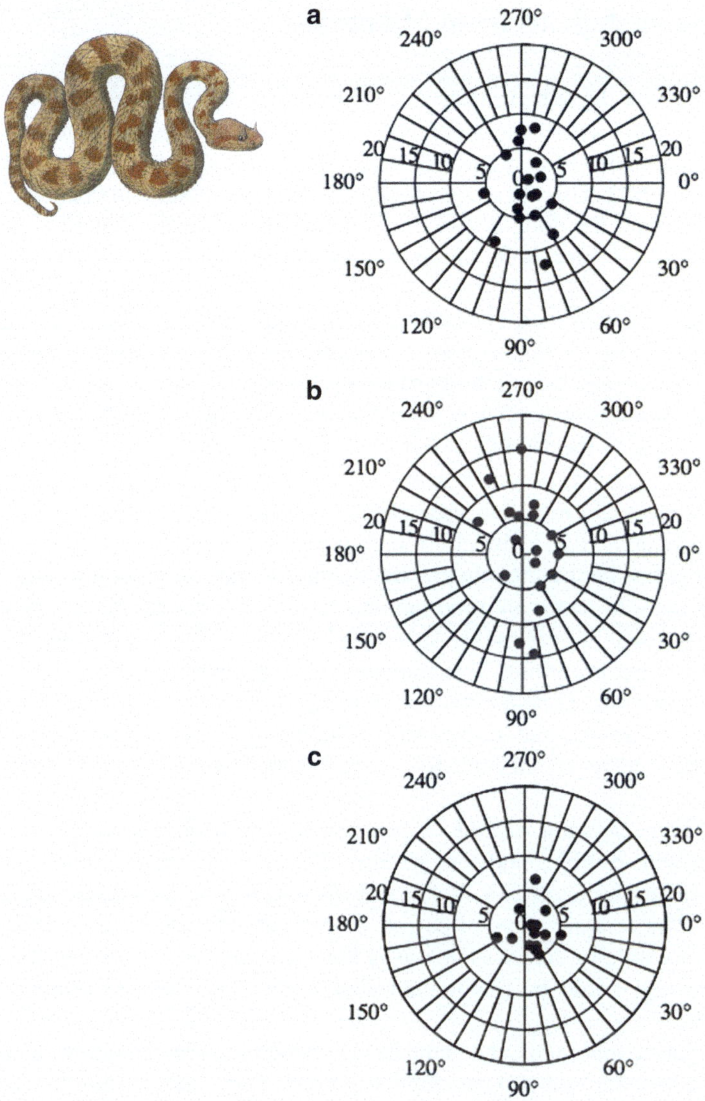

Fig. 8.1 Experimental manipulation of Saharan sand viper (*Cerastes cerastes*): polar coordinates of predatory strike and distance from the prey. (**a**) Control strike. (**b**) Strikes after denervation of olfactory and vomeronasal nerves. (**c**) Strikes after denervation as in (**b**) and bilateral occlusion of the eyes. The nonsignificant difference between trials (**a**) and (**c**) demonstrates that this species can direct strikes using other sensory reception (Reproduced with permission from Young and Moran 2002)

8.3 Seismic Signals from Mammals

Considering that the subterranean environment is not favorable to the diffusion of airborne sounds, seismic communication has been proved in blind subterranean and fossorial mammals (Mason and Narins 2001), as in the case of *Spalax ehrenbergi* (Nevo et al. 1991). This superspecies has a somatosensorship independent from the auditory system for long-distance communication. Head thumping along the excavated tunnel represents the vibration-borne mechanism for long-distance communication.

Seismic vibrations have been found in the mole rat (*Georychus capensis*) and in the spalacine mole rat *Nannospalax*. Recently the golden mole (*Eremitalpa granti namibiensis*) has also been hypothesized to use seismic vibrations to locate food.

Utilization of the seismic waves produced by clumps of grasses scattered among the sand dunes of the Namib desert have been shown to be used by the Namib golden mole (*Eremitalpa granti namibensis*). The grasses under a light wind emit low-amplitude vibrations that are used by this mole to orient itself toward such clumps of plants, rich in invertebrates, when it is searching for food by sand-swimming a few centimeters below the surface of the sand (Narins et al. 2006).

Soil percussion seems an important system for communication in many animals from insects to large mammals such as elephants. Vibrational signaling can be produced by foot drumming, as in kangaroo rats (genus *Dipodomys*). In the banner-tailed kangaroo rat (*Dipodoomys spectabilis*), foot drumming is produced during courtship or when encountering a snake. Ehrenberg's mole rat (*Spalax ehrenbergi*) thumps the head to signal but also to receive back the seismic echo to localize obstacles.

The Asian elephant (*Elaphus maximum*) has been shown to use signals associated with rumbles and foot stomps. The acoustic signal travels at 309 m/s and the seismic signals at 248–264 m/s. Both signals have a predominant frequency of 20 Hz. The seismic waveforms can be detected at a distance of 16 km. If the elephants are able to detect such waves the communication between the different familiar groups could be maintained also at long distance (O'Connell-Rodwell et al. 2000).

As in marine mammals, the elephant's foot cushion can be used to facilitate the impedance between soil and the elephant's body. For a review about the communication tools adopted by elephants, see O'Connell-Rodwell et al. (2000) and O'Connell-Rodwell (2007).

Seismic sensors have been applied to detect elephants and other large mammals. Using a buried geophone, Wood et al. (2005) have demonstrated it is possible to use this technique to estimate the energy produced by passing elephants (Fig. 8.2) and to discriminate species using the spectral content of their footfall with an accuracy of 82 %. This technique has been recommended to estimate species and number of individuals of large African mammals.

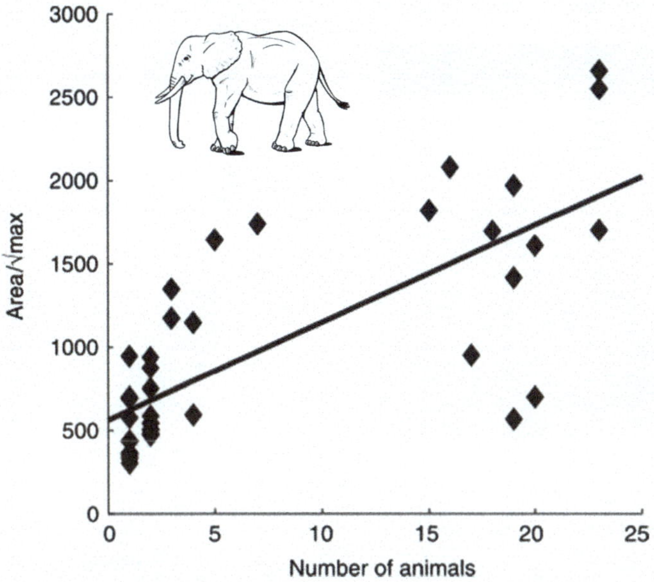

Fig. 8.2 Relationship between number of elephants and Area/$\sqrt{\text{maximum}}$ (amplitude envelope) (Reproduced with permission from Wood et al. 2005)

8.4 Seismic Signals from Arthropods

Seismic vibrations cause the *Diplocardia* earthworm to emerge from soil. This phenomenon is well known and is utilized to harvest earthworms. This practice is called "worm grunting" and is used in the Southeast United States by local families (worm grunters) and is well diffused around Florida's Apalachicola National Forest. The earthworms are forced to emerge from the soil by the use of a very peculiar technique named grunting, which consists of scraping a wooden stack driven into the soil with a flat metal object. The sound when produced at regular intervals causes the worm *Diplocardia mississippiensis* to emerge in a few minutes from the soil. A broadband frequency is produced by this grunting; the number of emerging worms decreases according to increasing distance from the origin of the vibration. Strokes produced for a duration of approximately 760 ms and repeated every 1.2 s produce the emergence of worms. The seismic signals are broadband, below 500 Hz with a dominant frequency of 97 Hz. The worms emerge 54–131 s after the onset of the vibrations. The number of emerging worms is positively correlated with the signal strength.

Some animals use the technique of grunting to capture worms. It has been observed in wood turtles (*Clemmys insculpta*), which stomp their front feet with a frequency of once per second, but also birds such as gulls have a paddling behavior, as observed by Tinbergen (1960).

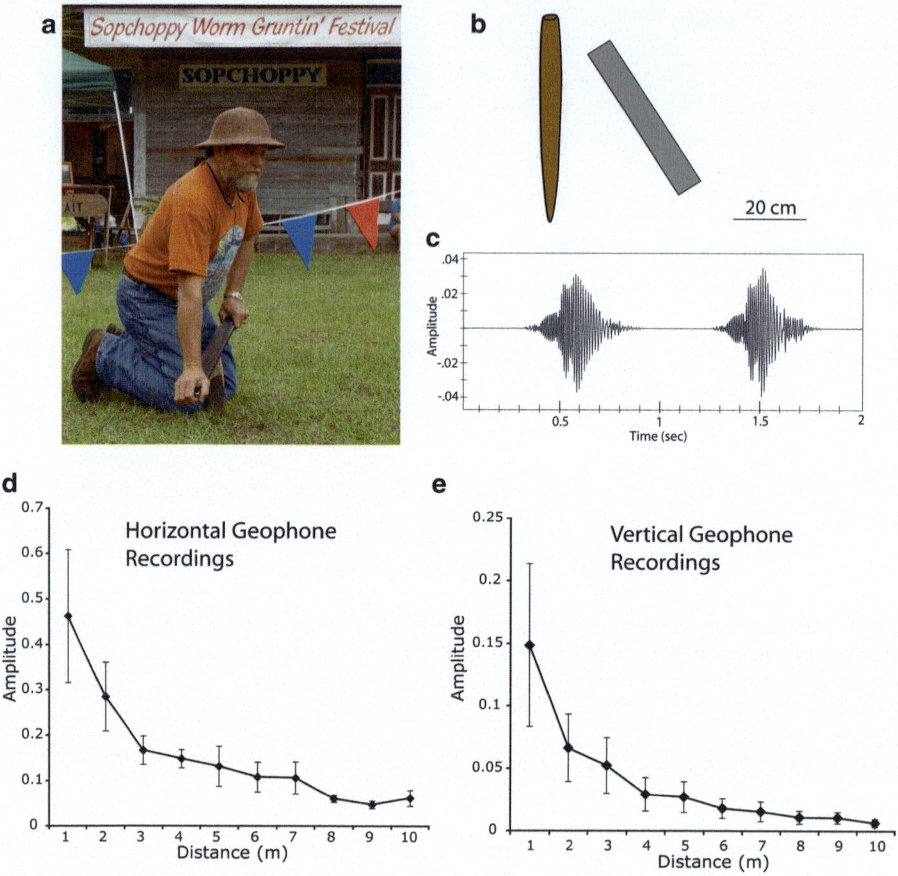

Fig. 8.3 The grunting technique used to capture *Diplocardia* earthworms. In (**a**), the authors demonstrate the use of this technique. (**b**) The tools used to produce vibration into the soil. (**c**) Oscillogram and spectrogram produced by the seismic vibration recorded at 1.8 m from the stake. (**d**, **e**) Horizontal and vertical geophone recordings, respectively (From Catania 2008)

Two opposed hypotheses have been discussed by Mitra et al. (2009) to explain the grunting: worms emerge from the soil in natural conditions to escaping the vibrations caused by digging foragers, or emerge as result of the vibrations produced by falling rain.

Catania (2008) has excluded the idea that worms emerge after human grunting because the vibrations copy the vibration of rainfall (Fig. 8.3). This author ascribes the emergence of worms from soil to vibrations resembling mole-like digging in the attempt to escape mole predation, considering that humans are playing the role of "rare predators."

Vibration is extensively used by social insects such as bees and ants, but other social insects use vibration in food recruitment, mating, and antipredatory signals. Food recruitment is well documented in treehoppers, in sawflies, and in tortoise beetles, as reviewed by Cocroft and Hamel (2010), but the different uses of vibration mechanisms in related species remain quite obscure.

The jumping spider *Hebronattus dossenus*, which during courtship uses seismic signals, seems to do more signaling in leaf litter than in sandy soil or rocks (Elias et al. 2004).

Stridulatory signals are used by ants although chemical signals seem more common. It is questioned if stridulatory signals are airborne sounds or are vibratorial signals across the substrate. It is reasonable to admit that ants are unresponsive to airborne sound at a distance of a meter or more. *Trichoid sensilla* probably use their antennae as receptors of airborne sound.

Selenopsis richteri is a species of fire ant that has a sting which can kill insects, birds, and small mammals. For instance, in humans the bite of this species produces a painful allergic reaction that in some cases may be lethal. When Hichling and Brown (2000) posed a microphone in a mound of this species, the microphone was immediately attacked, accompanied by different types of acoustic signals. The rapidity of the response suggests an acoustic and a nonchemical communication between individuals (Fig. 8.4).

Despite the assumption by these authors about the hearing capacity of this group, other authors (Roces and Tautz 2001) consider ant stridulation as a substrate-borne component, admitting that ants are definitively deaf.

The African fungus-growing termites of the genus *Macrotermes* utilize a complex system of vibrational long-distance alarm communication (Röhrig et al. 1999). When the nest is disturbed the males of *Macrotermes bellicousus* and *M. subhyalinus* produce vibrational signals by drumming their heads against the substratum. The frequency with which they emit such vibration is species specific: *M. bellicousus* uses a pulse frequency of 26 Hz and *M. subhyalinus* a pulse frequency of 13 Hz.

The drumming executed by males is transmitted by soldiers and intercepted by the workers, which enter into an alarm status and escape toward the nest. This system rapidly propagates where there are soldiers and represents a long-distance communication produced by individual preamplification.

According to the intensity of the drumming, the reaction of the workers can produce a mass retreat toward the nest.

Substrate-born vibration is used extensively by animals to locate mates, prey, and predators. Antlion larvae build funnel-shaped traps to capture ants. Recently Fertin and Casas (2007) have experimentally demonstrated that antlion larvae use the vibrations produced by ants to locate their prey.

Japanese rhinoceros beetle (*Trypoxylus dichotoma*) larvae have been shown experimentally to avoid pupal cells harboring live pupae, possible because pupae produce a specific vibration of 3–7 pulses at 1.3 s of interval by beating the pronotum against the inner wall of the pupal cell (Kojima et al. 2012). Such vibration reduces the risk that a larva of the same species living in the same patch of humus can damage the pupal cell. This behavior seems strongly correlated with an ancestral antipredatory mechanism (Fig. 8.5).

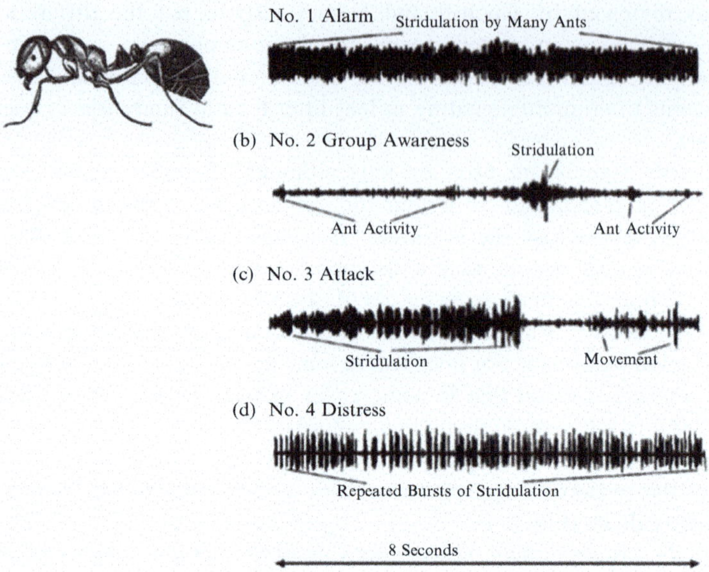

Fig. 8.4 Pressure–time spectra of stridulating signals from *Solenopsis richteri*. *1* Alarms (**a**), *2* Group Awareness (**b**), *3* Attack (**c**), *4* Distress (**d**) (Reproduced with permission from Hickling and Brown 2000)

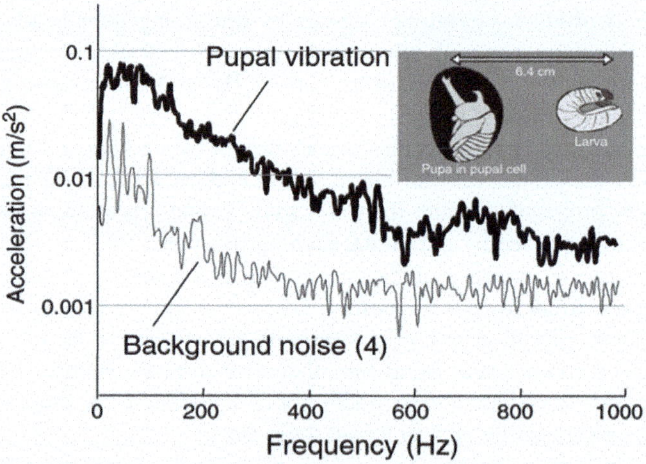

Fig. 8.5 Averaged power spectra of pupal vibration from natural pupal cells harboring a male pupa of Japanese rhinoceros beetle (*Tryopoxylus dichotoma*) and background noise. The larvae usually is 6.4 cm on average distant from the pupa (Reproduced with permission from Kojima et al. 2012)

The male prairie mole cricket (*Gryllotalpa major*) of the tallgrass prairies of the south-central United States builds a specialized acoustical burrow to attract females using airborne signals. Hill and Shadley (2001) have observed that contemporary seismic signals are produced in the soil from unknown mechanisms, which are hypothesized to be a mechanism to spacing males away from each other

8.5 Soil Pests, Sound Detection, and Biodiversity Assessment

Analysis of soil sonic ambience allows monitoring soil-dwelling insect pests and also evaluating soil biodiversity. The first aspect has been extensively studied in applied agroecology but the second remains a quite unexplored field.

Acoustic evaluation of soil pests requires specific recording techniques. Collection of soil information requires not only standard microphones, but also piezoelectric microphones, geophones, and accelerometers to capture soilborne vibrations (Mankin et al. 2011).

For instance, the vibrations of soil insects considered agricultural pests, such as *Phyllophaga* white grubs and *Diaprepes* citrus rootweevils (*Diaprepes abbreviatus* L.) (Coleoptera: Curculionidae), have been studied by Mankin et al. (1998) adopting a portable accelerometer coupled to a microphone system.

To solve a practical problem such as that presented by a termite infestation in trees and in houses, Mankin et al. (2002) proposed the use of an accelerometer connected to an amplifier (Mankin et al. 1998). These authors were able to distinguish between background noise and three species of termites (*Reticulitermes virginicus*, *Reticulitermes flavipes*, and *Coptotermes formosanus*), and they found a correlation between the activity rate and the number of termites present.

The quality of soils is important in every ecosystem because soils are the home for an important biodiversity. Soil health is coincident with the health of fungi, plants, and animals. Monitoring soils represents an important practice to survey the quality of the environment. The use of acoustic methodologies to survey the animal content of a soil appears a very promising technique, encouraged by sophisticated digital probes that can "listen" with great accuracy to the soil-borne sounds.

For instance, the mapping of soil invertebrates carried out by Brandhorst-Hubbard et al. (2001) in hayfields of Alabama using electronic microphones has demonstrated a good correlation between destructive traditional soil investigation and acoustic analysis (Fig. 8.6).

Summary
Soil is a medium rich in acoustic information. Most of the sounds present in the soil are originated by stridulating animals or are the result of the digging activity of fossorial organisms. Acoustic information from the soil is an important tool to monitor soil biodiversity in a nondestructive way.

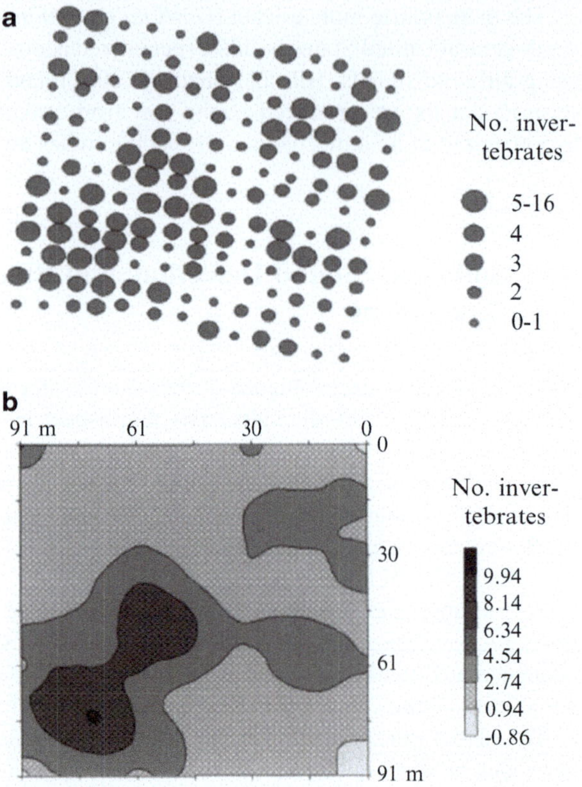

Fig. 8.6 Comparison between a map of counts of soil invertebrates (**a**) and a contour map of estimated invertebrates (**b**) by using an acoustic method (Reproduced with permission from Brandhorst-Hubbard et al. 2001)

Some reptiles use vibrations produced by their prey for hunting. Vegetation vibration has been proved to orient fossorial moles toward vegetation clumps rich in insect food. Soil percussion is also a common mechanism used by elephants to communicate at great distance.

Insects and arthropods use vibrations extensively to communicate or to escape predators. Well documented is the case of *Diplocardia mississippiensis*, an earthworm that emerges from the soil after vibration is produced in the soil using a "grunting" device. This behavior is explained as a strategy utilized by the worms to escape hunting by moles.

It is disputed if ants have hearing systems or body vibrational sensing, but their capacity to react to soil vibrations is surprisingly rapid. Substrate-borne vibrations are extensively used by animals to locate mate, prey, and predators, as in the well-documented case of antlion larvae.

Despite the scarcity of studies, soil monitoring appears a promising field of research. Sound prospecting of the soil requires that standard microphones are integrated with piezoelectric microphones, geophones, and accelerometers to capture airborne and soilborne vibrations.

References

Brandhorst-Hubbard JL, Flanders K, Mankin RW, Guertal EA, Crocker RL (2001) Mapping of soil insect infestations sampled by excavation and acoustic methods. J Econ Entomol 94(6):1452–1458

Catania KC (2008) Worm grunting, fiddling, and charming: humans unknowingly mimic a predator to harvest bait. PLoS One 3(10):e3472

Cocroft RB, Hamel JA (2010) Vibrational communication in the "other insect societies:" a diversity of ecology signals, and signal functions. In: O'Connell-Rodwell CE (ed) The use of vibration in communication: properties, mechanisms and function across taxa. Transworld Research Network, Trivandrum, Kerala, India

Cocroft RB, Rodriguez RL (2005) The behavioral ecology of insect vibrational communication. Bioscience 323:323–334

Elias DO, Mason AC, Hoy RR (2004) The effect of substrate on the efficacy of seismic courtship signal transmission in the jumping spider *Hobromattus dossenus* (Araneae: Salticidae). J Exp Biol 207:4105–4110

Fertin A, Casas J (2007) Orientation towards prey in antlions: efficient use of wave propagation in sand. J Exp Biol 210:3337–3343

Hartline PH (1971) Physiological basis for detection of sound and vibration in snakes. J Exp Biol 54:349–371

Hickling R, Brown RL (2000) Analysis of acoustic communication by ants. J Acoust Soc Am 108 (4):1920–1929

Hill PSM, Shadley JR (2001) Talking back: sending soil vibration signals to lekking prairie mole cricket males. Am Zool 41:1200–1214

Kojima W, Takanashi T, Ishikawa Y (2012) Vibratory communication in the soil: pupal signals deter larval intrusion in a group-living beetle *Trypoxylus dichotoma*. Behav Ecol Sociobiol 66:171–179

Mankin RW, McCoy CW, Flanders KL, Brandhorst J, Crocker RL, Shapiro JP (1998) Methods for acoustic detection of insect pests in soil. In: Proceedings of the Soil Science Society of America conference on Agroecoacoustics, Buoyoucos, MS

Mankin RW, Osbrink WL, Oi FM, Anderson JB (2002) Acoustic detection of termite infestation in urban trees. J Econ Entomol 95(5):981–988

Mankin RW, Hagstrum DW, Smith MT, Roda AL, Kairo MTK (2011) Perspective and promise: a century of insect acoustic detection and monitoring. Am Entomol 57(1):30–44

Mason MJ, Narins PM (2001) Seismic signal use by fossorial mammals. Am Zool 41:1171–1184

Mitra O, Callaham MA Jr, Smith ML, Yack JE (2009) Grunting for worms: seismic vibrations cause *Diplocardia* earthworms to emerge from the soil. Biol Lett 5:16–19

Narins PM, Jarvis JUM, Bronner G, Mason MJ (2006) Preliminary evidence for the use of microseismic cues for navigation by the Namib golden mole. J Acoust Soc Am 119 (2):1260–1268

Nevo E, Heth G, Pratt H (1991) Seismic communication in a blind subterranean mammal: a major somatosensory mechanism in adaptive evolution underground. Proc Natl Acad Sci USA 88:1256–1260

O'Connell-Rodwell CE (2007) Keeping an "ear" to the ground: seismic communication in elephants. Physiology 22:287–294

O'Connell-Rodwell CE, Arnason BT, Hart LA (2000) Seismic properties of Asian elephant (*Elaphas maximus*) vocalization and locomotion. J Acoust Soc Am 108:3066–3072

Roces F, Tautz J (2001) Ants are deaf. J Acoust Soc Am 109(6):3080–3082

Röhrig A, Kirchner WH, Leuthold RH (1999) Vibrational alarm communication in the African fungus-growing termite genus *Macrotermes* (Isoptera, Termitidae). Insect Soc 46:71–77

Tinbergen N (1960) The herring gull's world. A study of the social behaviour of birds. Basic Books, New York

Wood JD, O'Connell-Rodwell CE, Klemperer SL (2005) Using seismic sensors to detect elephants and other large mammals: a potential census technique. J Appl Ecol 42:587–594

Young BA, Moran M (2002) The use of ground-borne vibration for prey localization in the Saharan sand vipers (*Cerastes*). J Exp Biol 205:661–665

Chapter 9
Methods

9.1 Introduction

Soundscape ecology is a young science with a broad range of field and laboratory approaches from bioacoustics to psychoacoustics and truly heterogeneous methods to collect, process, and interpret sonic data. Two principal fields of investigation are direct investigation of the physical structure of the sound (amount of energy, distribution of such energy in the different frequency domains) and a semantic approach to sound description as culturally perceived by humans. The two approaches can be combined when, for instance, the goal is to investigate the physical and psychological consequences of noise on human physiology and behavior.

Physical and biological dimensions of the soundscape are collected with hardware. The human dimension of the sonic environment is more often investigated by using interviews or ad hoc forms. In some cases the interviews or the questionnaires are coupled to the presentation of images because the psychological components of sound perception in many cases are dependent on simple acoustic perception. Several approaches thus are available to collect information from the sonic environment:

(a) Automatic Recording Procedure (ARP)
(b) Direct Aural Procedure (DAP)
(c) Direct Aural Interpretation Procedure (DAIP)
(d) Public Interviews Procedure (PIP)

ARP can be used to collect sonic information for all occasions.

DAP is a procedure utilized by ecologists to evaluate richness, diversity, and dynamics of vocal animals directly by ears.

DAIP is a procedure utilized for soundscape evaluation or assessment for human use.

PIP is a psychoacoustic procedure to collect human feelings about the soundscape.

A. Farina, *Soundscape Ecology: Principles, Patterns, Methods and Applications*,
DOI 10.1007/978-94-007-7374-5_9,

Capturing the physical sound means to collect under a digital format the numerical value of the sound at a precise temporal interval.

In terms of sound measurements we have to distinguish between

(a) sound pressure investigation and
(b) spectral investigation.

Sound pressure is measured using a calibrated sound meter or phonometer that measures the sound pressure level (SPL). This measure is commonly used for the quantification of different kinds of noise/sounds, especially for industrial and environmental noise.

The spectral measure is obtained by recording sounds with microphones, storing the data in digital format in memory mass storage devices (e.g., external hard disk drive), and processing the data by applying, for instance, the Fourier transform procedure.

9.2 Basic Elements of Acoustics

To allow the reader to fully understand the terminology adopted in this book, we have considered it essential to report a summary of the fundamentals of acoustics necessary to develop our reasoning before describing some methodologies in soundscape analysis. This fundamentals do not represent an exhaustive treatment of acoustic terms but some essential synthetic and incomplete references.

9.2.1 The Sound

Sound is produced by a rapid variation in the average density or pressure of air molecules above and below the current atmospheric pressure. Sound is defined by The American Heritage Dictionary of the English Language (fourth edition, 2000) "*as a vibration transmitted through an elastic solid or a liquid or gas, with frequencies in the approximate range from 20 to 20,000 hertz, capable of being detected by human organs of hearing*".

Sound can be also defined as a sequence of waves of pressure that propagates through compressible media such as air, water, or soil in a process of compression and rarefaction. The changes in atmospheric pressure originated by a vibrating device produce a sound pressure, and the fluctuations create sound waves.

Humans perceive sound from approximately a minimum threshold of 20 μPa (20×10^{-6} Pa) to 100 Pa (the threshold of pain) (~140 dB). The smallest perceptible change in sound is about 1 dB.

Every sound can be distinguished according to these different properties.

- Frequency
- Pitch
- Period
- Wavelength
- Sound speed
- Wavenumber
- Amplitude
- Sound Pressure
- Sound pressure level (SPL)
- Sound power
- Sound power level
- Sound intensity
- Sound intensity level
- Sound exposure level
- Loudness
- Acoustic impedance

9.2.1.1 Frequency

Frequency (f) is the number of occurrences of a repeating event per unit time and is the inverse of the period T where $f = 1/T$ and is measured in hertz (Hz) or in cycles per second (CPS). For instance, if we have 50 events that occur in 25 s the frequency $f = 50/25 = 2$ Hz.

The hertz (Hz) is a measure of the frequency of sounds expressed in cycles per second for a regular phenomenon (Fig. 9.1). This measure can be applied to every cyclic phenomenon, such as heartbeat (an infant heart has 120 beats per minute, which equals 2 Hz; an adult heart has 70 beats per minute, equal to approximately 1.17 Hz). The commonest frequency ranges in acoustics are abbreviated as kilohertz (kHz) (10^3 Hz), megahertz (MHz) (10^6 Hz), and gigahertz (GHz) (10^9 Hz.)

Light, as an electromagnetic radiation, has a frequency ranging from tens (infrared) to thousands (ultraviolet) of terahertz (10^{12} Hz). Radiofrequency radiation is usually measured in kilohertz (kHz), megahertz (MHz), or gigahertz (GHz).

Humans perceive sound frequency as pitch (a subjective perception of an objective frequency), with differences according to age: infants have a full range of perception from 20 to 20,000 Hz, adult humans can hear sounds between 20 and 16,000 Hz, and in the aged human the high frequencies are dramatically reduced.

We consider infrasound and ultrasound sounds those with frequencies below and above human audibility. For instance, whales and other marine mammals are sensitive to infrared frequency from 20 Hz down to 0.001 Hz. This type of energy covers long distances and passes over obstacles with little dissipation. This frequency range is generally used for monitoring earthquakes and in oil prospecting.

The relationship between frequency and wavelength is represented by the equation $\lambda = c/f$, where $c =$ sound speed and λ is wavelength.

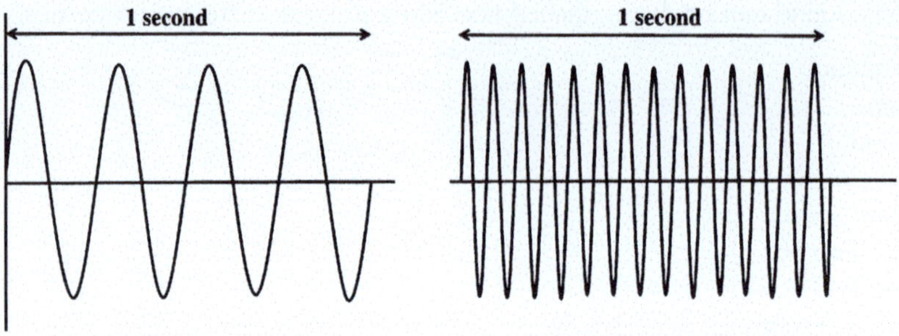

Fig. 9.1 Example of sinusoidal curves of different hertz (Hz): 4 and 14 Hz

9.2.1.2 Pitch

Pitch is our subjective interpretation of frequency.

9.2.1.3 Period

Period (T) is the time necessary to complete a cycle of a wave for a given point: $T = 1/f$. For example, for $f = 550$ $T = 1/550$ s $= 1.81$ ms (milliseconds).

9.2.1.4 Wavelength

Wavelength (λ) is the spatial period of a sinusoidal wave and is measured in a homogeneous medium as the distance between two points with the same phase (crest to crest), or at the intercept with zero: it results from the relationship between speed and frequency $\lambda = c/f$ or $\lambda = cT$ where c is the speed of the sound and T is the period (Fig. 9.2).

9.2.1.5 Sound Speed

Sound speed (c) is the distance that sound covers in a given unit of time. Sound propagates at different speeds according to the media and is proportional to the square root of the ratio of the elastic modulus (stiffness) [k] of the medium to its density [ρ]. $c = \sqrt{K/\rho}$, which can be calculated as $c = \lambda f$ or $c = \lambda/T$.

The ambient conditions modify sound propagation. In dry (0 % humidity) air at 20 °C at sea level, the speed of sound is approximately 343 m/s (1,230 km/h; 767 mph). This number is derived by the formula $c_{air} = (331.3 + 0.606\vartheta)$ m/s where ϑ is the temperature in degrees Celsius (°C).

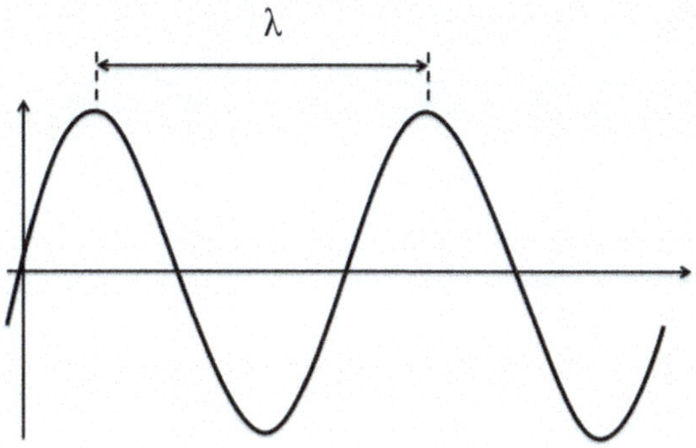

Fig. 9.2 The wavelength λ is the distance between two crests or at the intercept with zero of a sinusoidal wave

Table 9.1 Examples of sound speed according to different materials

Media	Speed of sound (m/s)
PVC soft	80
Air	343 at 20 °C
Cork	500
Helium	1,020
Water	1,480
Polystirol	1,800
Plexiglas	1,840
Concrete	3,100
Tile	3,600
Pine wood	3,600
Granite	3,950
Oak wood	4,100
Iron	5,000
Steel	5,050
Aluminum	5,200
Quartz glass	5,400

Sound propagation in air depends on air pressure, humidity, and the presence of wind. Wind creates a sort of downwind refraction, an acoustic shadow, so wind at 4 m/s/km creates the effect of a temperature of 7.5 °C (Table 9.1).

In freshwater, at 20 °C, the speed of sound is approximately 1.482 m/s (5,335 km/h, or 3.315 mph). In saltwater that is free of air bubbles or suspended sediment, sound travels at about 1,560 m/s.

The specific impedance, or resistance, in water is about 3,500 times greater than in air and is defined by the equation $Z = \rho c$,

where $\rho = \text{kg/m}^3$ is the density of the medium and $c = \text{m/s}$ is the speed of sound in the medium.

The speed of sound in water depends also on temperature, salinity, and depth. This is the empirical equation:

$$c = 1449 + 4.6t - 0.055t^2 + 1.4(S - 35) + 0.017D$$

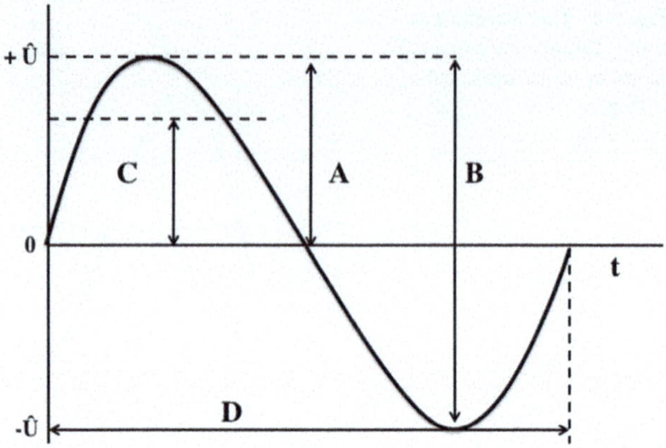

Fig. 9.3 Representation of a sinusoidal curve and the different way to measure amplitude: *A* peak amplitude ($\hat{U}$), *B* peak-to-peak amplitude ($2\hat{U}$), C = RMS (root mean square) amplitude $\hat{U}/\sqrt{2}$, *D* wave period *t*

where c = m/s is the speed of sound, t = temperature in °C, S (PSU) = salinity, and D (m) = depth.

The speed of sound in seawater increases with the increase of pressure (hence depth), temperature (a change of 1 °C ~ 4 m/s), and salinity (a change of 1 % ~ 1 m/s).

The speed of sound in water is about 4.5 times greater than in air for every equifrequent sound.

9.2.1.6 Wavenumber

Wavenumber, defined as the reciprocal of the wavelength or also a spatial frequency, is calculated by the formula $\overline{v} = 1/\widetilde{\lambda}$.

9.2.1.7 Amplitude

Amplitude is the maximum amount by which the instantaneous sound pressure differs from the ambient pressure. It is the dimension of the change in the oscillating wave and can be measured as peak to peak, that is, the distance between two peaks of a sinusoidal wave. Other ways to measure amplitude are *peak amplitude* (Û) when a signal is not sinusoidal and the *root mean square* (RMS) amplitude, which is the square root of the mean over time of the square of the vertical distance of the graph from the rest state (Fig. 9.3). The form of variation of amplitude is called the envelope of the wave.

For a sine wave the relationship between peak-to-peak and RMS is Peak-to-peak $= 2\sqrt{2} - \text{RMS}$

Peak - to - peak $= 2 \times \hat{U}$ substituting in the above equation $2 \times \hat{U} = 2\sqrt{2} \times \text{RMS}$ then RMS $= (\hat{U}/\sqrt{2})$.

9.2.1.8 Sound Pressure

Sound pressure or acoustic pressure is the deviation from ambient atmospheric pressure caused by a sound wave. The SI unit of sound pressure, p, is the pascal (Pa). A pascal Pa $= 1 \times \text{N/m}^2 = 1 \times \text{kg/m} \times \text{s}^2$ (N = Newton, that is, the force that produces an acceleration of 1 m/s when exerted on a mass of 1 kg). Considering that the smallest audible sound pressure in air is of order of 10^{-6} Pa, sound pressure is usually expressed in μPa.

9.2.1.9 Sound Pressure Level

Sound pressure level (SPL) or sound level (L_{p}) is a logarithmic measure of the effective sound pressure of a sound relative to a reference value. It is measured in decibels (dB) above a standard reference level. The commonly used "zero" reference sound pressure in air is 20 μPa RMS, which is usually considered the threshold of human hearing (at 1 kHz).

$$L_{\mathrm{p}} = 20 \log_{10}\left(\frac{P}{P_{\mathrm{ref}}}\right) \mathrm{dB}$$

In water this reference pressure = 1 μPa.

9.2.1.10 Sound Power

Sound power or acoustic power, P_{ac}, is a measure of sonic energy per time t unit $P_{\mathrm{ac}} = I \times A$, where I = sound intensity and A = unit area.

The unit of measurement for power is the *watt* (W): 1 W = 1 N of work per second.

9.2.1.11 Sound Power Level

The sound power level (SWL, Lw, $L_{P_{\mathrm{ac}}}$) is the difference between two P_{ac} expressed in dB according the formula

$$\mathrm{Lw} = 10 \log_{10}\left(\frac{P_1}{P_0}\right) \mathrm{dB}$$

9.2.1.12 Sound Intensity

Sound intensity or acoustic intensity (I) is defined as the *sound power*, P_{ac}, (ac = acoustic) per unit area A: $I = P_{\mathrm{ac}}/A$.

Sound intensity is related to the inverse square law, which states the amount of reduction of a power source with the distance. In fact, in ideal conditions sound power from a point source is spread over the surface area of a sphere where

$$I = \frac{P_{ac}}{4\pi r^2}$$

where r is the distance from the source or the radius of the sphere.

Considering that the power of the source is measured in W (watts), then the intensity of power for unit area at distance r is $I = W/4\pi r^2$.

9.2.1.13 Sound Intensity Level

Sound intensity level (IL or L_i) is defined as the difference between *sound* or *acoustic intensity* and a reference value of 10^{-12} W/m^2. This value, which represents the sound intensity level (IL), can be converted to dB:

$$Lw = 10 \log_{10}\left(\frac{I}{I_0}\right) dB$$

where I = *acoustic intensity* and I_0 = the reference value (10^{-12} W/m^2)

If for instance the power source is = 0.5 W, at 5 m of distance the intensity would be

$I = 0.5/4 \times 3.14 \times 5^2 = 0.00159$ W/m^2 and the

$$IL(dB) = 10 \log_{10}\left(\frac{1.59 \times 10^{-3}}{10^{-12}}\right)$$

9.2.1.14 Sound Exposure Level

Sound exposure level (SEL) is the measure of the energy of a sound and is the time-integral of the instantaneous squared sound pressure normalized to a square reference measure in a 1-s period.

$$SEL = \sqrt{\int_0^T |x(t)^2| dt}$$

This formula can be expressed also as SEL = SPL (rms) + 10*log (duration).

9.2.1.15 Loudness

Loudness is an important characteristic of a sound as it is perceived by the human ear. Loudness is correlated to the amplitude of a sound and is defined as "that attribute of auditory sensation in terms of which sounds can be ordered on a scale extending from quiet to loud."

Loudness is a subjective measure of sound strength and must not be confused with objective measures of sound energy such as the sound pressure, the sound intensity, or the sound power. The A-weighting method to measure sound power is an attempt to adjust sound measurement to human acoustic perception. Loudness depends on either sound pressure or sound duration. Actually, the human hearing system averages the effect of sound pressure level (SPL) over an interval of 600–1,000 ms. A sound with a constant SPL seems to increase in loudness after a duration of 20, 50, 100, or 200 ms up to the duration of 1 s; after that, loudness is stabilized. For sounds persisting longer than 1 s the loudness is fixed after an averaging of 600–1,000 ms.

9.2.1.16 Acoustic Impedance

Acoustic impedance indicates the part of sound pressure generated by the vibration of molecules of a specific medium at a given frequency. The acoustic impedance Z (or sound impedance) is frequency (f) dependent and is very useful for describing the behavior of musical wind instruments, for example.

The acoustic impedance is calculated as $Z = P/vS$ where P = sound pressure, v = particle velocity, and S is the surface through which an acoustic wave of frequency f propagates. The specific acoustic impedance Z (also called shock impedance) is the ratio of sound pressure P to particle velocity v at a single frequency and is expressed in rayls (Rayleighs).

9.2.2 The Decibel

In acoustics, the decibel is an important unit of measure of the sound power.

The unit dB, which is one tenth of a bel, is a logarithmic (to base 10) unit used to measure a ratio that may be represented by power, sound pressure, voltage, or intensity.

The equation that describes the difference in decibels between two measurements is

$$\Delta X(\text{dB}) = 10 \log \frac{X_2}{X_1}$$

where ΔX is the difference between two quantities expressed in dB. Considering that the bel unit is transformed for better use as a decibel (1/10 bel), in the logarithmic formula it is sufficient to multiply the logarithmic ratio by 10. In acoustics the sound intensity that is measured by microphones is generally expressed as variation of sound pressure. To do this a microphone registers the

Fig. 9.4 The isophonic curves of human audibility according to dB and frequencies: the system appears not linear and is strongly influenced by age class and individual auditive capacity (From ISO 226:2003)

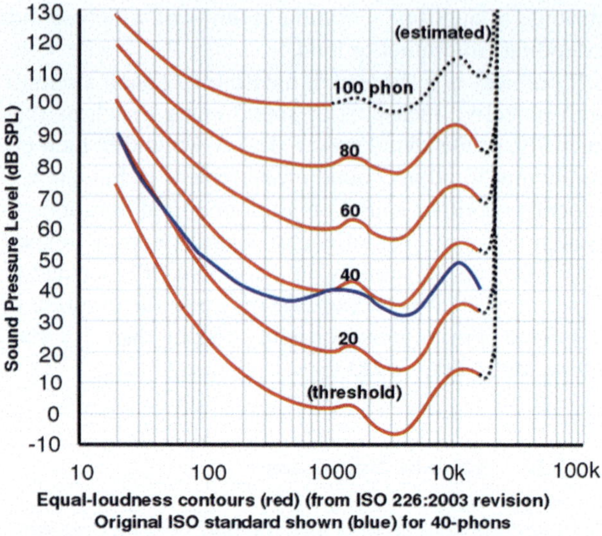

variation of voltage of a transducer, and this voltage is proportional to the change of sound pressure P. The power carried by a pressure wave is proportional to the square amplitude, and the formula becomes

$$\Delta I(\mathrm{dB}) = 10\log\frac{I_2}{I_1} = 10\log\frac{P_2^2}{P_1^2} = 10\log\left(\frac{P_2}{P_1}\right)^2 = 20\log\left(\frac{P_2}{P_1}\right)$$

The factor of 2 is added to the equation because the logarithm of the square of a quantity is equal to twice the logarithm of that quantity.

Considering that a microphone or a transducer converts the acoustic pressure into voltage, the formula can be rewritten:

$$\Delta I(\mathrm{dB}) = 20\,\log\frac{V_2}{V_1}$$

In acoustics is quite hard to measure the sound in terms of power (W, watt), intensity (W/m^2), and acoustic pressure (Pa, pascal) because the variation between the lowest and the highest value create difficulties for understanding the phenomenon. In fact, the acoustic pressure for humans varies between 20 μPa and 63.2 Pa, which is the limit for the pain threshold in humans with variation of 10^6. In Fig. 9.4 are reported the isophonic curves that represent the level of audibility by humans according to dB and frequencies. The system appears not linear and is strongly influenced by age class and individual auditory capacity.

The logarithmic scale is more convenient because sound magnitude has a range that can be of eight orders of magnitude (in submarine condition) and the human internal ear system works in acoustic perception similarly to a logarithmic scale.

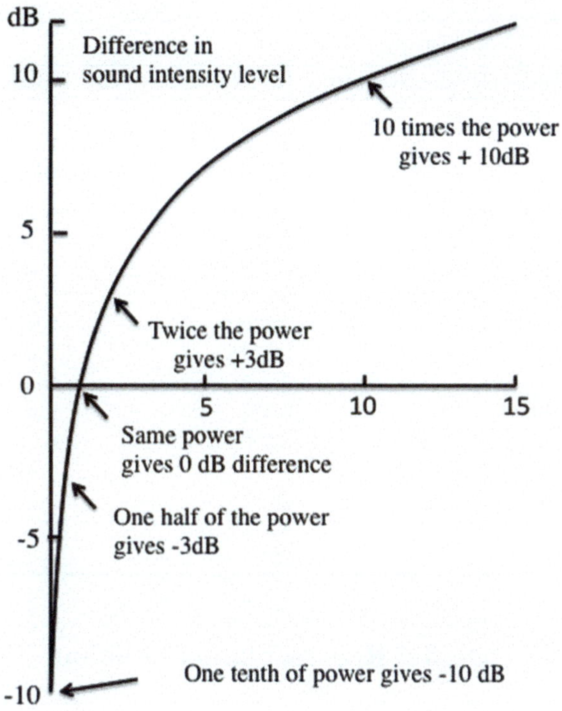

Fig. 9.5 Example of variation of dB according to the difference in sound intensity level. For instance, doubling the power of the sound signal corresponds to a gain of +3 dB

The logarithmic scale has been adopted also because a doubling of acoustic power does not correspond to a doubling of hearing perception. For instance, the difference in decibels between a signal with a power of 1 W and another of 2 W is 10 log (2/1) = 3 dB. If the difference in signal between 1 and 2 V, 20 log (2/1) = 6 dB. Figure 9.5 indicates how the decibel scale varies according the values of sound pressure.

Doubling a power gives a rise of 3 dB and doubling a voltage produce a rise of 6 dB. In Table 9.2 are reported the dB level and the amount of acoustic pressure of some common sounds and the noise level caused by anthropogenic causes.

A value in decibels has no meaning if there is no reference measure. In some cases the reference measure is implicit; for example, the sound pressure level (SPL) has a reference level defined worldwide as 2×10^{-5} N/m^2 (20 μPa). Often a symbol is added after dB (V, u, v, m, A) that means a reference level:

dBV: 1 V
dBu: 0.775 V (Europe)
dBv: 0.775 V (USA)
dBm: 1 milliwatt
dbA: dB SPL, A-weighted response

The decibel is used also to describe the voltage gain of a device. For example, when a microphone amplifier has a gain of 50 dB, this means that the input voltage has been multiplied by a factor 317 because 20 log 317/1 = 50 dB.

Table 9.2 Sound pressure vevel and sound pressure of the most common anthropogenic noises and level of hearing risk

Sound pressure level (dB)	Sound pressure (mPa)	Examples	
140	200.000.000	Jet engine	Harmful condition
130	63.245.555	Jackhammer	
120	20.000.000	Airplane	
Pain threshold			
110	6.324.555	Disco	Critical condition
100	2.000.000	Industrial machinery	
90	632.455	Heavy truck	
80	200.000	Intense traffic	Safety condition
70	63.245	Vacuum cleaner	
60	20.000	Offices	
50	6.324	Low-volume music	
40	2.000	Library	
30	632	Steps on leaves	
20	200	Home at night	
10	63	Clock tick-tock	
0	20	Hearing threshold	

9.3 Digital Representation of Sound

To manage a sound on a digital computer, it must be transformed from an analog format to a digital format (A/D), which is done by a hardware device called an analog-to-digital converter or digitizer.

The digitizer usually samples the instantaneous voltage amplitude of a continuously varying analog input signal that results from a microphone output where the time-varying voltage is proportional to the sound pressure. The digital representation of a signal produced by a digitizer is represented by a sequence of numeric values, evenly distributed, that represents the amplitude of the original waveform (Fig. 9.6).

Two parameters are involved in the A/D process: the sampling rate or sampling frequency (SR) and the number of bits used to represent each amplitude (sample size or bit depth). Commercial digital audio applications usually utilize sample rates of 44.1 kHz. The sampling rate at which a signal is stored in a magnetic device (e.g., AIFF or WAVE format) is saved in the file.

The sampling rate that must be selected to obtain a reasonable digital representation must be more than twice the highest frequency present in the signal to avoid producing the phenomenon of aliasing. Aliasing is a phantom frequency that is not present in the original signal but is the result of an inadequate sample rate (Fig. 9.7). To prevent this effect the signal can be passed to a low-pass filter (anti-aliasing filter) that removes, if present, every frequency higher than the Nyquist frequency. The Nyquist frequency is the highest frequency of a digital signal without aliasing and is equal to half the frequency at which an analog signal is digitized

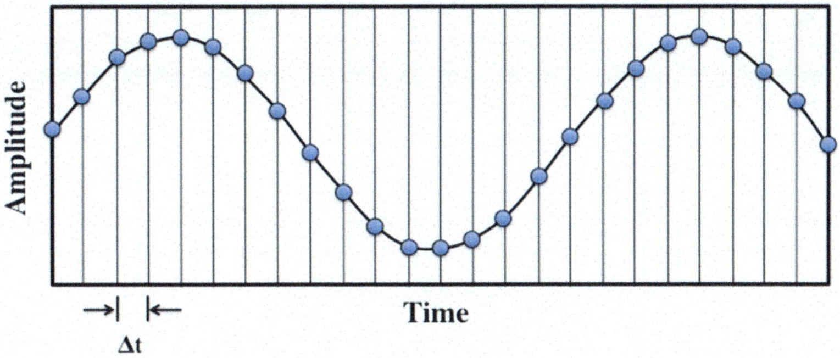

Fig. 9.6 A sinusoidal curve that represents the continuous analog waveform sampled at a Δt interval: every sampled point represents the digital conversion of the signal

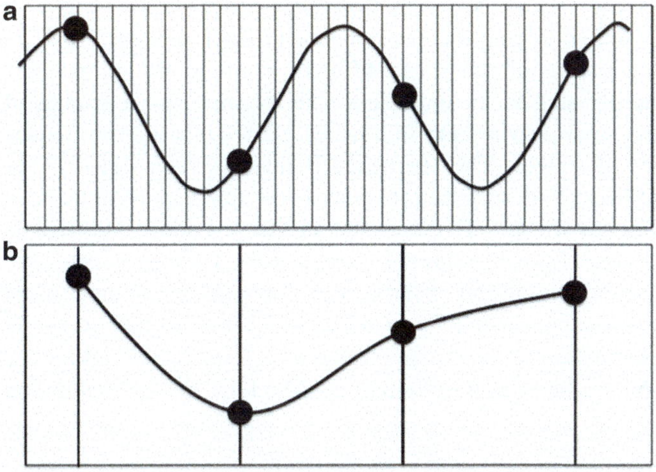

Fig. 9.7 Aliasing is the result of a poor sample rate of a sinusoidal curve (**a**). (**b**) The reconstruction of a digitalized curve that is not correspondent with the original analog curve. In this scheme the sampling in (**b**) has to be nine times larger (in Hz) than the original

$$f_{\max} = \frac{\mathrm{SR}}{2}$$

where:

$f_{\max}$ = Nyquist limit frequency (Hz)
SR = sampling rate (samples/s)

So if the digital sampling rate (SR) is 44.1 kHz, the signal is digitized to the frequency of 22.05 kHz.

The precision at which the analog instantaneous amplitude is converted into a digital format depends on the sample size or number of bits (bit depth) utilized in the binary representation of the amplitude value. The bit depth depends on the software; for instance, an 8-bit sample can resolve 256 (2^8) different amplitude values, and a 16-bit sample can resolve 65,536 (2^{16}) values. A high value of bit depth reduces the quantization noise introduced when the approximation between analog value and digitized version is too coarse.

9.4 Time Domain and Frequency Domain of Sound

Sound can be represented in a time domain and in a frequency domain. In the time domain the instantaneous pressure is represented as a function of the time (Fig. 9.8). In a frequency domain the pressure expressed as amplitude and phase is plotted against frequency. The pressure is converted by a microphone in volts and the amplitude is expressed in voltage. The frequency domain of a pure tone is represented by a vertical line with length corresponding to the amplitude. Any sound can be represented by a series of pure tones different in amplitude and phase. The spectral representation of sounds is more intuitive because in mammals the cochlea transforms the eardrum vibration into neural pulses that are the frequency domain representation of sounds. The frequency domain is represented by a magnitude spectrum and by a phase spectrum that has a few applications in bioacoustics.

The Fourier transform (FT) is a mathematical procedure that transforms a short segment of sound from a temporal domain to a frequency domain. A specific version of the FT, the discrete Fourier transform (DFT), is used when the signal and spectrum are represented by a sequence of discrete digital values. The obtained output of this transformation is represented by a sequence of values evenly spaced from 0 Hz to half the sampling frequency. The DFT is implemented applying the

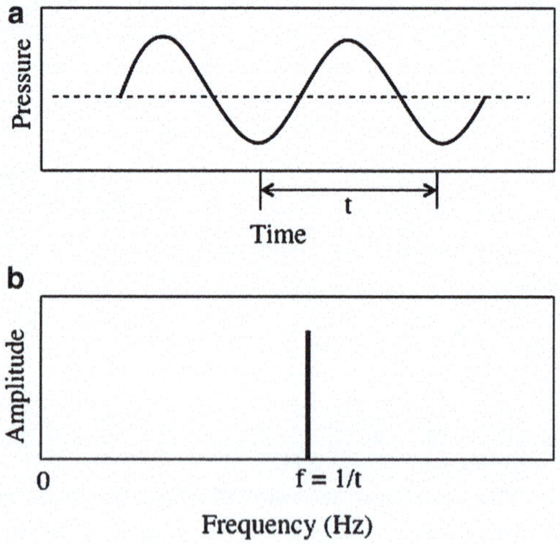

Fig. 9.8 Representation of the time domain (**a**) and frequency domain (**b**) of a pure tone. In (**a**) the instantaneous pressure is represented with time; in (**b**) the pressure is represented as amplitude along the frequency axes. In this case a pure tone belongs only to a frequency class

fast Fourier transform (FFT). The FT is reversible without losing information. The time domain and the frequency domain are equivalent.

9.5 The Spectrogram

Most signals of biological origin have a change in frequency with time, and to understanding the dynamics of a so complex signal is necessary to analyze the spectrum with the time because an individual magnitude spectrum of signal has less information.

The spectrogram is the visual representation of the sound in a space where the horizontal dimension corresponds to time (reading from left to right), and the vertical dimension represents the frequency (or pitch) measured in hertz (Hz), or cycles per second. The amplitude of the sound at any particular time and frequency is indicated by the color of the spectrogram at that point.

The spectrogram is the result of a procedure known as short-time Fourier transform (STFT).

This procedure divides the entire signal into a series of successive short time segments, or records (frames). Each record is used as input of a DFT that generates a series of spectra (one for each record). To represent the entire series, spectra are plotted size by size.

The length of each record of a STFT determines the time analysis resolution (Δt) of the spectrogram. The bandwidth of an STFT determines the frequency analysis resolution (Δf) of the spectrogram. A STFT with a wide bandwidth will have a poorer frequency analysis resolution than a narrower bandwidth.

It is not possible to have a very fine time and frequency simultaneously: this status is incompatible.

The record length and the filter bandwidth of a STFT are inversely proportional and are not independent. When we adopt a short record length, this produces a

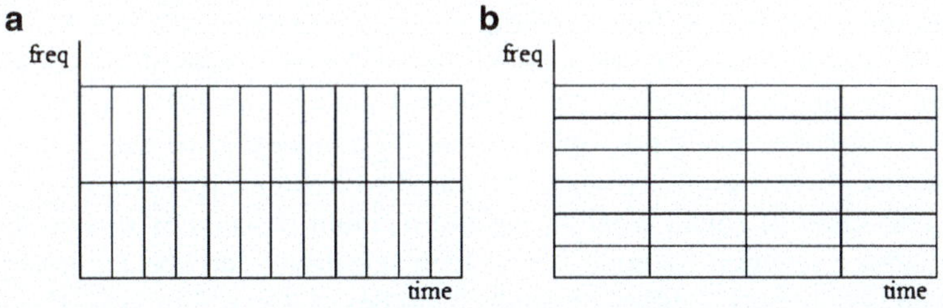

Fig. 9.9 One of the limits of the short-time Fourier transform (STFT) is that it has a fixed resolution. A wide window gives better frequency (*freq*) resolution (narrowband) but poor time resolution. A narrower window gives good time resolution but poor frequency resolution (wide-band). In (**a**) there is a fine time resolution (12 steps) but only two classes of frequency ($12 \times 2 = 24$). In (**b**) there is a coarse time resolution (four steps) with six classes of frequency ($6 \times 4 = 24$)

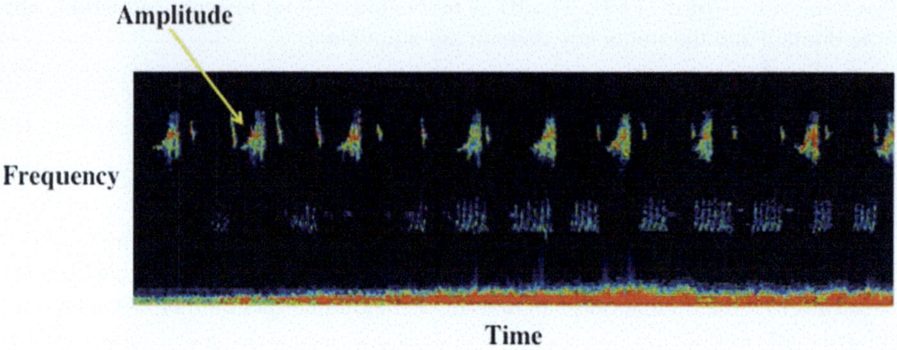

Fig. 9.10 A spectrogram represents the distribution of frequencies with time and according to the value of amplitude (*in colors*)

spectrogram with a finer time analysis resolution but correspondingly a coarse frequency resolution (Fig. 9.9). Generally the selection of a record of 512 points is a reasonable compromise to have a satisfactory resolution in time and frequency, but the choice of resolution depends on the goal of the spectral analysis (Fig. 9.10). Otherwise it is possible to process two spectrograms, one at high-frequency resolution and the second at high time resolution. Several types of software offer a spectrographic analysis with different graphical resolution and editing routines and the possibility to banding the sound data into equally spaced fractional octave bands (1-, 1/3-, 1/12-, 1/24-).

9.6 Field Recording Procedures

Today technological improvements in the field of digital sound offer a great potentiality in term of recording apparatus. After the period of magnetic tape recording now we are in the era of digital recording (Rumsey and McCormick 2009). The broad variety of commercial digital devices available is sufficient to guarantee the quality of the recordings (Mennill and Fristrup 2012), as is especially true for the airborne sound microphones that for a modest price can guarantee the same accuracy with frequency when compared with more expensive devices.

The collection of sound from an aquatic medium remains expensive. For this application are required hydrophones at high sensitivity and special apparatus to place such instruments at different depths in bodies of water. Similarly, the devices required to collect information from soil where geophones and piezoelectric sensors coupled with signal amplifiers and filters are required appear quite sophisticated.

Recording in the field can be done using a single microphone (mono/stereo), a set of coupled oriented microphones (mono/stereo), or an array of microphones (mono/stereo) Fig. 9.11.

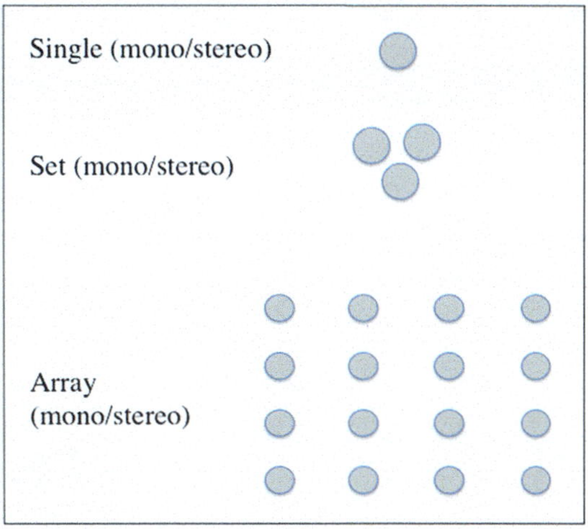

Fig. 9.11 Schematic representation of recording procedures by using a single microphone, or a set or an array of microphones

The single microphone collects sounds without associating information about the origin of sound, whereas a set of oriented microphones allows discovering the precise (geographic) origin of the acoustic source. Finally, the array allows us to cover the sonic information of a wide area, which in turn depends on the distance at which the microphones are placed.

For instance, to estimate bird species abundance, richness, and composition, Cellis-Murillo et al. (2009) have designed a soundscape recording system (SRS) based on four microphones placed at 90°, shielded with plastic dividers, and oriented 30° from the horizontal plane, connected to two Sony DAT TCD-D8s. This system offers the advantage of listening to the sound files in the laboratory in playback and locating the direction of the sound with precision. With this device the authors have compared data collected directly in the field with data collected in a post-processing procedure.

In another study, a wireless sensor network composed of four nodes was located in a neotropical rainforest of Chiapas, Mexico, by Collier et al. (2010) to perform an acoustic localization. Each node had a subarray of four microphones spaced 16.97 cm apart with a tetrahedral configuration. The system was equipped with a GPS, an acoustic emitter, a computer, a wireless 802.11 network interface, and a smart lithium ion battery pack; all the devices were packed into a Pelican 1,150 case. The entire system was remotely handled with a laptop or a smart phone. The system was able to localize antbirds using the sound in playbacks with an accuracy of 0.199 and 0.445 m, respectively.

An array of eight microphones at a distance of 75 m was used by Mennill et al. (2006) to locate acoustically the position of duetting rufous-and-white wrens (*Thryothorus rufalbus*) in a Costa Rica evergreen forest using playback songs. The position of the loudspeakers was estimated with an accuracy of 2.82 ± 0.26 m, and it was possible to calculate the "male" and "female" loudspeakers with an error of 2.12 ± 0.42 m, confirming the validity of the method. Minor errors have been a consequence of imperfect functioning of the GPS under a thick vegetation cover.

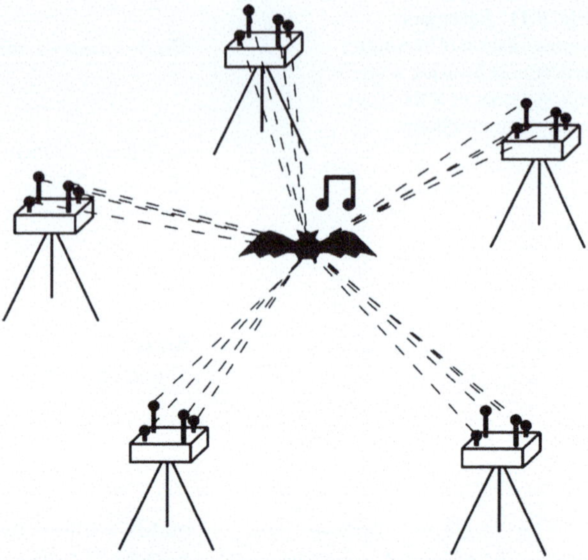

Fig. 9.12 Example of a microphone array to collect information on location of vocalizing bats. Each node is composed of four microphones that create a 3D representation of the sonic environment (Reproduced with permission from Blumstein et al. 2011)

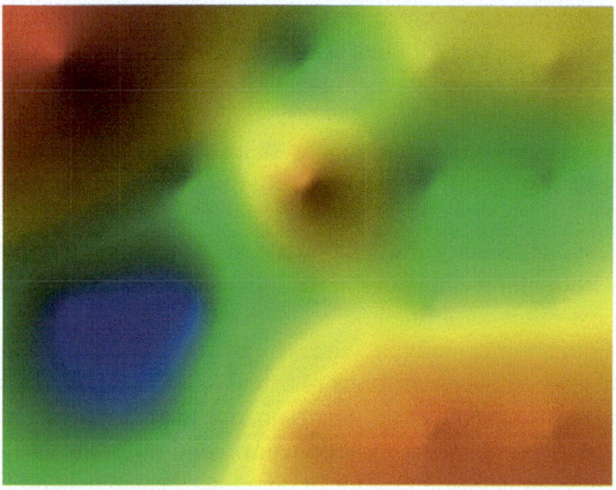

Fig. 9.13 Example of sonic distribution resulting from the interpolation of the acoustic complexity index (ACI) from a 4 × 5 array of microphones displaced at a distance of 25 m to each other in a dense Mediterranean maqui (From Farina and Pieretti, Ecological Informatics 2014, in press)

The use of arrays of microphones allows many advantages such as knowing the spatial disposition of the acoustic sources and the rate of interactions between a sender and a respondent, but at the same time the great amount of data to be processed requires powerful computation tools (for a review, see Blumstein et al. 2011) (Fig. 9.12).

In particular, a set of spatially arranged microphones seems the most efficient way to investigate the relationship between landscape patterns and the sonic environment. In the literature there are not many examples of this technique (e.g., Pieretti and Farina, Ecological Informatics 2014, in press) that offer the possibility of measuring the influence on local characters of the landscape (vegetation composition and structure, patch size and shape, ecotone distribution, heterogeneity and fragmentation levels, etc.), the sonic aggregations or sonotopes, and the behavioral effects that generate the soundtopes (Fig. 9.13).

Employment of digital recorders in the field poses some constraints such as the necessity to download the mass storage memory when filled and to periodically renew the power supply. To solve in part these technical problems the remote download of data by using smart phones seems a promising solution, coupled to independent powering with solar cells. In fact, the innovative use of smart phones to record the sonic environment is just at the beginning but looks very promising. There are different examples in the literature. For instance, a system of smart phones has been tested around the Brisbane airport in Australia by Cai et al. (2007), but see also Zhang et al. (2010), creating a sensor network for monitoring.

9.7 Metrics in Soundscape Ecology

The processing of acoustic data is a very complex action when the identification of a species is requested, but the main goal of soundscape ecology is to process data to extract the emerging patterns in terms of complexity/information of the sonic environment. In other words, automatic processing means to extract information using the sound pressure approach or a more sophisticated spectral analysis of frequency (Fig. 9.14). The two approaches are quite different from each other and also the goals are different. The first approach measures the amount of sonic energy (sound pressure/power) with particular attention to the effect of noise on human life and generally collects this information using phonometers. The second approach to the study of the spectral patterns created by sounds uses digital recorders and requires a complex post-processing of the recorded sound, made in the majority of the cases by adopting the Fourier transform algorithms.

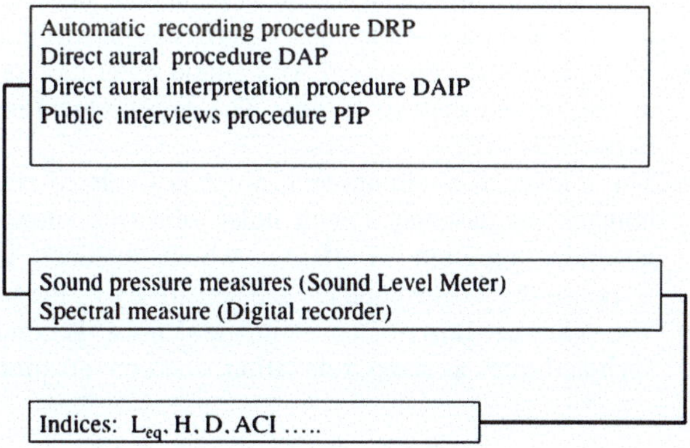

Fig. 9.14 Schematic representation of the methodological approaches to sonic environment analysis

9.7.1 Sound Pressure Measurements

Sound pressure measurements are largely used to assess the environmental "noise" and represent an application to sound analysis largely devoted to industrial and urban indoor and outdoor evaluation of the quality of the sonic environment These measurements are required by law to be carried out by the Environment Protection Policy on Noise. This approach is very popular for evaluating the effects of noise on human life from different environmental conditions and is regulated by national and international standards. Sound pressure measurement allows collecting the amount of sound energy that is produced in a specific site at a precise instant or during a time interval.

Here are reported a basic conceptual model of sound emission, propagation, and potential annoyance as results from the *Noise Measurement Procedures Manual* (2nd Edition, 2008; Environment Division, Department of Environment, Parks, Heritage and the Arts, Australia).

- Acoustic energy is emitted from specific items of equipment or machinery.
- The energy radiation pattern can generally be approximated by assuming that all sound energy from a particular item is emitted from a point source located at about the centre of the item. Some variation to this may be necessary for extended sources at close listening positions.
- The sound energy radiation pattern can generally be regarded as isotropic (i.e. the energy radiates equally in all directions) unless there is evidence to the contrary.
- The sound energy reduces with distance from the source in proportion to the inverse square of the distance between the effective point source and any listening position.
- The sound propagates through the air in relatively straight paths and can be subjected to various modifications such as atmospheric absorption and refraction, which become increasingly more significant with increasing distance from the source.
- The noise may be modified by the presence of sufficiently large objects that can reflect, absorb or screen sound. This can lead to associated increases or decreases in the sound pressure level at a particular listening position.
- Noise levels are measured with sound level meters that provide various signal averaging and analysis options. The measurement results are normally displayed in decibels (dB).
- The impact of environmental noise is likely to be greatest at locations where humans are associated with noise sensitive situations. Thus noise levels are generally reported in dB(A), which includes the A-weighting frequency response that corresponds to the sensitivity of the human ear.
- The main concerns of environmental noise generally relate to interference of such activities as sleep, relaxation, and conversation.

- Various features of a particular noise, generally identified as tonality, impulsiveness, modulation and/or low-frequency components, can increase its potential interference. These features tend to increase the information content of the particular noise and thus raise its audibility. The audibility of such a noise is also influenced by the level and characteristics of other ambient noise in the area.
- Non-acoustic factors that can influence the subjective audibility and annoyance potential of a particular noise include the subjective value of the activity generating the noise and the effectiveness or otherwise of any attempt to control the noise.
- Adaptation to the noise, past experience, individual differences, and personality can influence the response of a particular listener.

Only a few measures that process this information are reported in this book because most of the metrics are utilized in industrial and civil engineering and could be of less interest for our readers.

Considering that this approach focuses on the effect of noise on human life, however, it is necessary to add some details before the description of the metrics.

The human ear has different sensitivity according to frequency range: the major sensitivity is between 500 Hz and 8 kHz, with less sensitivity to very low and very high frequencies. To account for this physiological constraint, the recorded sound pressure level (SPL) is corrected accordingly with an *A-Weighting*, *C-Weighting*, or *Z-Weighting* procedure to compensate for the nonlinear hearing.

A-Weighting is the most common weighting used in noise measurement. As does the human ear, this method cuts off the lower and higher frequencies that the average person cannot hear. The dB is represented as dBA or dB(A).

C-Weighting. At the higher levels sounds produce a flat response in the human ear, and a measurement considering this limitation may be useful for specific applications. The dB is represented as dBC or dB(C).

Z-Weighting. In this case, the measurement is independent of frequency perception.

9.7.1.1 Equivalent Continuous Noise Level

The equivalent continuous noise level ($L_{\text{eq},\ T}$) metric stores the amount of sonic energy produced by the environment for a specific time lag and then produces an averaged value (Fig. 9.15).

$$L_{\text{eq},T} = 10\log_{10}\left(\frac{1}{T}\int_{t-T}^{T}\frac{p^2(t)}{p_0^2}\,\mathrm{d}t\right)$$

where

$T = t_2 - t_1$ is the time interval considered
$p^2(t)$ is the squared instant pressure
p_0^2 is the square of reference pressure

When the L_{eq} is averaged, the frequency is indicated as LA_{eq}, LC_{eq}.

Fig. 9.15 Representation of the pressure level equivalent with time. L_p = pressure amplitude, L_{eq} = pressure level equivalent

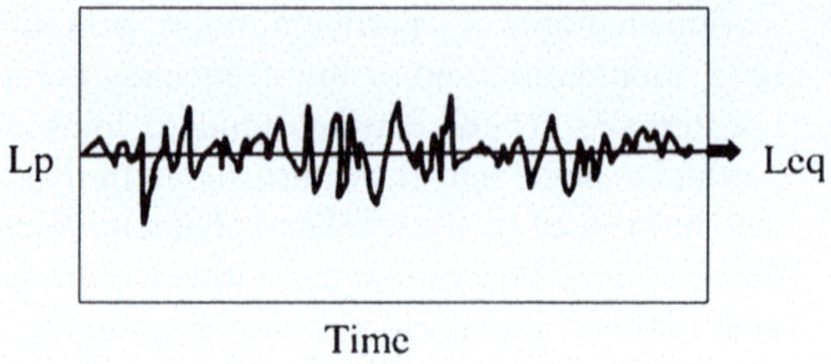

From this metric is possible to add partial $L_{eq\ e,T}$ according to the following formula:

$$L_{eq} = 10\log_{10}\sum_{1}^{n}\frac{T_i}{T}10^{\frac{L_{eq,1}}{10}}$$

where $T = \sum_{i}^{n} T_i$

9.7.1.2 Sound Exposure Level

Sound exposure level (SEL, L_{AE}) is defined as the total noise energy produced from a single noise event such as an aircraft takeoff.

The SEL is calculated in a precise temporal interval $t_1 - t_2$.

$$L_{(E,T)} = 10\,log_{10}\left(\int_{t1}^{t2}\frac{p^2(t)}{p_0^2 T}\right) = L_{eq,T} + 10\log_{10}\frac{T}{T_0}$$

where

$T = t_2 - t_1$, the time interval considered
$p^2(t)$ = the squared instant pressure
$p_0^2 T$ = the square of reference pressure
$L_{eq,T}$ = equivalent continuous noise level
$T_0 = 1$ s

This parameter is expressed in decibels and can be weighted for frequency with an *A-Weighting*.

9.7.2 *Spectral Analysis*

This approach, very popular in bioacoustic and environmental studies, is based on analysis of the spectral representation of sound when resolution of different frequencies the relative intensity of the sound is considered with a specific time lag (e.g. Joo et al. 2011).

The spectral analysis allows in some cases the automated identification of the species, but this is possible only in favorable context conditions (no wind, no rain, no anthropogenic noise, no choruses, etc.). The difficulties in analyzing aurally acoustic data can be in part overcome with a hybrid processing that consists of the identification by experts of some key features of the acoustic data bank and then the use of this key information to match the data processed automatically.

Here we describe a few selected metrics that are specifically devoted to the identification of the (ecological) characters of a soundscape independently from the source of the sounds.

9.7.2.1 Acoustic Entropy Index (*H*)

This index, based on the Shannon theory, which measures the information present in a sonogram, was first developed by Sueur et al. (2008) and is composed of two subindices:

The temporal entropy index H_t and the spectral entropy index H_f,
where $H = H_t \times H_f$ with $0 \leq H \leq 1$.

$$H_t = -\sum_{t=1}^{n} A(t) \times \log_2(A(t)) \times \log_2(n)^{-1}$$

with $H_t \in [0,1]$

$$H_f = -\sum_{f=1}^{N} S(f) \times \log_2(S(f)) \times \log_2(N)^{-1}$$

with $H_f \in [0,1]$

where n = length of the signal in number of digitized points

$A(t)$ = probability mass function of the amplitude envelope

$S(f)$ = probability mass function of the mean spectrum calculated using a short-time Fourier transform (STFT), with a non-overlapping Hanning window of $N = 512$ points.

This index has been tested with success in a tropical forest, but when applied to temperate conditions where the diversity of species is lower and the background noise is high, some biases result in the H_f component as reported by these authors.

9.7.2.2 Median of amplitude envelope (*M*)

This index, proposed by Depraetere et al. (2012), is calculated as

$$M = \text{median}\, A(t) \times 2^{(1-\text{depth})} \quad \text{with } 0 \leq M \leq 1$$

where $A(t)$ = amplitude envelope and depth = signal digitization depth (8 or 16 bits). To obtain M values varying from 0 to 1, M has been standardized as $M/A(t)$ max.

9.7.2.3 Acoustic Richness (AR)

This index is the combination of the indices described in the section 9.7.2.1 and 9.7.2.2 (Sueur et al. 2008)

$$\mathrm{AR} = \frac{(\mathrm{rank}(H_t) \times \mathrm{rank}(M))}{n^2}$$

with $0 \leq M \leq 1$

9.7.2.4 The Acoustic Dissimilarity Index (*D*)

This index measures the β diversity between two acoustic communities (Sueur et al. 2008). D is composed of two subindices: D_t, the temporal dissimilarity index and D_f, the spectral dissimilarity index: $D = D_t \times D_f$

for $D \in [0,1]$

$$D_t = 0.5 \times \sum_{t=1}^{n} |A_1(t) - A_2(t)|$$

with $D_t \in [0,1]$

$$D_f = 0.5 \times \sum_{f=1}^{N} |S_1(f) - S_2(f)|$$

with $D_f \in [0,1]$

where $A_1(t)$ and $A_2(t)$ = probability mass function of the amplitude envelope and $S_1(f)$ and $S_2(f)$ = probability mass function of the mean spectrum.

9.7.2.5 The Acoustic Complexity Index (ACI)

This index, elaborated by Farina and Morri (2008) and by Pieretti et al. (2011), is applied to a spectral representation of the sound signals registered in audio files and produces a direct and quick quantification of the soundscape structure (Figs. 9.16 and 9.17). The computation, starting from a *.wav* file, is carried out by the SoundscapeMeter (Farina et al. 2012), a plug-in of WaveSurfer (Sjölander and Beskow 2000).

The hypothesis on which the ACI algorithm has been created is based on the observation that many biophonies, such as bird songs, are characterized by an

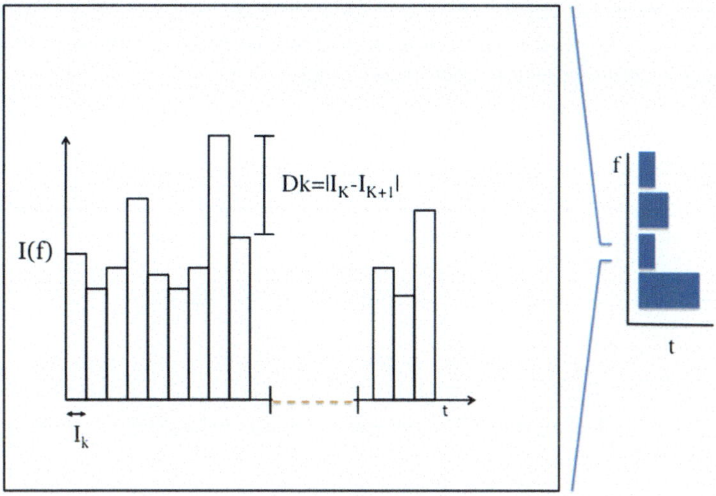

Fig. 9.16 Schematic representation on the way the ACI index is calculated. For each frequency band is calculated the difference between each temporal datum of the spectrogram amplitude (here, I). ACI can be calculated according to different temporal resolution to better fit the spectral heterogeneity (Reproduced with permission from Pieretti et al. 2011)

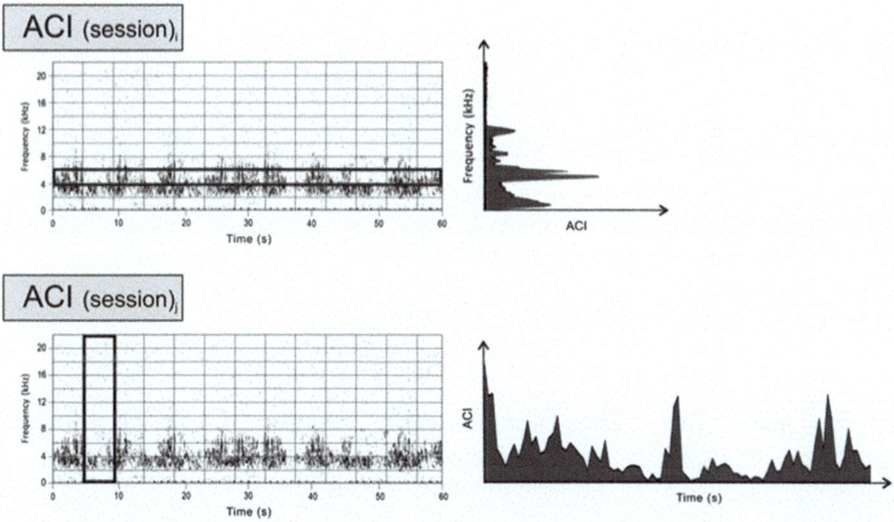

Fig. 9.17 ACI (session)$_i$ and ACI (session)$_j$ are, respectively, the vectorial transformation of the matrix Aij according to the frequency (f) and the temporal (t) axes. In our case, the Aij is composed of 157 classes of frequency and 60 temporal subsets (Reproduced from Farina et al. 2011)

intrinsic temporal variability of amplitude, whereas some types of geophonies (e.g., wind, flowing waters, rain) and anthrophonies (e.g., car passing or airplane transit) present more constant amplitude values.

When the acoustic variations of amplitude are converted from a time domain into a frequency domain by a Fourier transform, such variations appear along every frequency category, resembling an indented border. More different are the values of amplitude along a temporal step, and more information is contained in the spectral representation. The ACI index measures indirectly such information.

The ACI calculates the absolute difference (Δk) between two adjacent values of spectral amplitude (I_k and $I_{(k+1)}$) in a single frequency bin (Δf_l):

$$\Delta k = |I_k - I_{(k+1)}| \tag{9.1}$$

and then adds together all the Δk encompassed in a temporal step J of the recording (e.g., J may be 1 s, 5 s, 30 s, 60 s, etc.):

$$D = \sum_1^n \Delta k \quad \text{for } j = \sum_1^n \Delta t_k; \quad n = \text{number of } \Delta t_k \text{ in } j \tag{9.2}$$

where D is the sum of all the Δk contained in j. To obtain the relative intensity and to reduce the effect of the distance of the animals from the recording microphone, this result is then divided by the total sum of the amplitude values registered during the time interval j:

$$\text{ACI} = \frac{D}{\sum_1^n I_k} \tag{9.3}$$

where the ACI is calculated in a single temporal interval (j) and in a single frequency bin (Δfl). Thereafter, the ACI, which was worked out on all the temporal steps encompassed in the recording, is calculated:

$$\text{ACI}(\Delta fl) = \sum_1^m \text{ACI} \tag{9.4}$$

for m = number of j intervals in the entire recording, where the ACI(Δfl) corresponds to the ACI of an entire frequency bin.

Finally, the total ACI for all the frequency bins is calculated:

$$\text{ACI(tot)} = \sum_1^q \text{ACI}(\Delta fl) \quad \text{for}: \Delta f = \sum_1^q \Delta fl \quad q = \text{number of } \Delta fl \tag{9.5}$$

where the ACI(tot) is the total value of the index for the entire recording (for more details, see Pieretti et al. 2011).

This algorithm can also be applied to compare the differences between two magnitudes along the axes of the frequencies comparing the magnitude of a frequency bin with the magnitude of a successive frequency bin at the same time.

The combination of these two indices ACIt and ACIf becomes a better indicator of the acoustic complexity inside a spectrogram.

9.7.2.6 ACI Evenness

This index measures the distribution of ACI along each frequency class

$$\mathrm{ACI_e} = \frac{1}{\sum pi^2}$$

where *pi* is the relative abundance of the ACI(Δfl) index along the different *i* frequency classes (Farina et al. 2012).

This index describes the level of saturation of a spectrogram to which all the frequencies belong.

9.8 Pattern Recognition of Vocalizations

The identification of species has primarily a bioacoustics conservation valence, but the identification of species from a sound file may also find several applications in the field of soundscape ecology. In particular, the identification may be extremely useful to validate synthetic (blind) indexes.

Vocal animals have species-specific signatures that can be extracted for their identification. Today this is relatively easy for species that have stereotyped vocalizations, adopting an ad hoc training set and after adequate filtering, but when a complete soundscape is under investigation background noise produced by wind or falling waters and the biophonic cacophonies produced by a low acoustic partitioning create problems of identification and require great care (Depraetere et al. 2012).

The automated identifications of animal vocalization using, for example, the hidden Markov model (HMM), dynamic time warping (DTW), or Gaussian mixture model (GMM), are procedures in quick improvement (Anderson et al. 1996; Kwan et al. 2004; Chen and Maher 2006; Deecke and Janik 2006; Sumervuo et al. 2006; Trifa et al. 2008), although at the present time these procedures can be used only for a few species that have a simple sound structure and for records of quality without the masking effects of geophonies (rain, wind) or anthrophonies (airplane noise, traffic noise, etc.). For instance, Obrist et al. (2004) have applied a pattern recognition approach to the automated identification of most of the Swiss bat species using "training-calls" (but see also Skowronski and Harris 2006).

Bardeli et al. (2010) have suggested methods based on pattern recognition working on the vocalization of the Eurasian bittern (*Botaurus stellaris*) and of Savi's warbler (*Locustella luscinioides*). This approach is particularly useful to detect rare and endangered species using remote monitoring. Ranjard and Ross (2008) have proposed an unsupervised bird song syllable classification using neural networks applied to the song of the white-crowned sparrow (*Zonotrichia leucophrys pugetensis*).

An automatic call recognition (ACR) has been discussed by Brandes et al. (2006) for classifying narrow-band cricket and frog calls. These authors have used image processing techniques applied on spectrogram images to detect and classify constant-frequency cricket and frog calls. A blur filter was used with a thresholding filter to isolate calling events.

Gunasekaran and Revathy (2011) have proposed an automatic recognition of wild animal vocalizations using a fractal method that offers the advantage of selecting the right part of the signal to extract the features. Fractal approach was proposed also by Maragos and Potamianos (1999) for speech recognition where fractal dimension of speech signals was based on a multiscale morphological filter.

Automating identification based on Gabor transform has been recently proposed by Connor et al. (2012). These authors have utilized the Gabor frequency-amplitude spectrum (GFAS) and Gabor time-amplitude profile (GTAP). These two parameters have been combined and compared using a distance metric with a library of known birdsongs/calls. The identification of 67 bird species on 568 digital recordings of songs/calls was successful, although songs or calls of species dominated by low frequencies were more difficult to identify than species characterized by higher frequencies.

A processing software such as Song Scope (Wildlife Acoustics), able to recognize selected sound from a spectrogram after comparison with training sets, seems a very promising approach to study sound in both natural and human-modified landscapes. For instance, its application to study the long-term dynamics of frogs has been tried by Waddle et al. (2009), processing 200 h of recordings of three species of frogs: green treefrog (*Hyla cinerea*), American bullfrog (*Lithobates catesbeianus*), and bronze frog (*Lithobates clamitans*). Identification of the different species of frogs was biased by false positives (2.7–15.8 % per species) and failed to identify species (false negatives) in 45–51 % of recordings.

This method requires a prudent approach to be used for a large monitoring network.

In fact, despite new automated digital recording systems (ADRS) such as the Soundmeter™ (Wildlife Acoustics) or similar automatic devices (e.g., Zoom H4™ coupled to a timer), that also allow collecting sonic information for a long time, the processing procedure and species identification remain a challenge.

However, instead of identifying each species, if the goal of the soundscape approach is to assess the acoustic complexity of a site, it is possible to adopt new powerful and relatively simple metrics.

The ADRS approach can be distinct in two separate steps: in situ recording and ex situ processing (Sueur et al. 2012) (Fig. 9.18). In situ processing consists of the

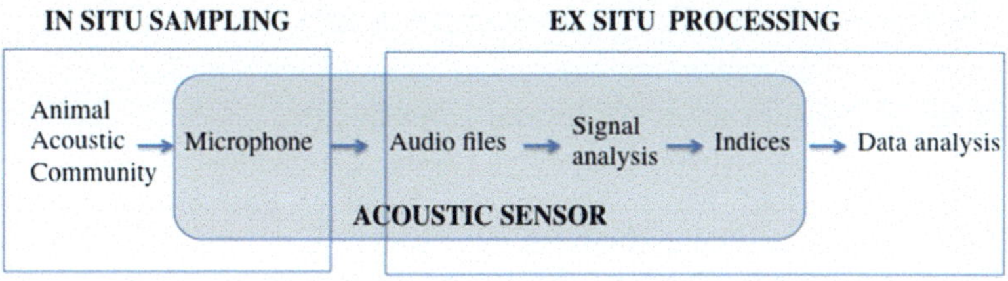

Fig. 9.18 Schematic view of the method to collect and process audio data. In *in situ sampling*, the animal acoustic community is recorded, then in *ex situ processing* the audio files are processed and signal analysis is performed applying metrics (Modified with permission from Sueur et al. 2012)

sampling/collection of field data using a single or a combination/array of microphones. The ex situ phase encompasses all the processing of audio files, their labeling, and insertion in a database. Then, indices are applied to extract the required information. This procedure in the future could be done in a portable all-in-one system where the entire process could be obtained from a specific application.

9.9 Functional Diversity Indices

Biodiversity is a well-known concept but despite its universal use remains quite difficult to quantify appropriately (Kaennel 1998; Petchey and Gaston 2002; Duelli and Obrist 2003). According to Wagner and Danchin (2010) the information that is contained in the acoustic repertoire of vocal animals can be considered a social information with two subrepartitions: public information (accessible to others) and private data (inaccessible to others).

Acoustic complexity of biophonies is a relevant part of biodiversity, although the contribution that every vocal species can hear remains undetermined without an aural species-specific determination (Obrist et al. 2010).

The complexity that emerges from a spectrogram of a community of vocalizing animals can be measured by applying functional diversity indices in which at every frequency category is assigned a value of functionality (Schleuter et al. 2010). This procedure can be used by transferring the algorithms designed for calculating the functional richness (FR) to the spectral domain where each frequency category could be considered a "vocal trait." An example of application of this approach to acoustic complexity has been posed by Pekin et al. (2012) when modeling acoustic diversity for a comparison with LIDAR methodology and by Gasc et al. (2013).

These last authors have considered the acoustic distance between species and used four different dissimilarity indices: spectral dissimilarity index (Df) (Sueur et al. 2008), Kolmorov–Smirbìnov distance (KS) (Rachev 1991), symmetric Kullback–Leibler (KL) distance (Kullback and Leibler 1951), and the similarity RV correlation coefficient (Josse et al. 2008).

9.10 Software Tools for Soundscape Analysis

In the field of acoustics there are several softwares that process audio data and produce sophisticated statistical analyses. Some commercial software can edit sounds: CoolEdit, Gold Wave, Spectrafoo, Amadeus Pro, Free Audio Editor, WavePad Sound editor, Sound Studio 2.

Others are more oriented on bioacoustics analysis, such as Raven, Song Scope, Avisoft, Wildspectra, Canary, Spectra Pro, Syrinx, Sonobat, Pamguard, Soundruler, Wavesurfer, SoundscapeMeter, Sound Analysis Pro, SeaWave, and Ishmael.

Some software is dedicated to signal analysis: Spectrogram, Syrinx pc, Osprey, Signalize, Sigview, SpectraPlus.

This is not absolutely a complete list but just an indication for a first orientation. According our knowledge only two free-domain softwares (Seewave http://rug.mnhn.fr/seewave and SoundscapeMeter www.disbef.uniurb.it/biomia/soundscapemeter/) have specific algorithms to processes the soundscape applying the metrics described in the previous section.

9.11 Browsing the Audio Database

The autonomous recording units are able to store a giant amount of data at different sampling intervals, but handling and inspecting audio files requires a massive amount of work energy and time. A new web-based application that archives, manages, retrieves, analyzes, visualizes, and allows listening to the audio files has been recently designed by Villanueva-Rivera and Pijanowski (2012). This system is called Pumilio and runs on the Ubuntu (Canonical Ltd. 2010) distribution of the Linus operating system. Pumilio uses three moduli: a tool to import audio files, an interface for browsing the archive, and a sound visualization and analysis tool. Data can be imported in one click from a directory of the server (all files) or imported a file at a time (Fig. 9.19).

Browsing data can be carried out by different approaches:

- A first method uses a Google map via spatial interface if the files are spatially associated to a map.
- A second method consists of a form-based query or database, for instance by selecting time or data ranges and site.
- A third method allows to browsing the archive using the label of a specific collection.
- A fourth method consists in browsing files according to tags assigned a specific file.

Sound is visualized, and is possible to select a part of the file and apply a bandpass filter. The visualization allows marking specific signals, labeling them, and then saving to the database. The selected data can be annotated with a specific

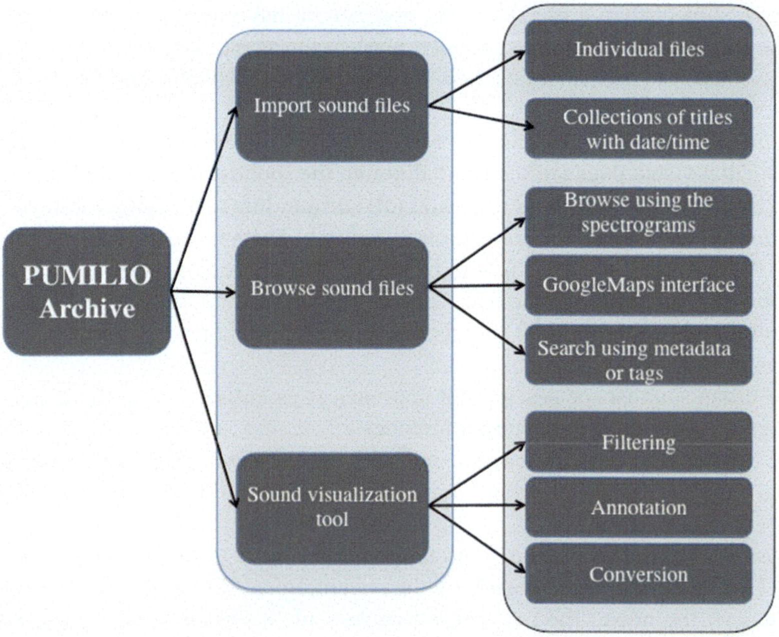

Fig. 9.19 Schematic representation of the Pumilio modules (Import sound files, Browse sound files, and Sound visualization tool). In each module different functions are associated (Reproduced with permission from Villanueva-Rivera and Pijanowski 2012)

marker and exported as comma-separated files for statistical processing. Parallel computation is possible, saving in this way a large amount of time. The advantages to using Pumilio are quick browsing, analysis in the browser, public data sharing, user accounts, and flexible parallel computing.

9.12 Methods to Evaluate/Rate the Quality of the Sonic Ambient

The urban soundscape is "overshadowed by continuous mechanical noise" (Yang and Kang 2005), where the quality of the urban soundscape represents an important component of human well-being.

Sound perception varies according to different urban open spaces because social, demographic, and behavioral factors (including age, gender, occupation, education, and residential status) have an important effect on sound level evaluation (Yu and Kang 2008). During these last years people are surrounded by a growing presence of unwanted sound (noise), especially in urban areas and along infrastructures such as railways, highways, airports, and ports. The sound quality becomes more a cognitive subject than a physical one, and people have the capacity to select and rate the different subjects that produce unpleasant or pleasant sounds. The use of sound semantics to investigate how people perceive sounds and noise is quite common.

Visual and aural perceptions differ in terms of physical facilitating mechanisms, meaning that visual cues are sectorial although the soundscape is ubiquitous. Sound has no barrier: it has a circular diffusion (all surrounding). The aural harmonization is temporal, the visual harmonization is spatial. The visual object is a physical entity, the acoustic subject is a field of energy that expresses meaningful information. Sound perception may be less rich in information than the visuals but is richer in emotion, and an intentional soundscape is an important part of design and urban planning.

The urban soundscape assessment approach is essentially an aesthetic approach that regards individual or group preferences.

The information contained in a sound, the context in which a sound is perceived, and the sound level are elements that people utilize in the sound rating in a city soundscape.

Physiological and psychological components concur to the acoustic annoyance rating. In other words, this depends on the physiological skill of the ear, the physical attributes of the noise, the cultural component of the listener, etc. A plethora of experiments have demonstrated how important is the psychological component of the hearing process, more than the physical attributes of the noise per itself. Long-term environmental experience and cultural background may rate differently the perceived sounds.

In conclusion, the rating in sound appreciation largely depends on the single listener attitude. For instance, expert ornithologists can judge the song of a bird as having great value when non-expert listeners can ignore/exclude that sound in the rating process of a soundscape.

Yang and Kang (2005) have investigated the soundscape of two squares in the city of Sheffield carrying out an intensive questionnaire in which sound level and comfort were both analyzed. The questionnaire was quite general on the environment of the two squares, just to reduce the bias that can be introduced when too specific questions or themes are proposed. During the compilation of the questionnaire (two pages), A-weighted L_{eq} (equivalent continuous noise) and microclimate indices were collected.

Despite the expectation, people identify first not the loudest sound but the soundmarks as the noticed sound. The natural sounds were preferred in the two urban squares selected by the researchers. Young people showed a tolerance toward music and technological sound, while aged people showed more tolerance to natural sounds and to sound related to culture or human activities. Only slight difference was found between the sexes in sound appreciation.

Investigation about the sound quality of urban areas in many cases is the only tool that certifies the noise exposure of visitors to urban noise when they spend time in urban parks. Urban parks can be considered in this case real ecological traps because they produce a visible attraction for people but they often do not offer the therapeutic effect that people hope to find in it. The simple measure of the sound pressure level (SPL) is not sufficient to evaluate the quality of the environment, and acoustic comfort is a parameter that is not sufficiently popular in urban monitoring. Often the acoustic comfort, and more in general the benefits that visitors hope to find, in a green urban space can be misinterpreted by other co-factors. For instance, prostitution, vagrancy, or water pollution are factors not directly related to the sonic environment but that decrease the total rating, as found by Szeremeta and Zannin (2009) after interviews on the state of noise level in a public urban park in Curitiba (Brazil). From this study emerges the realization that the collection of sound levels or the level of human-perceived comfort are not sufficient to assess the real comfort of an area: to make a realistic assessment it is necessary to combine acoustic pressure level with human perception.

For instance, Gustavino (2006) submitted a questionnaire composed of five main questions:

1. According to you, what would be the ideal urban soundscape?
2. In your urban sonic environment, what do you find pleasant/unpleasant?
3. In your urban sonic environment, are there high-pitched /low-pitched sounds? If so, describe them.
4. Do you perceive background noise in urban environments? Is so, under which circumstances? How would you describe it?
5. In urban areas, are you sensitive to transportation noise? Describe its characteristics?

This questionnaire proposed with questions scored according to most general to most particular, was posted to a group of inhabitants of Paris, Lyon, and Nantes.

The reaction of people using the semantic of sounds was positioned on the positive site for most identified sounds with the exception of mechanical sounds such as traffic, cars, construction work, car horns, and unidentified noise (Fig. 9.20).

Four sound quality descriptors emerged: variety (38 %), tranquility (30 %), animation (20 %), and non-aggressiveness (12 %). Variety means the contribution of a great variety of sounds that can be located inside a well-structured environment. Tranquility does not mean the absence of sound but the absence of noise. Animation refers to the human activity in a lively city where people have different functions blended into a public space. Finally, non-aggressiveness means a sonic environment without social tension or individual anomalous behavior.

About the second question, most of the judgment was expressed in a negative way (cars, motorcycles, other vehicles, car horns, background noise). Positive perception was focused on other people, nature, birds, and music.

The sources of sound/noise are mediated by personal judgment. Thus, public transportation is considered less harmful than private personal vehicles, and this creates a cognitive representation of the sonic ambience that reflects personal culture and personal expectations for a public space. The qualitative evaluation is not directly related to the physical properties of the heard sounds but to the cognitive template that elaborates and processes the sound.

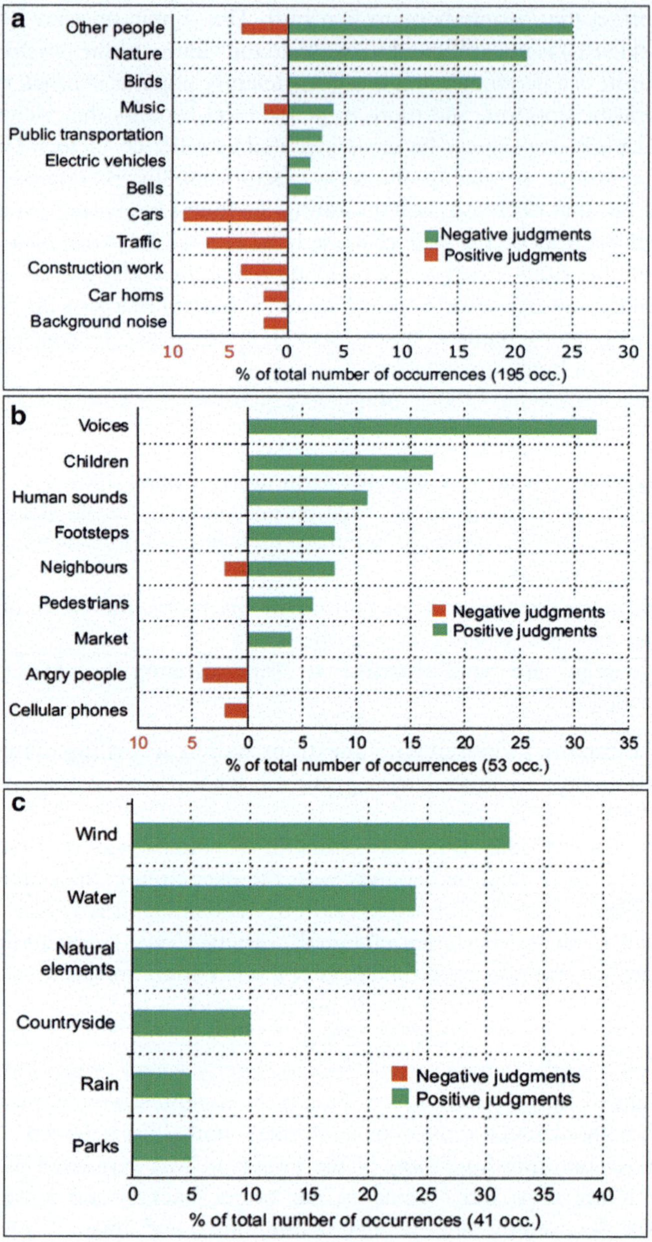

Fig. 9.20 Percentage of positive and negative judgments of the main categories of sonic sources from participants for spontaneous descriptions of the ideal urban sonic environment. (**a**) Main categories of sources. (**b**) Subcategories of sources within the main categories "Other people." (**c**) Subcategories of sources within the main category "Nature" (Reproduced with permission from Gustavino 2006)

In evaluating the sounds, uniform low-frequency sounds are considered more as background noise than pitched discontinuous sounds that are much better recognized.

The cognitive representation of a sound reflects more than the physical recognition of the sound the perceived soundscape. So, definitely, at least for human subjects, there are specific eco-fields for every recognized category of sounds.

The urban soundscape reflects the life of people and other organisms, and people require a diverse soundscape in a city. The apparent conflict between physical versus psychological measurement and assessment of the sound in urban areas can be easily overcome because the two components are strictly connected. Loud sounds produce a more convergent attitude in their assessment using either a psychological or a physical approach.

9.13 Method of Soundwalking

Soundwalks are direct experience of the sonic environment conducted in isolation or in groups to collect, via recorders or via annotation, the sonic environment encountered.

As reported in detail by Westerkamp [1974 (2007)]:

> A soundwalk is any excursion whose main purpose is listening to the environment. It is exposing our ears to every sound around us no matter where we are. We may be at home, we may be walking across a downtown street, through a park, along the beach; we may be sitting in a doctor's office, in a hotel lobby, in a bank; we may be shopping in a supermarket, a department store, or a Chinese grocery store; we may be standing at the airport, the train station, the bus-stop. Wherever we go we will give our ears priority. They have been neglected by us for a long time and, as a result, we have done little to develop an acoustic environment of good quality.

Soundwalks collect the personal experience about the sonic environment, and this experience can be conducted in every place you wish from the more neglected periphery of a metropolitan area to a remote cloud forest of an equatorial mountain range. The exercise to collect field information about the sonic environment is called *phonography*.

A soundwalk is an educational tool to explore the sonic ambience, to appreciate nature music, to reconstruct the confidence with nature lost after several years of sedentary life and urbanization.

A soundwalk is an exercise for which you must be educated by steps.

Soundwalking could be introduced as an educational tool in the schools and represents an important approach to consider the importance of the sonic environment for human well-being and for social aggregation also. Polli (2012) has recently outlined the political and social dimension of soundscape composition and of soundwalking and has addressed the different attitudes between music and soundscape.

9.14 Soundscape Customer Satisfaction

The auditory landscape or soundscape has many positive aspects, for instance, flowing waters from urban fountains may have a soothing effect masking irritating sounds. Traffic noise can produce annoyance, but the sound from a single vehicle can be considered an important signal for pedestrian safety when crossing a road. Human sounds of talking can be interpreted as irritating by people reading a book or engaged in conversation, but could be pleasant during shopping activity. All these elements are extremely useful to engineers and architects for planning or restyling urban sonic spaces.

The soundscape can be an inconvenient and highly complex intermediate step in the process of customer satisfaction. Customer satisfaction is a theory of product development and customer satisfaction developed in the 1980s by Noriaki Kano that classified customer preferences into five categories: Attractive, One-Dimensional, Must-Be, Indifferent, Reverse. We briefly describe these quality categories.

Attractive Quality. This attribute contributes to satisfaction when achieved fully, but does not cause dissatisfaction when not fulfilled. Normally is not expected.

One-Dimensional Quality. This attribute contributes to satisfaction when fulfilled and dissatisfaction when not fulfilled because it is expected.

Must-be Quality. This attribute is taken for granted when fulfilled but result in dissatisfaction when not fulfilled.

Indifferent Quality. This attribute refers to aspects that are neither good nor bad, and they do not result in either customer satisfaction or customer dissatisfaction.

Reverse Quality. This attribute refers to a high degree of achievement resulting in dissatisfaction and to the fact that not all customers are alike.

Intervention in a space modifies the position of the sound source, the reflective surfaces, and in general the physical environment in which the sound is produced and diffused. Changes in space can create new more interesting configurations, increase of comfort, or vibrant, spooky, obtrusive, or dull feelings.

In the creation of an attractive soundscape the application of a Kano model seems a good solution. In Fig. 9.21 is shown a typical Kano plot in which the horizontal axis indicates the degree at which an aspect of product or service is present. The vertical axis shows the level of customer satisfaction. These aspects are split into three categories: *Performance, Basic, Excitement*. *Performance* works at one dimensionality: "more is better" – "less is better."

The *basic* requirement considers absence as a dissatisfaction. For instance, a car starter: a poor starter leads to dissatisfaction, but a good starter does not increase the car satisfaction.

Excitement or attractive requirement, a usually unexpected aspect of a product, and its absence does not produce dissatisfaction.

The Kano model allows locating the order at which a soundscape evaluation must be carried out.

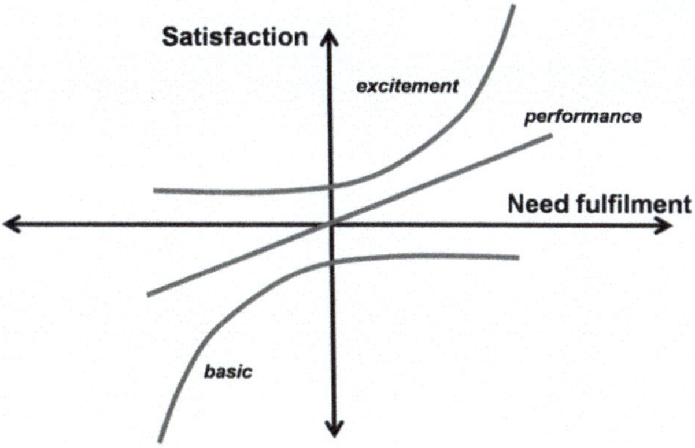

Fig. 9.21 A typical Kano plot, where the horizontal axis indicates the degree of expectation of a product or of a service. The vertical axis expresses the level of satisfaction of the customer that is split into three components: performance or satisfier, basic or must-be requirements, and excitement or attractive requirements (Reproduced with permission from Jennings and Cain 2012)

9.15 The Assessment of the Human-Perceived Soundscape

Managers, decision makers, and policymakers feel growing concern for noise pollution. The universality and severity of noise penetration mean a noise impact assessment should be considered an important procedure both in developing areas and in protected areas.

The soundscape approach highlights the necessity to integrate the sonic ambience from different perspectives that should not be limited to only physical attributes of the sound.

In fact, study of the soundscape requires diverse fields of practices and competencies: sound quality, human acoustic comfort in buildings, music, and relationships with wilderness and recreation management, urban and house design, landscape planning, and management.

The soundscape analysis represents a complex exercise that is carried out to improve the quality of the human sonic environment. In fact, the quality of a soundscape largely depends on the cultural filter used to separate the different components of the sonic environment (Table 9.3).

The context in which the information is collected affects the final assessment. The context is represented by place, dimension of the physical environment, personal activity in the place, dimension of the social environment, and personal dimension.

The spatial dimension of a soundscape allows distinguishing a background from a foreground and classifying a sound as intrusive or acceptable; this largely depends on the role of the context.

To create a standardization for an assessment, a sonic procedure is necessary to create categories of places (indoor or outdoor) where outdoors must further distinguish urban, rural, wilderness, and underwater areas and adopt a detailed semantic of sound classification.

Table 9.3 List of outcomes useful to appreciate the quality of a soundscape

Acceptability	Identification of place	Relaxation
Appropriateness	Importance	Safety
Clarity	Information	Satisfaction
Comfort	Liveliness	Sense of control
Communication	Naturalness	Solitude
Enjoyment	Nature appreciation	Tranquility
Excitement	Nostalgic attachment	Uniqueness
Happiness	Peacefulness	Variety
Harmony	Place attachment	Well-being

Source: Adapted with permission from Brown et al. (2011)

Summary
The soundscape analysis at present moment can be distinguished in two main fields: the measurement of the physical sound (sound pressure and its repartition in a frequency domain) and a semantic approach that describes how the sound is perceived by humans and what part culture has in the evaluating process.

Recognizing the difficulty to represent in a synthetic way fields of research that until now have evolved independently, the different procedures to approach the sonic phenomenon are summarized as four main actions: automatic recording procedure; direct aural procedure; direct aural interpretation procedure; and public interviews procedure.

The sound is a complex phenomenon that copies the environmental context in which is generated. For this reason many descriptors are requested to approaching its behavior and its effects on physical and biological objects. frequency, pitch, period, wavelength, sound speed, wavenumber, amplitude, sound pressure, sound power, sound intensity, and loudness are some of the distinctive attributes.

The decibel (dB) that represents a ratio between two signals is a very popular conversion of the sound pressure at a logarithmic scale. The decibel with reference has at the denominator a value of reference such as like 1 V or 20 μPa.

In modern usage the sounds are converted after the passage in volts by the microphone transducer into a digital format. To avoid the aliasing effect, the sample rate, in hertz, of the analog signal must respect the Nyquist frequency, which is half of the sample rate. The level of accuracy in the analog-to-digital conversion depends also by the bit depth (2^8 or 2^{16}): the higher the bit depth, the lower is the quantization noise.

Sound can be represented in terms of frequency domains. The reversible conversion between the two domains is made generally adopting the Fourier transform (FT). For a digital signal the adoption of discrete Fourier transform (DFT) can be implemented by the fast Fourier transform (FFT). To evaluate the variation of frequency with time, the entire signal is divided into discrete windows (frames), and to these windows is applied the short-time Fourier transform (STFT).

The spectrogram is the visual representation of the sound in a space where the horizontal dimension corresponds to time and the vertical dimension represents the frequency (or pitch), with higher sounds shown higher on the display. The color of the spectrogram represents the magnitude of the signal.

A field record offers a great variety of recording devices that can sample at different rates and on different types of digital memories. Single microphones, sets of adjacent microphones, and arrays of microphones are some of the strategies adopted to collect sound from the environment.

The main goal of soundscape ecology is to process data to extract the emerging patterns in terms of complexity/information. A sound pressure measurement and a spectral-frequency analysis are the two distinct approaches to day adopted. The sound pressure measurement, adopted by engineering and acousticians, is obtained with at least three different types of weighting to correct the different sensitivity of human hear to frequencies. Equivalent continuous noise level and sound exposure level are two examples of this approach applied in the study of sound/noise transmission across a medium.

The spectral analysis adopted by bioacousticians offers at the moment very few metrics to evaluate the complexity of the sonic environment. The acoustic entropy index (H), median of amplitude envelope (M), acoustic richness (AR), acoustic dissimilarity index (D), the acoustic complexity index (ACI), and ACI evenness are some of the available metrics.

The automated recognition of environmental sounds has been pursued by bioacousticians. Most animal vocalizations represent acoustic signatures that can be recognized and classified. Today, computing facilities and new algorithms allow a satisfactory processing of sound data, although several problems still remain, especially when long-time recordings introduce unwanted sounds that mask the biophonies.

Acoustic diversity as results from a spectral analysis represents a significant component of biodiversity because this diversity represent some functional traits.

Several softwares are today offered for processing sounds. Some software is dedicated to sound editing, such as CoolEdit, Gold Wave, Spectrafoo, Amadeus Pro, Free Audio Editor, WavePad Sound editor, Sound Studio 2; other software is dedicated to bioacoustic analysis such as Raven, Song Scope, Avisoft, Wildspectra, Canary, Spectra Pro, Syrinx, Sonobat, Pamguard, Soundruler, Wavesurfer, SoundscapeMeter, Sound Analysis Pro, SeaWave, and Ishmael, and finally some software is dedicated to the physical analysis of the spectral representation of sounds such as Spectrogram, Syrinx pc, Osprey, Signalize, Sigview, SpectraPlus.

Because of the great amount of data that automated recording stations can collect, it is urgent to create data banks easily browsed. A new software, Pumilio, is proposed characterized by quick browsing and by flexible parallel computing.

The soundwalking method is suggested as an efficient approach to soundscape description from a psychological point of view. The procedure to describe the sonic environment also represents an important educational tool.

References

Anderson SE, Dave AS, Margoliash D (1996) Template-based automatic recognition of birdsong syllables from continuous recordings. J Acoust Soc Am 100(2):1209–1219
Bardeli R, Wolff D, Kurth F, Koch M, Tauchert K-H, Frommolt K-H (2010) Detecting bird sounds in a complex acoustic environment and application to bioacoustic monitoring. Pattern Recogn Lett 31:1524–1534
Blumstein DT, Mennill DJ, Clemins P, Girod L, Yao K, Patricelli G, Deppe JL, Krakauer AH, Clark C, Cortopassi KA, Hanser SF, McCowan B, Ali AM, Kirschel ANG (2011) Acoustic monitoring in terrestrial environments using microphone arrays: applications, technological considerations and prospectus. J Appl Ecol 48:758–767
Brandes TS, Naskrecki P, Figueroa HK (2006) Using image processing to detect and classify narrow-band cricket and frog calls. J Acoust Soc Am 120(5):2950–2957
Brown AL, Kang J, Gjestland T (2011) Towards standardization in soundscape preference assessment. Appl Acoust 72:387–392
Cai J, Ee D, Pham B, Roe P, Zhng J (2007) Sensor network for the monitoring of ecosystems: bird species recognition. ISSNIP conference, Melbourne
Cellis-Murillo A, Deppe JL, Allen MF (2009) Using soundscape recording to estimate bird species abundance, richness, and composition. J Field Ornithol 80(1):64–78
Chen Z, Maher RC (2006) Semi-automatic classification of bird vocalizations using spectral peak tracks. J Acoust Soc Am 120(5):2974–2984
Collier TC, Kirschel NG, Taylor CE (2010) Acoustic localization of antbirds in a Mexican rainforest using a wireless sensor network. J Acoust Soc Am 128(1):182–189
Connor EF, Li S, Li S (2012) Automating identification of avian vocalizations using time-frequency information extracted from the Gabor transform. J Acoust Soc Am 132(1):507–517
Deecke VB, Janik VM (2006) Automated categorization of bioacoustics signals: avoiding perceptual pitfalls. J Acoust Soc Am 119(1):645–653
Depraetere M, Pavoine S, Jiguet F, Gasc A, Duvail S, Sueur J (2012) Monitoring animal diversity using acoustic indices: implementation in a temperate woodland. Ecol Indic 13:46–54
Duelli P, Obrist M (2003) Biodiversity indicators: the choice of values and measures. Agric Ecosyst Environ 98:87–98
Farina A, Morri D (2008) Source-sink e eco-field: ipotesi ed evidenze sperimentali. Atti X congresso nazionale della SIE-IALE. Ecologia e Governance del paesaggio: esperienze e prospettive. Bari, pp 365–372
Farina A, Pieretti N, Piccioli L (2011) The soundscape methodology for long-term bird monitoring: a Mediterranean Europe case-study. Ecol Inform 6:354–363
Farina A, Lattanzi E, Piccioli L, Pieretti N (2012) SoundscapeMeter. Bioacoustic Software. Version 1.0 Beta. DISBEF, Urbino University
Gasc A, Sueur J, Jiguet F, Devictor V, Grandcolas P, Burrow C, Depraetere M, Pavoine S (2013) Assessing biodiversity with sound: do acoustic diversity indices reflect phylogenetic and functional diversities of bird communities? Ecol Indic 25:279–287
Gunasekaran S, Revathy K (2011) Automatic recognition and retrieval of wild animal vocalization. Int J Comput Theor Eng 3(1):136–140
Gustavino C (2006) The ideal urban soundscape: investigating the sound quality of French cities. Acta Acustica 92:945–951
Jennings P, Cain R (2012) A framework for improving urban soundscapes. Appl Acoust 2013:293–299
Joo W, Gage SH, Kasten EP (2011) Analysis and interpretation of variability in soundscapes along an urban–rural gradient. Landsc Urban Plan 103:259–276
Josse J, Pages J, Husson F (2008) Testing the significance of the RV coefficient. Comput Stat Data Anal 53:82–91

Kaennel M (1998) Biodiversity: a diversity in definition. In: Bachmann P, Köhl M, Päivinen R (eds) Assessment of biodiversity for improved forest planning. Kluwer, Dordrecht

Kullback S, Leibler RA (1951) On information and sufficiency. Ann Math Stat 22:79–86

Kwan C, Mei G, Zhao X, Ren Z, Xu R, Stanford V, Rochet C, Aube J, Ho KC (2004) Bird classification algorithms: theory and experimental results. In: Proceedings of the IEEE international conference on acoustics, speech and signal processing, vol V, pp 289–292

Maragos P, Potamianos A (1999) Fractal dimension of speech sounds: computation and application to automatic speech recognition. J Acoust Soc Am 105(3):1925–1932

Mennill DJ, Fristrup KM (2012) Obtaining calibrated sound pressure levels for consumer digital audio recorded. Appl Acoust 73:1138–1145

Mennill DJ, Burt JM, Fristrup KM, Vahrencamp SL (2006) Accuracy of an acoustic location system for monitoring the position of duetting songbirds in tropical forest. J Acoust Soc Am 119(5):2832–2839

Obrist MK, Boesch R, Flückiger PF (2004) Variability in echolocation call design of 26 Swiss bat species: consequences, limits and options for automated field identification with a synergetic pattern recognition approach. Mammalia 68(4):307–322

Obrist MK, Pavan G, Sueur J, Riede K, Llusia D, Márquez R (2010) Bioacoustic approaches in biodiversity inventories. In: Manual on field recording techniques and protocols for all taxa biodiversity inventories. Abc Taxa 8:68–99

Pekin BK, Jung J, Villanueva-Rivera LJ, Pijanowski BC, Ahumada JA (2012) Modeling acoustic diversity using soundscape recordings and LIDAR-derived metrics of vertical forest structure in a neotropical rainforest. Landsc Ecol 27:1513–1522

Petchey OL, Gaston KJ (2002) Functional diversity (FD), species richness and community composition. Ecol Lett 5:402–411

Pieretti N, Farina A. Application of a recently introduced index for acoustic complexity to an avian soundscape with traffic noise. J Acoust Soc Am 1

Pieretti N, Farina A, Morri D (2011) A new methodology to infer the singing activity of an avian community: the Acoustic Complexity Index (ACI). Ecol Indic 11(3):868–873

Polli A (2012) Soundscape, sonification, and sound activism. AI Soc 27:257–268

Rachev ST (1991) Probability metrics and stability of stochastic models. Wiley, New York

Ranjard L, Ross HA (2008) Unsupervised birds song syllable classification using evolving neural networks. J Acoust Soc Am 123(6):4358–4368

Rumsey F, McCormick T (2009) Sound and recording, 6th edn. Focal Press, Burlington

Schleuter D, Daufrensne M, Massol F, Argillier C (2010) A user's guide to functional diversity indices. Ecol Monogr 80(3):469–484

Sjölander K, Beskow J (2000) WaveSurfer – an open source speech tool. In: Proceedings of the ICSLP 2000, IV, pp 464–467

Skowronski MD, Harris JG (2006) Acoustic detection and classification of microchiroptera using machine learning: lessons learned from automatic speech recognition. J Acoust Soc Am 119 (3):1817–1833

Somervuo P, Harma A, Fagerlund S (2006) Parametric representations of bird sounds for automatic species recognition. IEEE Trans Audio Speech Lang Process 14(6):2252–2263

Szeremeta B, Zannin PHT (2009) Analysis and evaluation of soundscapes in public parks through interviews and measurement noise. Sci Total Environ 407:6143–6149

Sueur J, Pavoine S, Amerlynck O, Duvail S (2008) Rapid acoustic survey for biodiversity appraisal. PLoS One 3(12):e4065

Sueur J, Gasc A, Grandcolas P, Pavoine S (2012) Global estimation of animal diversity using automatic acoustic sensors. In: Le Galliard JF, Guarini JM, Gaill F (eds) Sensors for ecology: towards integrated knowledge of ecosystems. CNRS Editions, Paris, pp 101–119

Trifa VM, Kirshel ANG, Taylor CE (2008) Automated species recognition of antbirds in a Mexican rainforest using hidden Markov models. J Acoust Soc Am 123(4):2424–2431

Villanueva-Rivera LJ, Pijanowski BC (2012) Pumilio: a web-based management system for ecological recordings. Bull Ecol Soc Am 91:71–81

Waddle JH, Thigpen TF, Glorioso BM (2009) Efficacy of automatic vocalization recognition software for anuran monitoring. Herpetol Conserv Biol 4(3):384–388

Wagner RH, Danchin E (2010) A taxonomy of biological information. Oikos 119:203–209

Westerkamp H (2007) Soundwalking. In: Carlyle A, Entendre D (eds) Autumn leaves, sound and the environment in artistic practice. Association Double-Entendre in association with CRISAP, Paris, CRISAP - University of the Arts London, p 49

Yang W, Kang J (2005) Soundscape and sound preferences in urban squares: a case study in Sheffield. J Urban Design 10:61–80

Yu L, Kang J (2008) Effects of social, demographical and behavioural factors on the sound level evaluation in urban open spaces. J Acoust Soc Am 123(2):772–783

Zhang J, Roe P, Pham B, Mason R, Towsey M, Sumitomo J (2010) Archiving nature's heartbeat using smart phones. In: Wachowicz M (ed) Movement-aware applications for sustainable mobility: technologies and approaches, IGI Global, Hershey. pp 19. doi: 10.4018/978-1-61520-769-5.ch008

Chapter 10
Applications

10.1 Introduction

Today there is growing evidence that the human impact on landscapes is not only restricted to habitat destruction, fragmentation, and chemical pollution but that a subtle degradation of the sonic environment represents a new dramatic escalation of human dominance of the Earth ecosystem.

The soundscape theory seems a very promising approach to a sustainable design in urban and rural areas and a powerful pedagogic tool to be utilized by students of design (e.g., Schafer 1994, Fowler 2012).

Soundscape is defined as a semiotic dimension of sonic information when perceived as noise can be considered not simply a negative character of a place but also an important component of the living space. For instance, the soundscape can greatly concur to create the sense of place, improving human well-being and representing an important landmark.

Sonic intrusion can be observed not only in urban areas or in rural landscapes but also in remote ranges and in shallow and deep seas.

Although other factors such as habitat isolation, habitat loss, land transformation, chemical pollution, hunting activity, food supply, and tourism safaris have a great influence on animal communication, noise effects occupy a high rank.

10.2 Threats to Animal Communication

In the past bioacoustics research was primarily focused on the characteristics and properties of acoustic signals, but recently this discipline has become more aware of the importance of investigations on the effects of anthropogenic intrusion on the sonic environment of animals. Laiolo (2010) has discussed in detail this aspect, confirming an exponential growth of interest for researchers in animal communication in relationship to anthropogenic impact (Fig. 10.1).

A. Farina, *Soundscape Ecology: Principles, Patterns, Methods and Applications*,
DOI 10.1007/978-94-007-7374-5_10,

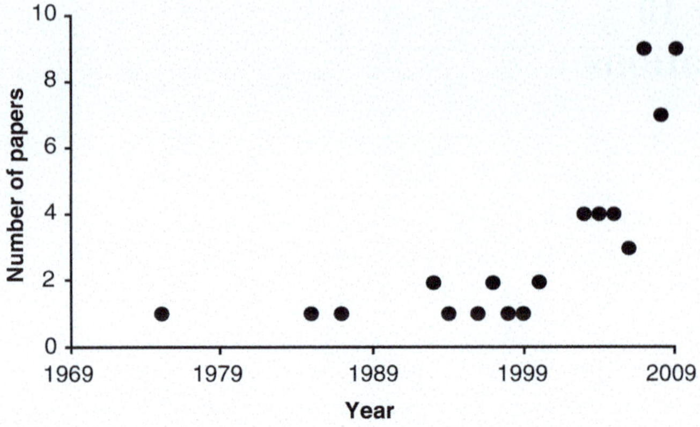

Fig. 10.1 Distribution of the recent number of papers regarding "Conservation Bioacoustics:" the exponential growth since the year 2000 is evident (Reproduced with permission from Laiolo 2010)

Several animals, such as humans, bony fishes, amphibians, reptiles, birds, and mammals, and invertebrates, such as insects, spiders, crustaceans, and nematodes, use acoustic mechanisms for courtship, territorial defense, and social behavior. In some groups the acoustic repertoire is simply innate, but in others as in humans or in some groups of birds, it is learned.

Communication is the most powerful semiotic process by which organisms provide information on individual status, mood, and intention. Humans use the interspecific communication process to catch resources. For instance, the great sensitivity of fishes to acoustic cues has been used in fisheries for some time. Fish schools can be driven into nets using the sound of Risso's dolphin (*Grampus griseus*), and the "Damburi" or "boko" is a traditional Japanese fishing methods to capture red sea bream (*Pagrus major*). This device is composed of a conically shaped lead (Fig. 10.2) suspended by a rope 7 m long that is immersed in the water, creating a column of tiny bubbles that attract fish to the nets (Yan et al. 2010).

In urban areas where the noise impact is strongest, some birds to avoid the masking produced by traffic noise have shifted song frequencies, duration, amplitude, and timing of singing. The shift of frequency probably represents a trade-off between noise masking and quality of the acoustic message. A high pitch increases sound propagation but a low-frequency sound is more informative for the quality of the signaler than the higher frequencies to which species have to adapt. In several cases, large birds that cannot change the frequency of their sounds prefer to abandon or avoid noisy areas to reduce the impact of noise.

In the marine environment, noise pollution has reached dramatic levels because of the greater propagation of acoustic waves in waters. Ship traffic, work in the harbors, and the use of military and commercial sonar affects pelagic fishes, whales, dolphins, and pinnipeds.

Experiments conducted in a marine protected area (Miramare, Italy) by Codarin et al. (2009) have demonstrated how boating and shipping negatively affect the

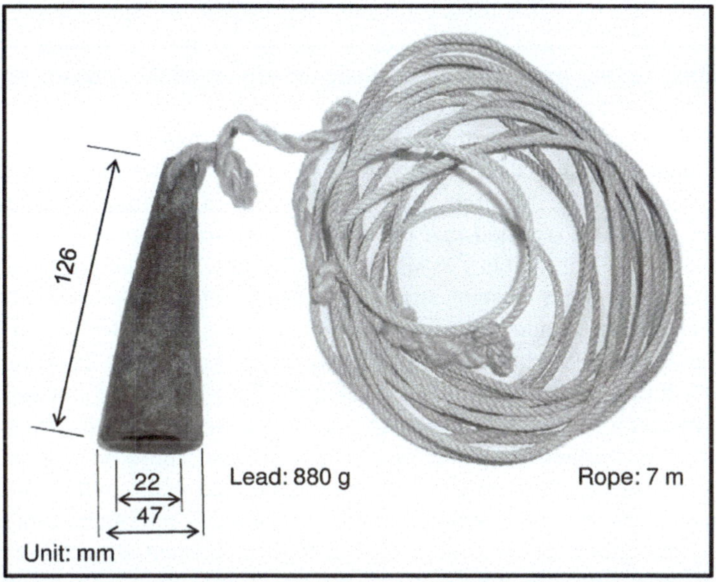

Fig. 10.2 Domburi, a fish attraction device used in Shibushi area by Japanese fishermen (Reproduced with permission from Yan et al. 2010)

hearing capacity of local fishes. To evaluate the effect of such sources of noise on some fishes (*Chromis chromis, Sciaena umbra*, and *Gobius cruentatus*) these authors played back in the laboratory the recording of a ship noise and the sound pressure level. The auditory sensitivity was evaluated recording the auditory evoked potential recording. For all species, the masking effect of boat noise clearly resulted in interference with the acoustic communication.

Habitat isolation is an important driver in communication system, especially when isolation means difficulties to transmit cultural traits of an acoustic repertoire. Dialects in birds are example of the presence of barriers that restrict the circulation of individuals. Dialects, although this is a controversial argument, represent the first step of genetic isolation.

Habitat loss is a further menace to the persistence of species, populations, and behavioral traits. Small areas support small populations that can experience a decreasing variety of acoustic traits, and signal modification has a direct effect on the reproductive results. In fact, if the acoustic signal is an "honest signal" the deterioration of this trait could open the way to non-honest competing systems with high risk to lose some of the genetic integrity. Deterioration of signals can have effects not only on reproductive success but also in efficiency in intercepting resources or maintaining social cohesion.

Land transformation, with the changes of the spatial aspects of the landscape by changing density and position of trees (from savanna like to linear hedgerows, for instance), modify the sonic environment. Consequently, this change can cause the disappearance of some vocal species even if other resources such as nesting sites or food remain unaltered. This aspect has not been considered until the present, but soundscape ecology could in perspective operate in this promising sector by adopting bioacoustics procedures to evaluate the effects on the acoustic broadcasting of species and the population dynamics.

Chemical pollution is a further consequence of human-based degrading processes that can affect the acoustic repertoire of species. It has been demonstrated that some birds reduce their repertoire when living in a strongly polluted area. Gorissen et al. (2005) reported that the singing repertoire of birds living in a site very polluted by heavy metals was significantly smaller and the birds emitted less song than singers 4 km distant from the polluted area. Alteration of singing behavior is a powerful indicator of environmental stress at the population level.

In a population of starlings (*Sturnus vulgaris*) living close to 20 sewage treatment plants in the southwest UK, Markman et al. (2008) observed changes in their acoustic behavior as a consequence of ingestion of earthworms highly contaminated by synthetic and natural estrogens. In fact, these birds were singing longer and performing a more complex song compared with control males. This change was observed to be a major attraction for females, although in these males an increase of immune function was experimentally confirmed. It was observed also that the key brain area that controls the male song complexity is significantly larger in the contaminated subjects. We can conclude that these animals are exposed to pollution that reduces their health but that paradoxically these animals are more attractive for females, changing the significance of the "honest signal" attributed by several observations and experiments to the song function. This trait, modified by altered physiology caused by environmental pollution, distorts sexual selection, producing long-term negative effects in the population.

Hunting activity has direct numerical effects on populations of wildlife but other aspects have to be considered. For instance, the rule to remove the older individuals from one population, justified by the model that the older individual is at the end of the life cycle and that young individuals have more possibilities to recruit is only in part confirmed by evidence. Often the older exemplars have in their memory all the experience accumulated and their removal reduces the chance for a group to better survive without the use of this experience. For example, family groups of elephants are guided by old matriarchs that have the capacity to discriminate between several other families. The personal knowledge of the other individuals encountered during the year is important for social cohesion, territory delimitation, and social exchanges for all. Experiments conducted by McComb et al. (2001) on the vocal recognition of different individuals have verified this hypothesis. Families with older matriarchs were able to better discriminate familiar and unfamiliar females when experimentally submitted to the playback of different groups.

In Gunnison's prairie dogs (*Cynomys gunnisoni*), a species distributed across the southwestern United States, Slobodchikoff et al. (1998) have found a change in alarm repertoire (local dialects) with changes in syllable length, number of syllables, and length of calls. In this species the alarm calls vary according to environmental conditions and the type of predators. Variations have been found both at regional scale and at local scale. The differences at the local scale may depend on local environmental conditions but also on the different human hunting pressure.

Wildlife safaris are a relevant practice in touristic offering. Large animals attract the curiosity of visitors, who often use aircraft facilities for a short survey, as is the case of the Barren-ground caribou (*Rangifer tarandus*), a species that lives in great herds in the extreme northern ranges. This species attracts tourists who generally approach the largest herds by renting small airplanes. The observation from the air creates several problems in this species, which depending on the height of the airplane may show a panic response, strong escape response, stationary response, or no visible response. Observations conducted to censuses the herds were collected by Calef et al. (1976), who have investigated 736 groups of caribou. The variety of reactions has been largely correlated with the height of the airplane. At an altitude of 60 m the caribou had a panic reaction, with animals out of control, stumbling and colliding with one another and crossing obstacles such as willow patches and trees. When the airplane flight is at an altitude of 150 m during spring and fall, and 300 m at other periods, this is sufficient to prevent the reactions of the caribou. The noise produced by the airplane probably is the major cause of the panic reaction. This study is important to set the height at which tourists may watch caribou without negative consequences.

Also, actions motivated by the noble goal of feeding "wild pets" may represent a serious threat affecting communication mechanisms. For instance, supplementing birds can produce a negative effect on their song repertoire. In fact, Zanette et al. (2009) have demonstrated as young song sparrow (*Melospiza melodia*) fed by food-supplemented parents sang a less diverse song than young fed by non-food-supplemented adults, because fed parents can produce more offspring, but with poorer quality, increasing the risk of local population decline. This evidence indicates that the manipulation of populations for "conservation" purposes in some cases could produce negative effects on the future generations.

10.3 Acoustic Monitoring

Monitoring means to evaluate changes over time using a repeated standardized sampling procedure. Monitoring is particularly required to track the dramatic changes that modern human societies are producing at the planetary scale. In particular, ecosystem dynamics represent the focus of most monitoring programs, where cycling of water, cycling of nutrients, energy transfer, and species relationships are the four main processes to be considered. Despite the relative simplicity of using automated remote procedures and instruments to capture the parameter described in the first three processes, the last requires a more sophisticated approach.

Nature under energy constraints is rich with sounds and each sound can explain in detail the magnitude, occurrence, and dynamics of a single process.

Unfortunately, natural cycles and trends of the biological and ecological systems are rarely considered from the sonic perspective and few data are yet available on the variation of sound amplitude, frequency, and typology at the long term. The requested and recommended actions for sonic monitoring is not a short, intense, and scientifically sound investigation but consists of a long-term evaluation of the natural soundscape able to cover hours, days, and seasons over the years.

Krause (1993) reported that in old growth forest the repartition of frequencies are quite continuous, and the concept of acoustic niche was strongly supported by this author. In secondary forests, however, the gaps along this spectrographic sequence indicate defaunation. The study of frequency distribution and of the frequency gaps could help to better understand the effects on ecosystems and communities of human intrusion.

Environmental monitoring using acoustic cues is an innovative and nonintrusive methodology to collect biological and nonbiological information from the sonic environment (André et al. 2011; Bassett et al. 2012). In this direction is oriented the Perennial Acoustic Observatory in the Antarctic Ocean (PALAOA), also the Hawaiian word for "whale," constructed in austral summer 2005–2006 on the Ekström ice shelf 15 km north of Germany's Neumayer Base. This system collects acoustic information 365 days per year, 24 h per day, real-time data access with full frequency and dynamic coverage (Boebel et al. 2006).

An acoustic environmental observatory (Mason et al. 2008) can represent an important tool for monitoring changes at a broad range of temporal and spatial scales and to realize an ecosystem-based management.

Acoustics coupled with other data can produce a very efficient tool to respond to questions pertaining to the effects of global warming, the changes in land use, and the effect of agriculture practices on natural systems.

Acoustic procedures can measure and recognize phenomena such as different species and species density, species identification, species behavior, and localization of individuals.

Ecosystem-based management (EBM) requires more information than single-species management, and synthetic indices are necessary for efficient evaluation. Acoustic methods seem to have a great potential to contribute to EBM, as recently stressed by Trenkel et al. (2011).

For instance, recent recording methodologies coupled to ad hoc metrics have allowed distinguishing individuals in populations of endangered species. In the European eagle owl (*Bubo bubo*), an endangered species that suffers from human presence, Grava et al. (2008) have identified a vocal signature for every individual contacted in the field. This approach opens new perspectives to study population composition, territorial fidelity, and definitively indicates the availability of environmental resources for this owl.

Monitoring the sounds uttered by marine animals and in particular mammals seems a further promising perspective because animal vocalizations are audible at a long distance and represent a more efficient methodology to estimate abundance, activity, and dynamics of such pelagic animals. For this are required robust methods of surveying abundance and migration, and good knowledge of behavioral traits and knowledge of the acoustic propagation and ambient noise, as suggested by Cato et al. (2006).

10.4 A Sonic Observatory

Many natural phenomena produce direct or indirect sounds and the analysis of such sounds can be extended not only to animal and human vocalizations but also to precipitation intensity, air turbulence and wind, and soil wetting. From biophonies it is possible to extract much information such as community complexity, species composition, acoustic activity, masking and overlap phenomena correlation, etc. A sonic observatory can be located in natural environments, at ecotones, and in urban areas as well. The sonic recording could be scaled according to hours, days, weeks, months, seasons, and years. It is possible to select the daily time (dawn, dusk, midday, night, etc.) and adopt instruments at different sensitivities, operating different sampling techniques and processing methods.

A sonic observatory may have as target global warming, changes in land use, the effect of agriculture, the way pests are spreading, and the value of an environmental bank.

Acoustic cues can be used to identify different species and species density, species behavior and localization of individuals.

Acoustic monitoring does not mean only collecting information of the vocal activity of birds, frogs, insects, and some mammals but can be used to measure road traffic, weather, soil moisture, streams, and a river hydraulic regime.

An observatory can record sudden changes in the environment such as the quantitative variation of frog choruses in a pond, and data can be shared in the web.

Different methods are available today to collect and store acoustic data from nature via data loggers (e.g., SM1/2, Wildlife Acoustics) or with autonomous wireless sensors. The first approach is simple but requires fieldwork, maintenance of the device, changes of batteries, and downloading of data. The use of autonomous wireless sensors when powered adequately (e.g., solar panels) allow handling data for longer time. The weak point of these methods is their security, because such devices that remain for a long time in the field can be damaged by extreme weather events and wild animals, stolen, or simply destroyed (Planitz et al. 2009).

10.5 Mapping Quiet Areas

The representation of a soundscape means to fix on maps most of the elements that produce sounds (Papadimitriou et al. 2009; Mazaris et al. 2009).

If we consider the sound as composed of local sonic information, the distribution of sonic sources can be used to design sonic maps. The actual sonic maps are based

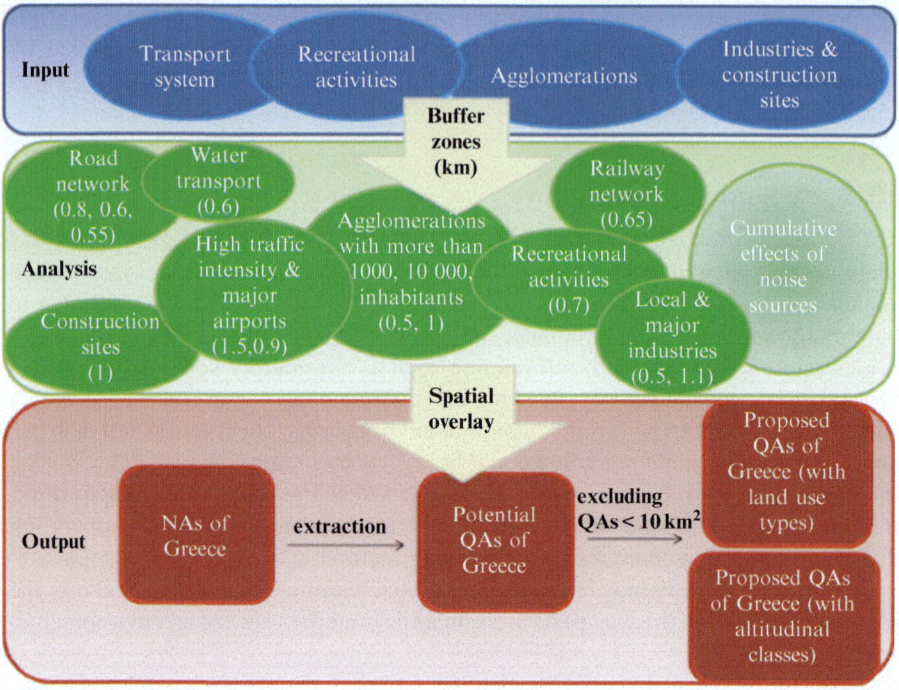

Fig. 10.3 GIS processing of noise sources across Greece to calculate the Quiet Areas (Reproduced with permission from Votsi et al. 2012)

on instrumental recording of the sound pressure, but depicting a soundscape requires a more integrated approach not easily transferred into visual maps.

The assessment of quiet areas at a country scale is important for future planning and management of natural resources. Votsi et al. (2012) (Fig. 10.3), according to the EU's Environmental Noise Directive (END, 2002/49/EC), reported a map of Quiet Areas in open country areas of Greece using a distance-based criteria where the quiet areas (QAs) were calculated as distance from noise sources (the categories considered were road traffic, railway noise, residential areas, industrial noise, aircraft noise, ports, construction sites, recreational activities). This approach is less accurate than the direct noise measurement but the results are useful to locate the critical areas in which to collect true noise data in further investigation.

10.6 Valuing Noises in Recreation Areas

Recreation noise is becoming an important issue in many national parks and natural areas and poses management problems. In particular in accessible front-country areas the concentration of loud people in some periods of the year can create problems for the natural quiet.

Noise effect management is important for wildlife species in protected areas (Cessford 2000), but noise per se becomes a complex and intricate issue that crosses several areas in nature conservation, social and economic development, and local development expectations. Natural quiet for visitor enjoyment is an important focus that does not diverge from wildlife conservation.

Parks have important tangible components: geological features, water, animals, plants, historical buildings, archeological and religious sites. A park has also intangible qualities: scenery, space, solitude, clear days and night skies, sound of nature, and natural quiet.

Natural quiet does not mean simply silence: it may represent the sound of nature without human technical sounds. Natural quiet is becoming more and more important because the cacophony of the modern life is growing: 90 % of park visitors consider the natural quiet an important source of well-being.

Noise can impact on visitor enjoyment and can represent an important source of negative rating.

The evaluation of visitor enjoyment is a complex process. For instance, mountaineers have a more precise idea of the sounds either in positive or in negative directions, quite different from roadside campers. Mountaineers rate with a high score the sounds from nature and feel annoyance for technologically borne sounds. In conclusion although the noise is the same, the valuation differs with the subjects, and people along a valley-floor trail are less annoyed by aircraft flights than those that are walking along rugged trails to high valley-wall viewpoints.

Safety concern represents another important component of subjective assessment of natural space. For some people the noise produced by a vehicle may represent a possible risk of crash but for others this sound can create a sense of reassurance in a "too wild" region. It is the same for the presence of the mobile phone, which represents a guaranty for a potential rescue but for others is an intolerable intrusion in the wild.

Noise impact in natural areas creates complex problems for managers, and better knowledge of noise sources can be extremely useful to reduce conflicts in management decisions.

Four possible categories of noises have been defined by Cessford (2000) in relationship to protected areas: external nonrecreational noise intrusions, external recreational noise intrusions, onsite interactivity noise intrusion, and onsite intraactivity noise intrusions.

10.6.1 External Nonrecreational Noise Intrusions

The sources of these noises, such as the low-altitude training of military aircraft, are outside the control of park managers. Commercial aircraft use routes that are established by other authorities, and in addition to the noise of their engines these flights produce distinctive stripes on the sky, reducing dramatically the sense of the range and integrity of nature. Other sources of noise are road and ship traffic, machinery use, industrial processes, and urban noise when the parks are close to large cities.

10.6.2 External Recreational Noise Intrusions

At the border of parks are concentrated many recreational activities such as music concerts, motorbike circuits*, and water-skiing that are considered as edge effects.

10.6.3 Onsite Interactivity Noise Intrusion

Motorized and nonmotorized recreational activities can be a source of concern when active inside the park borders, causing a conflict between people engaged in nonmotorized activity such as skiing, canoeing, walking, sailing, and swimming and on the other hand motorized activity such as snowmobiling, motorboats, jet skis and water-skiing, heli-skiing, motorbiking, and off-road driving.

10.6.4 Onsite Intra-activity Noise Intrusions

Often park visitors is engaged during their stay in different activities and can have different behavior accordingly.

Natural parks and protected areas in many regions of the Earth are under the attack of invading noise as the result of resource extraction, aircraft overflights, and roadway expansion, where in many cases the spatial scale of noise exposure is larger than the scale of protected areas and noise penetrates with increasing frequency inside the core areas. Barber et al. (2011) have investigated a case study in U.S. National Parks (Mesa Verde, Grand Teton, and Glacier) where was applied a spatially explicit land use change model to depict the intensity of human development at a sub-county resolution. They found, for instance, that the noise from roads is far greater than any protected areas.

Great attention has been paid in the recent years to evaluate and to enjoy the soundscape in U.S. national parks. For instance, an Educational Program Guide to Listening and Recording has been produced by the NPS (Krause 2002). The importance of the natural soundscape to visitors was stressed by introducing education about the natural soundscape and its importance to enjoy nature and to understand the healthy signals that nature transmit by sounds. In that guide, equipment and methods to listening and to recording sound and an essential glossary are offered.

To collect information about the visitor appreciation of sonic quality in protected areas, a standard practice is the distribution of a "listening exercise" to visitors with which the respondents identify natural and anthrophonic sounds heard in the park. Each sound is classified as "pleasing" or "annoying." For instance, a survey was conducted by Pilcher et al. (2009) at Muir Woods National Monument, California. Some visitor-caused sounds such as produced by a talking group were considered annoying, a strong indication that the human voice can be a good indicator of the sonic quality of a site, whereas wind and water were considered highly pleasing.

The aircraft overflights represent an important argument for wilderness recreationists. After an interview conducted with visitors from 12 wilderness areas, annoyance after exposure to overflight traffic by park visitors interviewed after 2 weeks from their visit demonstrated a value from 5 to 32 % (Fidell et al. 1996). This point indicates the importance of aircraft noise in the utilization of natural aesthetic resources.

10.7 Managing Sonic Context in Natural Areas

Noise is a good candidate to become the major actor of many social and land use conflicts of protected areas. In fact, natural sounds are considered an integral part of visitor experience in wild sites and also in archeological and cultural sites and caves, although the great sense of annoyance produced by anthropogenic noise in remote and wild areas can be the result of modest noise when compared with that in more intensively disturbed areas. In fact, noise produces annoyance that is the result of physical, psychological, and cultural filters, and to manage such annoyance is a real challenge for park managers.

The protection of natural sounds considered a physical component of the landscape cannot be achieved only by reducing the intrusion of mechanical noise. Noise that is a psychological intangible entity requires for its management social-at-large approaches as suggested by Cessford (2000) that distinguish three main strategies: manage the separation, reduce the noise effect, improve visitor expectations.

Managing separation means to reduce the overlap between conflicting activities, displacing in time or in space.

Reducing the noise effect can be achieved by regulating the use of vehicles or improving sonic insulation, shielding, and masking of the sources.

Improve visitor expectations means that visitors must be informed about the conditions that they will find inside a park. Thanks to this action people are more aware about the local conditions.

The integration of these three strategies is required to achieve a good trade-off between human enjoyment, nature dynamics, and local economies.

It is common to every conservation agenda to decide what type of sonic ambience may be appropriate for a park, which largely depends on several concurrent factors that are park specific.

Conflicts exist between the different strategies to preserve nature and associated values and at the same time to create all the necessary facilities to allow visitors to enjoy the aesthetics and beauty of park landscapes.

The National Parks try to find a trade-off between conservation of natural resources and limiting the impact of visitors. Within natural resources, the soundscape represents an important issue linked to the human concept of a quiet site, but with increased numbers of visitors in a park involuntary anthropogenic noise is introduced. Limiting the "audibility" of inappropriate sounds seems the major mission in conserving the wilderness of parks. The concept of audibility should be considered not only from a human point of view but also from the animals' point of view, and the threshold of audibility could be quite different and difficult to assess species by species.

In conclusion, sonic complexity in protected areas is an indicator of animal richness and habitat intactness, and this complexity is a strategic goal to pursue with priority.

The sustainable management of such areas requires the conservation of all the sonic attributes of the landscapes to guarantee the best functioning of ecological processes at the population, community, and ecosystem level.

The guidelines to preserve the sonic complexity in protected areas provide specific actions:

1. The reduction of anthropogenic sound intrusion
2. The reduction of every manmade disturbance that can fragment the habitats, reducing their extent
3. The reduction of motorized vehicle intrusions crossing the core areas
4. Reduction of airflights (commercial or touristic) in the perimeter of protected areas
5. The prevention of every pollutant
6. The education of visitors to respect the tranquility and the hot spots

10.8 Soundscape Protection in National Parks

Human noise can produce substantial changes in wildlife behavior in breeding and species success. The impact of noise on park visitors is an important issue to be addressed for every policy in natural areas.

Reconnaissance of the sonic quality of natural areas has introduced, for a long time, rules in the management of natural parks and protected areas, such as the Executive Order 10092 "establishing an Airspace Reservation Over Certain Areas of the Superior National Forest in Minnesota, US" signed by President Truman on December 17, 1949.

The natural sounds are considered by the National Park Service (NPS) Organic Act inherent components of "the scenery and the natural and historic objects and the wild life."

Intrusive sounds in natural areas become sources of concern for visitors (Gramman 1999). For instance, in the US the Natural Sounds Program Office operates inside the NPS system "to protect, maintain or restore natural, cultural, and other soundscape resources." This office "provides advice, guidance and technical support in characterizing and preserving park soundscapes."

The goal of this office is to heighten public and agency awareness of the value and character of park soundscapes. In this way the soundscape resources are preserved for the benefit of park visitors (Jensen and Thompson 2004).

10.9 Perceived Crowding in Natural Parks

A fundamental question that feeds major discussion about noise and protected areas is the following: "How much human-produced sound is appropriate in a National Park setting?" as advanced by Miller (2008) in a review on management of park soundscapes.

In the parks there are several sources of noise such as aircraft, snow machines, watercraft, road vehicles, and resource extraction (coal, oil, gas, minerals, and timber) that vary according to a seasonal dynamic. The measurement of source audibility and level requires standard procedures that are not easy to produce. Considering that different sounds are present in every park it becomes quite difficult to decide what sound is allowed in a specific park.

Crowding in national parks and reserves is the result of the recent boom of visitors in these areas. The increase of visitors poses management problems to parks in terms of impact on nature but also on the satisfaction of the visitors who may stay away as a result of low satisfaction (Fig. 10.4). So the problem is complex and

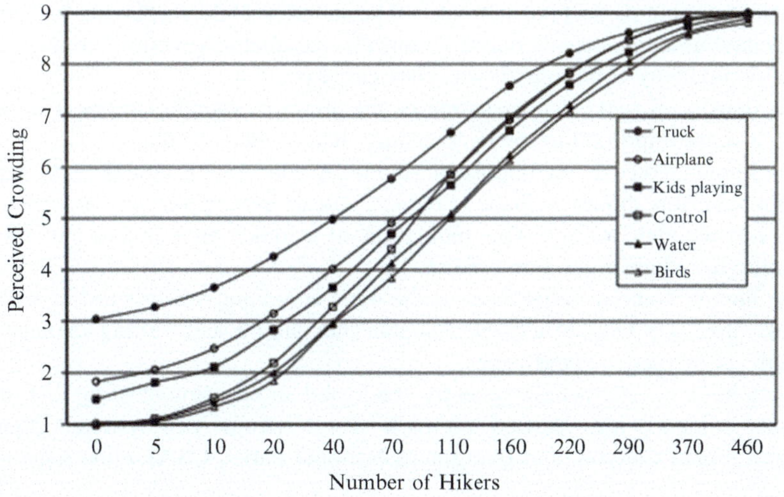

Fig. 10.4 Relationship between the number of hikers and perceived crowding (Reproduced with permission from Kim and Shelby 2011)

requires continuous monitoring. In a recent investigation carried out in the biosphere reserve of the “Schleswig-Holsteinisches Wattenmeer” on the island of Hallig Hooge (Germany), by Kalish and Klaphake (2007), only 20 % of visitors reported the presence of a large number of visitors. People more sensitive to crowding were non-guided day visitors; the guided visitors were more tolerant. This study represents a good example of monitoring of the tourism process and of the potential impact on the environment and on the visitors themselves.

Crowding produced by visitors in a natural park is of growing concern to planners. This phenomenon is not easy to evaluate because of the great variability of its assessment introduced by the different categories of park visitors. For this reason, crowding is often expressed as “perceived crowding” (PC).

Perceived crowding (PC) seems connected with perceived sounds and this effect is connected with the type of sound stimuli. Children playing, airplanes, truck engines, birds, water, and a control were the acoustic signals administered in an experiment conducted by (Kim and Shelby 2011). Topography, geography, and trip characteristics influenced perceived crowding too. Areas with low use are reported accurately by respondents of questionnaires but in areas with high use (high density of visitors) this value is underreported and more experienced users perceive the areas with more PC than normal people.

Wilderness users appreciate backpackers more than horseback riders and several small groups more than a few large groups. Bad behavior by visitors such as noise, yelling, and littering produces a feel of more crowding in respondents although the density is the same. Wilderness users prefer encounters near the road rather than in remote areas; the same is true for boaters close to the accesses compared to the middle of lakes.

Sounds have a role in the rating of perceived crowding and encounter norms. Natural sounds such as water and birds decrease PC and increase the tolerance to seeing other park visitors. An increase of PC and intolerance is produced by technological sounds (airplanes, trucks). The sound of children playing had ambiguous behavior: at low density of encounters this parameter was considered negative, but in high-density conditions it was considered positive.

The density of park visitors increases the effect of sound changes (Fig. 10.5). Sound is more important in the backcountry than in the forecountry.

The human voice is an important source of concern in natural parks and an educational effort should be made to reduce, at least in the proximity of very scenic spots, talking and yelling. The indication of a quiet area is just an efficient procedure to reduce voice-borne noise.

The human voice in some cases is the most annoying sound heard in a natural park because this was an expected sound and talking and yelling damages the aesthetic assessment of landscapes.

Benfield et al. (2010) have recently conducted an experiment using 251 undergraduate students who viewed 30 park scenes (from Yellowstone, Olympic, Saguaro, Grand Canyon, and Everglades National Parks, USA) while listening to

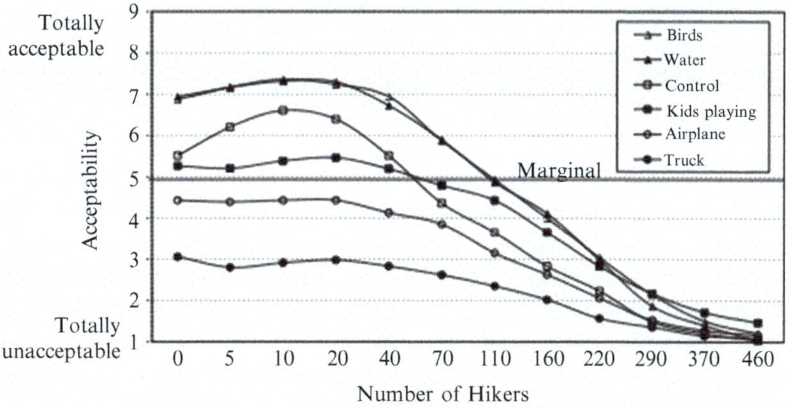

Fig. 10.5 Level of acceptability of different types of sounds for different numbers of hikers (With permission from Kim and Shelby 2011)

four soundtracks: natural sounds (birds and breeze), natural sounds with mechanical ground traffic added, and natural sounds with human voice added. The soundtrack was reproduced at three levels of loudness: control (40–45 dB(A), that was the background noise in the experimental room, soundtrack added at 40–45 dB(A), and soundtrack added at 60–65 dB(A). The Positive and Negative Affect Schedule Expanded Form (PANAS-X) were used to calibrate 60 emotion adjectives rated on a 5-point Lickert scale (Watson and Clark 1994). The results confirm that every added soundtrack of mechanical or human origin decreased the rating, and this effect was more evident at higher sound levels. In psychological terms, serenity was decreased and hostility increased. These effects were stronger for scenes of great beauty. Noise in parks is discussed in terms of conflict between preservation and recreation accessibility. In particular, noise produced by aircraft impacts in a direct way the rating in park evaluation procedures; aircraft noise has produced a larger variability in the responses than all the other administrated sounds. Ground traffic was found impacting independently from the volume at which it was added.

The incongruence between aircraft noise, ground traffic, and human voices and the expected sounds in natural parks decreases the rating in assessing quality.

Natural sound produces serenity and calming; anthropogenic sounds reduce serenity and create hostility. Natural sounds seem to have a minor impact on the natural assessment compared to anthropogenic sounds, which could be a good indication for future management. These sounds generate a greater reaction in listeners' shifting emotions.

Although many secondary aspects remain debated, the auditory soundscape has an important role in assessing the quality of natural and historical areas. Managing sounds in parks has as major an importance as managing scenic views.

10.10 Snowmobile Noise in National Parks

The visit of some mountain parks in winter time requires the use of snowmobiles and snowcoaches that allow the transportation of tourists along the best scenic views of parks while reducing the risk of casualties for unguided trekkers

For instance, visiting Yellowstone National Park (YNP) in wintertime has been a subject of controversies and discussion when visitors are encouraged to use snowmobiles and snowcoaches. Before 2004 the majority of vehicles were unguided snowmobiles. Since 2004 a new regulation has considered private snowmobile and snowcoach use as not allowed. The admitted snowmobiles and snowcoaches must be "Best Available Technology" to meet the NPS requirements. To verify and test the value of the sonic environment in wintertime at the Yellowstone National Park an investigation was conducted by Freimund et al. (2011) on 413 visitors in the Old Faithful area by using a questionnaire. The respondents were between 18 and 87 years old (average, 51). Personal information about the visit and modalities was collected as well.

The majority of respondents were in agreement with the statement that the YNP is a place for natural quiet, and "a place to hear natural sounds." Of the visitors, 99 % stated that "the opportunities to experience natural sounds were important to the overall value of the park." The respondents during the visit to the YNP used on different occasions different vehicles. According to the selection of vehicles there were differences in valuing the sonic environment: 81 % of respondents declared that natural sounds had a positive effect on their visit, and 19 % replied that the natural sound was not important for the quality of their visit (Fig. 10.6).

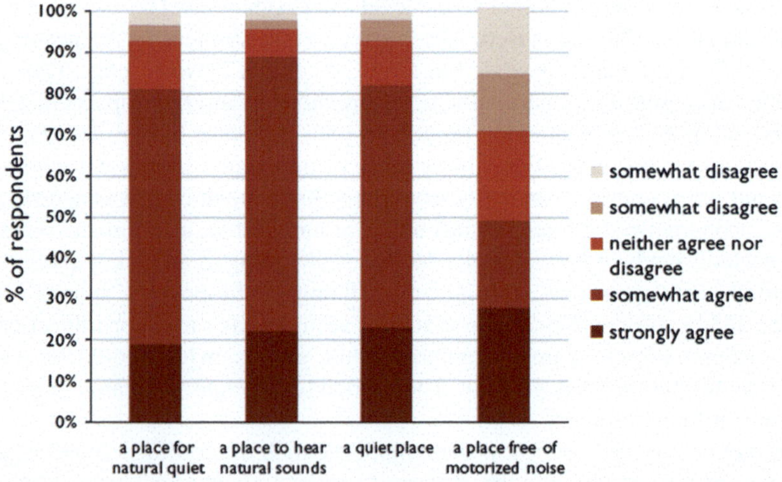

Fig. 10.6 Level of agreement of visitors about the different definitions of the Yellowstone National Park (Reproduced with permission from Freimund et al. 2011)

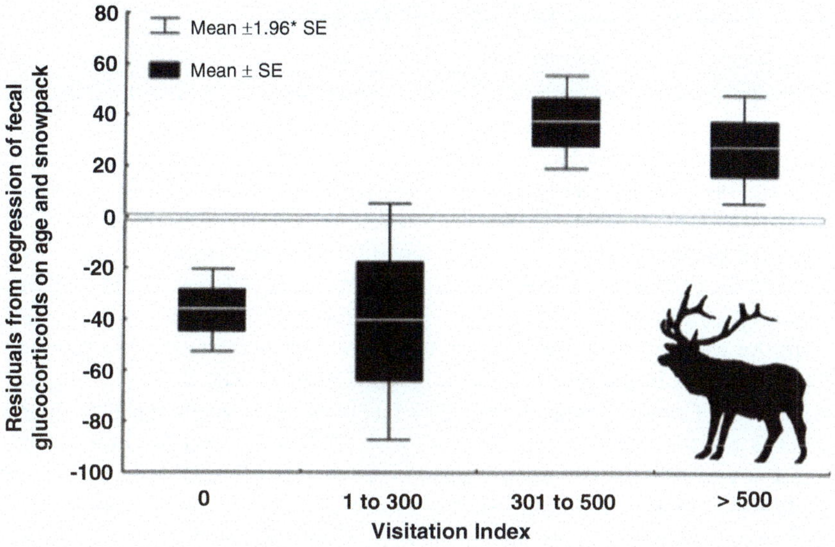

Fig. 10.7 Variation of fecal glucocorticoids (GC) in elk (*Cervus elaphus*) in Yellowstone National Park in relationship to the number of snowmobiles entering the protected area (Reproduced with permission from Creel et al. 2002)

This investigation clearly demonstrated that the last generation of visitors is aware of the consequences of vehicle use during wintertime in the parks and after regulation a significant decrease of vehicle use has been experienced by commercial guides.

Snowmobiles are a very intrusive source of noise in national parks and protected areas. The regulation of these vehicles is achieved in many countries, but great discussion persists against evidence that this device impacts animal life. Several studies demonstrate that snowmobiles are avoided by white-tailed deer (*Odocoileus virginianus*), mule deer (*Odocoileus hemionus*), reindeer (*Rangifer tarandus*), and moose (*Alces alces*). Other investigations have discovered an increase in heartrate in bighorn (*Ovis canadensis*) and white-tailed deer in the presence of human disturbances. The physiological and behavioral responses to stress causes an increase of glucocorticoids (GCs) secretion by the hypothalamic-pituitary-adrenal axis. A long-duration high concentration of GCs produces reproductive suppression, ulcers, muscle wasting, and immune system suppression.

Analysis of fecal GC content, a good noninvasive method to evaluate the level of stress, has been applied by Creel et al. (2002) on elk (*Cervus elaphus*) in Yellowstone National Park (Fig. 10.7) and on wolves (*Canis lupus*) in Yellowstone, Voyageurs, and Isle Royal National Parks. The results confirmed the increase of GC with the increase of snowmobiles when compared with other road vehicles. No evidence was collected about the effects of snowmobiles on population dynamics, however.

10.11 Aircraft Noise in National Parks

The noise produced by jet aircraft is becoming a growing concern around the world, and regulations have been provided most of the developed countries. We have already discussed the problem of aircraft noise in Chapter 6. In particular, natural parks in the backcountry are areas very sensitive to noise intrusion. In urban areas, the dose–response relationship between dwellers and noise sources, faced with efficient parameters and metrics (e.g., day–night average sound level; DNL, or L_{dn}), is well correlated with the percent of the surveyed residential population complaining of noise annoyance. This parameter is insufficient for park visitors who are not exposed to noise by aircraft or ground transportation machines for a prolonged time as in the proximity of airports or urban areas.

In the US a Congress mandate of 1987 has been received by the U.S. National Park Service in the 1990s in collaboration with the U.S. Federal Aviation Administration (FAA): 2,600 questionnaires when processed in different parks and reported by Anderson et al. (2011) found that the effects of aircraft noise on visitors largely depend on the characters of the visitors as well. The dose–response relationships are important but significant acoustic variables are also the total aircraft Leq during the visit, the energy percentage of that total from helicopters, and the energy percentage of that total from fixed-wing aircraft.

In such type of investigation resulted important also non-acoustic variables that are significant:

Visitors for whom natural quiet is very or extremely important are 10 to 50 dB more sensitive to aircraft noise than other visitors.

Visitors that have visited the site before are 20 to 35 dB more sensitive to aircraft noise than other visitors.

Visitors in adults-only groups are 10-to-20 dB more sensitive to aircraft noise than other visitors.

Visitors on short hikes are 35 to 70 dB more sensitive to aircraft noise than other visitors.

High-altitude jet aircraft produced annoyance for 16 % of visitors for 40 dB, but tour aircraft noise annoyed 36 % of visitor for 40 dB.

Monitoring the effect of aircraft noise on visitors in natural areas is becoming an urgent matter. The increase of the number of flights over natural areas exacerbates the relationships between ground-based recreationists and aircraft civilian and military air traffic, as reported by Booth et al. (1999). To reduce the impact of aircraft overflying, a monitoring scheme has been presented by Booth et al. (1999). The administration of a short questionnaire represents an efficient tool to investigate the status of aircraft-borne acoustic pollution. In remote areas of the parks the questionnaire was distributed for a self-pickup.

The questionnaire was regarding:

General likes and dislikes
Whether aircraft were noticed during the visit

Number of aircraft noticed during the visit
Experience compared with expectations
Estimate of aircraft threshold level
Reaction to aircraft: positive, negative, or neutral,
Extent of annoyance
Extent to which aircraft have affected total visit enjoyment

Some steps are requested to initiate the procedure for collecting the necessary information. The first step is to choose the monitoring area, prepare the questionnaire, choose the monitoring periods, select the monitoring days, and select and brief the survey interviewers.

The effects of aircraft overflights on recreationists are varied. Direct effects are visual and aural but the latter are more important. For some recreationists the noise of an aircraft or of an helicopter produces a feeling of security. In the case reported by these authors, aircraft overflight is a not strong indicator of thc quality of a park but there are other indicators that prevail, such as poor signage or conflict with other recreationists.

Expectation of aircraft noise reduces the annoyance effect and vice versa. The origin of the visitors makes the difference. European visitors are more tolerant of the presence of aircraft than natives.

The "Last Settler Syndrome" affects visitors with a past experience of satisfaction in the same place. When they visit the same site again they can make a comparison with the past experience, and considering that the aircraft flights are increasing the feeling is generally negative.

There are also differences between backcountry visitors and forecountry visitors, in that backcountry visitors are more sensitive to aircraft traffic.

Site location and activity can have a different feeling; for instance, visitors on track are more annoyed than when they are in the accommodation quarters.

The type of flights is perceived differently: rescues or service flight are highly tolerated, but scenic flights are considered "unnecessary."

Visitors disturbed by aircraft flights define this disturbance as annoying, irritating, intrusive, distracting, or disruptive, although helicopters create interest in the visitors.

Measuring the aircraft flights is not easy: it is possible to use the aircraft sound levels, length of time aircraft are audible, frequency of flights, and speed of onset. The most important effects are created by flight frequency and sound levels. For some people just the presence of an aircraft represents an impact.

The uniqueness of each wilderness area presents difficulties to refer concerns to other areas. Visitor crowding may reduce aircraft annoyance. Direct measure of noise-induced annoyance is a good indicator of the effect of aircraft flights rather than visit satisfaction or intent to revisit, as shown by the study of Fidell et al. (1996).

The increase of tourists in national parks has accompanied the intrusion of growing noise produced by vehicles such as snowmobiles, and scenic park overflights enter in direct conflict with the conservation goals. Severe consequences on wild fauna result, and a recent case was documented in a sky resort not far from Cortina d'Ampezzo (Italy) during the 2013 New Year of a brown bear that was probably awakened from its hibernation by fireworks.

Legislation when restrictive can create severe consequences both on the economic and on leisure and recreational offers. In a study conducted in the Grand Canyon National Parks, an air tour flying over Las Vegas has been estimated to contribute $504 million to the economy of southern Nevada, and restriction of such overflights would cut $249 million.

Noise is an ambient stressor. Landscape assessment is conducted using both visual and acoustic cues. Many researchers have demonstrated that people are escaping urban noise and as this is a priority for visitors of national parks, technological sounds in those conditions are considered highly negative, nonappropriate, and impacting. Motives (escaping urban noise) and expectations (natural sounds and quiet areas) are the two elements that drive the landscape assessment of park visitors.

In evaluating the effect of noise on landscape assessment, Mace et al. 1999 found that a airplane noise from 40 to 80 dB(A) was rated negative in landscape assessment. Other impacted sounds can decrease the rate as well.

Aircraft overflight noise produced by helicopter tours is frequently encountered in national parks such as Grand Canyon, Hawaii Volcanoes, Haleakala, and Bryce Canyon. Noise is considered an environmental stressor co-responsible with other disturbing sources to several physiological and psychological effects that can impair individual well-being for a long time. Although helicopters are associated with rescue activity and should not produce a hostile feeling, an experiment in which were utilized 200 undergraduates Mace et al. (2003) found that 60 dB (A) noise from a helicopter decreased the ratings of scenic beauty, solitude, tranquility, freedom, naturalness, and preference, and higher ratings of annoyance in every type of scenic context administrated.

The case reported by Bowlby et al. (1990) about the impact of noise produced by the activity of the local airport (Jackson Hole) on Grand Teton National Park quietness explains very well that monitoring the noise and appropriate regulations can improve the sonic environment. This study showed that the dose of noise for the year is not significant although a small difference appears between summer and winter; in winter, more noise was observed. Compliance was observed with the day–night level limit except when the wind direction requires departure of aircraft in the northern direction; then day–night level is not respected and is sanctioned. The new models of aircraft can solve in part the problem of preserving quietness in this remote area. When no aircraft are in flight this area is very quiet and so a distant airplane is sufficient to create an annoying event even far from the airport, an example of the relativity of sonic environment perception.

10.12 Valuing Sonic Heritage

Generally the multisensory nature of aesthetics does not consider the sonic ambience as a critical element and often the aesthetic process is limited to visual cues, as recently reported by O'Connor (2008) for an Australian case. According to Schafer (1977, p. 9–10) every place has a specific soundscape with at least three components:

1. 'Keynotes' – background sounds that form the base of the soundscape
2. 'Sound signals' – foreground sounds that are more distinct
3. 'Soundmarks' – sounds that have a particular meaning for local communities and that are analogous to landmarks in the structured environment

The acoustic value of a landscape is rarely considered in the evaluation of a heritage site, although there is a strong connection between sense of place and sound. The low attention to soundscape is the cause of poor acoustic management, so the acoustic quality of a place easily is involuntarily lost.

Recently the European Network of Universities for the Implementation of the European Landscape Convention (UNISCAPE) has produced an important declaration that is an example how to integrate the necessity to educate people in sonic evaluation under the framework of the implementation of the European Landscape Convention (Florence 2000). With this declaration, here reported in the integral way, the acoustic heritage of a place receives the same importance of other visual or historical components of the landscape, which opens new perspectives in valuing sonic heritage as an integral part of a holistic landscape evaluation.

CAREGGI LANDSCAPE DECLARATION ON SOUNDSCAPES – JUNE 2012

"Landscape" means an area, as perceived by people, whose character is the result of the action and interaction of natural and/or human factors . . .."

European Landscape Convention, 200.

Introduction

The soundscape can be defined as the acoustic property of every landscape according to a species' specific perception and is the result of physical (geophonies), biological (biophonies) and human (antrophonies) features and dynamics.

The soundscape is an important component of the terrestrial and aquatic environment especially from a human perspective, but it represents also an indispensable informative and communicative medium for every vocal animal. When lost or degraded it reduces the quality of a landscape and impacts on the human well-being and more in general on the biological activities of several organisms.

Modern human societies have strongly impacted on the soundscape, especially in urbanized areas, reducing the acoustic quality and functions of the surroundings. Noise pollution in urban areas is a problem for the health of billions of people at every latitude.

The quality of the soundscape is an important aspect not only in natural areas but also in cultural landscapes and in historical sites and monuments because it contributes to make residents and visitors fully benefit of such valuable heritage.

To preserve a soundscape and to mitigate the anthropogenic noises it is important to reduce unnecessary acoustic sources associated to mitigation planning strategies.

The protection and management of a native soundscape represents a qualified goal to maintain in situ the biodiversity and to guarantee the ecosystem's functionality as well.

Educating to listen is the obligate step to recognize the strategic value of the soundscapes and to support a protection and management policy. The creation of data banks of threatened soundscapes is one important action to preserve the memory of the past complexity that can be used in the future to establish more favorable acoustic conditions especially in urban areas.

To face the soundscape matter a multi-disciplinary and integrated approach is required because the soundscape is a focal point for many scholars (e.g. acousticians, bio-acousticians, engineers, planners, artists, psychologists etc.), therefore a world permanent forum must be encouraged to disseminate ideas, research projects, management.

1- Landscape strategy for your region. Characterizing soundscapes (but not only these…)

1.1- Policy makers should consider soundscapes as an important component of landscape planning and management since the sonic environment reflects human culture and natural heritage and effects our quality of life and health.

1.2- Main point of reference for valuating soundscapes is the well-being of the people, the local community.

1.3- Landscape should be lived and valued in a multisensorial way. Surveys on multidisciplinary basis among inhabitants and visitors are crucial to define the perceived values of the landscape, and the measure to which these characterize specific landscapes.

1.4-"Sound-marks" (including local dialects) may be defined just as well as landmarks. Soundscapes are by definition dynamic, and more characterized by rhythm (day-night, season) than by continuity.

1.5- Narratives describing typical landscapes and associated soundscapes are more powerful than maps.

1.6- It should be promoted that landscape and urban planning be based on multisensorial perception of the living environment.

1.7- Indicators for characterizing soundscapes are rather to be found in localized qualitative statements, to be related to more general quantified parameters on a regional level.

1.8- The disturbing effects of human activities in underwater and underground are to be considered.

2- Education and raising awareness

2.1- The first step in raising soundscape awareness is to improve listening skills. This can be achieved by a series of activities (for ex., soundwalks, sound recognition exercises, etc.) to be proposed to the general public including educational Institutions.

2.2- Sounds are perceived not only on a perceptive level but also on an emotional level, thus informed sound making and sound-art is relevant to enhancing awareness.

2.3- Raising awareness and at the same time preserve and manage soundscapes can be reached though specifically designed facilities such as sound libraries and sonic museums.

2.4- Education should ultimately aim at integrating soundscape awareness into a holistic perception of landscapes.

3- Scientific aspects for landscape conservation and management

3.1 Integration among the different approaches made by different field of expertize and multidisciplinary approach.

3.2 It is urgent to develop a research policy in soundscape ecology (SE) because the acoustic aspect of landscape is underestimated

3.3 The SE must be positioned under the umbrella of the landscape epistemology.

3.4 It is important to define the ecology of the Soundscape: limits and tools

3.5 It is recommended a standardization of sampling and processing methodologies.
3.6 The Soundscape of historical urban sites and its character should be considered a value (tourism, cultural heritage and quality of life).

10.13 Some Characteristics of the Urban Sonic Environment

In natural areas vegetation and geo-morphological structures act as a sponge that absorb certain frequencies. This effect is strongly modified by the position of the objects and the position of the listener.

In urban areas buildings function as screens, masking sounds, and the distance from sound sources is important. Elevated roads increase the noise effect at the upper level of buildings facing such roads but this solution significantly reduces the road noise for people living at the ground.

Squares fully surrounded by buildings can have a tranquil sonic environment: the presence of green roofs and balconies can increase the quiet façade effect. Roof slope can be another important factor of energy diffraction. Sound diffraction is reduced more by green covers than by traditional tile.

Streets in many cases function as an acoustic canyon in which sounds reverberate, creating a sense of annoyance. The sound is maintained longer along such streets, especially if the streets are narrow as in city centers. If the façade has the possibility for absorption this reduces the sound pressure, but unfortunately it is not possible to use such absorbing material because is not weather resistant. The use of green vegetation along the street façade is an important strategy to reduce noise and canyon effect.

Public spaces in an urban context offer different services: shopping, meeting, recreation, leisure, enjoyment, etc., and the paradigm that a silenced area may be the best is not the true goal.

Thus, noise remains an important factor in the quality of life for every urban dweller.

10.14 Urban Noise Management

One popular goal to manage the sonic environment in an urban context is to reduce noise and to create quiet areas. Sonic tranquility today is not seen as a necessity.

Tranquility of outdoor spaces is a positive environmental quality in both urban and rural landscapes. The assessment of tranquility remains a disputed argument and specific parameters are not available for such evaluation. Tranquility does not mean absence of noise but a combination of pleasant sounds with a scenic surrounding. Kaplan and Kaplan (1989) consider important elements for a tranquil place the "soft fascination" that should provide a pleasing level of sensory input.

Urban noise management means to reduce the level of unwanted noise or to mitigate the noise sources (planting trees, building sonic barriers, etc.).

Sonic barriers, either composed of plastic, concrete, or living materials, also reduce and degrade the scenic composition of a urban landscape. Technical solutions adopted on cars, trucks, and trains can reduce traffic noise. For instance, electrical engines can greatly reduce traffic noise.

Rolling noise is produced by tires and by the road surface, and a road well maintained reduces this rolling noise. Railway noise is mainly produced by the friction between metal surfaces of rail and wheel. Relevant improvements are possible especially for freight; noise barriers are significant solutions but not in urban areas because of space restrictions and visual impact.

The reaction of people to noise exposure depends on many factors. For instance, noise annoyance is higher for the same energetically averaged exposure for aircraft noise than for traffic noise. This is the same for industrial noise; a marble sawmill in Carrara City (Italy) is at the same time a soundmark for residents and a source of annoyance for visitors.

Indoor noise mitigation from outdoor noise sources is easily achieved using insulating materials and closing doors and windows, but these actions can create a sense of claustrophobia for dwellers and a negative effect on perceived annoyance.

10.15 Traffic Noise Mitigation Procedures and Urban Spaces

Orchestrating traffic noise represents a realistic opportunity to improve the sonic ambience in urban spaces. Traffic is the result of a flux of cars, trains, autobuses, and trucks. Direction, intensity and fluxes may be changed. For instance, the substitution of roundabouts for traffic lights can modify cyclic noise by reducing acoustic peaks after the lights turn green.

Traffic dynamics is under two main scaling constraints: a large scale that largely depends on the quantity of vehicles and their direction and speed and a microscale largely impacted by personal style in driving. The microscale modeling of traffic allows simulating every individual behavior that creates emergent phenomena such as traffic congestion. The noise emission coupled with individual behavior creates the conditions for overall traffic noise. Acceleration and deceleration near intersections are important components of traffic noise, and modeling close to intersections is an important component.

Noise redistribution and noise reduction are two different strategies for managing traffic. To create a quiet area in a city is possible by removing traffic, but such traffic may create congestion elsewhere, and increased noise, so in terms of overall noise impact on the city the redistribution of traffic is a local remediation but not a holistic solution strategy.

The effect of highway construction on the urban setting produces changes or "sonic fields migration." This effect was studied by Ramis et al. (2003) in the town

Motilla de Palancar in Spain when the A-3 motorway Madrid-Valencia was finished in 1999 and all the traffic load was removed from the center of the city. The displacement of traffic resulted a reduction of noise from 3 to 7.5 dB(A) with an average noise in the city of 60 dB(A) as recommended by the World Health Organization. This change has also produced a more satisfactory quality of human life, as seen from investigation of 100 people who had their homes in front of the principal road.

The noise of traffic is increasing in cities, with reduction of the extension of quiet areas, and this phenomenon is a continuous font of public concern for the annoyance produced and for the reduction of communication capacity of a densely aggregated people. The policy of removing traffic noise from the cities has exposed the outskirts to an increasing level of noise. In Europe 20 % of population is exposed to excessive road traffic noise, and this represents a severe health risk. In this issue the physical measurement of urban noise is not enough to allow architects, town planners, and landscape designers to initiate an action to revaluate and improving the sonic environment, in other words to create the changed sound in the right space, including in this process the subjective sonic experience of dwellers.

A way to escape or in part to solve this problem is to create green areas, urban parks far from the traffic congestion, but often in urban areas and specifically in urban parks reduction of traffic noise is not easily made.

A further possibility consists in masking the traffic noise, when it is at a lowest level, with wanted sounds such as water fountains or playback of bird songs. Waterfalls can represent a distracting sound that could help to alleviate the sonic annoyance. An experiment is described by De Coensel et al. (2011) asking 100 volunteers (45 women, 55 men) to evaluate in the laboratory the effect of waterfalls and bird song to mask road traffic by binaural acoustic files. The addition of fountain sound reduced the loudness of the traffic noise only when the traffic noise had a low temporal variability, less pulsing, but the addition of the sound of singing birds significantly enhanced soundscape pleasantness and eventfulness, more than the addition of water sounds.

This experiment clearly demonstrates that the loudness of unwanted sounds should not be the only direction in acoustic design but that the soundscape quality is strongly influenced by the meaning associated with the different sounds.

Jeon et al. (2010) have demonstrated that the water sounds such as "stream" and "waves of lakes" sounds can efficiently mask urban noise, when the level of the water sounds are similar or not less 3 dB below the level of the urban noise. This strategy seems an efficient method to reshape and improve outdoor urban spaces.

The manipulation of sound sources allows modifying the sonic environment. In this way the displacement of two annoying sounds does not change total annoyance, but is possible to overlap an annoying sound with a sound source of preferred sound quality and to create a masking effect.

Noise attenuation can be obtained by engineering barriers or by natural ones. In this second case the utilization of plants is very common because plants can produce many contemporary benefits such as soil cover, habitat for animals, atmospheric pollutant absorption, scenic view, and finally a sound barrier. In a recent investigation Fan et al. (2010) evaluated the level of noise absorption by six species of evergreen plants: arrowwood (*Viburnum odoratissimum*), oleander (*Nerium indicum*), Chinese photinia (*Photinia serrulata*), bamboo (*Oligostachyum lubricum*), red robin photinia (*Photinia fraseri*), and deodar cedar (*Cedrus deodara*). Every plant has demonstrated a specific spectrum response. The amount of noise reduction depends not only on the single plant species but also by the arrangement of different species according to the environmental noise specificity. These authors have presented some useful suggestions such as use of species that have a more similar noise reduction spectrum to environmental noise, and combining of different species is possible to improve the attenuation effect. They have found that deodar cedar reduces the low-frequency noise much better than the other five species considered and that some characters, such as leaf shape, leaf weight, and leaf tactility affect directly the noise attenuation value. Attenuation peaks are performed by the species but their arrangement is the major factor responsible for degree of attenuation.

10.16 Design the Soundscape of Urban Spaces and Parks

In urban planning sound was considered at the end of the projects, when structures and open spaces had been built, but in a new perspective sound should receive at least the same score of visual esthetics components because the soundscape contributes to the identity and specificity of this urban environment.

The acoustic design represents an important activity to improve sonic environment in urban spaces. High-frequency music and waterfalls of fountains can adequately mask traffic noise, transforming open spaces into a comfortable dimension. The inclusion of greenery in the urban context creates natural barriers to traffic noise and adds songs from birds and insects. In some cases playbacks by loudspeakers have been used to imitate natural sounds.

Designing a quiet area does not mean to exclude every sound; a natural quiet soundscape refers not to silence but to the presence of natural sound sources. Components of the natural quiet include the wind blowing through a forest, the babble of water in a stream, the howl of a wolf pack, the chirp of a bird, the rumble of an avalanche, thunder and a rainstorm, the crash of ocean waves, and the deafening roar of a waterfall, although in a urban context some sounds of the wild cannot be present.

The benefits of a good urban park to extend environmental benefits to the largest part of the population possible have created great interest in policymakers and decision makers. The majority of park design is based on visual components such as shape and position of vegetation, but a peaceful quiet place is characterized also by a comfortable sonic ambience. For this reason the soundscape is an important factor to rate the quality of an urban park. The sonic environment is the result of the contribution of many different types of sounds that occur either contemporarily or separately in time. For instance, bird twittering or falling water are pleasant sounds while mechanical sounds are considered unpleasant and not welcome in a urban park. In some cases, as when we are engaged in talking, most sounds such as the ring of a mobile telephone could be considered temporarily distasteful.

All these sounds can be pleasant or unpleasant with positive and negative effects according to the different categories of visitors (residents, tourists, young, aged, rich, poor, with different cultural background, etc.), and sound remains a mysterious matter because it is appreciated so differently by people. Furthermore there are differences in which a dweller evaluates the soundscape outdoors with respect to the indoor sonic environment.

Describing a soundscape is not an easy task. In fact, a sonic environment should be a match between expected sounds with less expected sounds. The expected sounds are the result of the context that a person expects according to the different signals in which he moves aside from other semiotic indicators. A high-quality soundscape should have several distinct and expected sounds. Sounds become the semiotic interface between the subject and the environment where sonic signals are both physical and personal. Physical signals are loudness, spectral, temporal, and information content of the sound, and visual aspects are movements of subjects, the landscape architecture, lighting, people's activity, temperature and humidity, and olfactory issues. The personal factors are related to personal traits (noise sensitivity, attitude to perceive other sources of information, the current activity of the subject, emotional status, personal goals, etc.).

Noise control engineering may have a negative impact on sonic environment: not all sounds are perceived as noise. Several sounds in a urban context represent landmarks and prevent people from feeling too extraneous to giant buildings and a huge plaza. Planning pleasant sonic space is important to our everyday life.

10.17 Sonification

Sonification is defined according Kramer et al. (1999) as "the use of nonspeech audio to convey information." "Sonification is the transformation of data relations into perceived relations in an acoustic signal for the purposes of facilitating communication or interpretation." This acoustic approach represents an important component of our communication network and in general presents a nonsecondary impact on the indoor and outdoor soundscape.

Sonification is a field of research that includes the psychological research in perception and cognition, the development of sonification, tools and sonification design and application.

Sonification becomes a germane discipline inside the constellation of acoustic/ sonic disciplines that can be grouped for their impact on the environment into the domain of soundscape ecology.

Examples of sonification are Geiger counters, sonar, auditory thermometers, and medical and cockpit auditory displays. The possibility offered by sonification includes the analysis of complex data sets such as seismic traces.

Sonification concurs in the complexity of the sonic offerings of this modern world, and its impact on the environment will increase in the future together with the "dematerialization" of modern devices.

Sound is a source of information. Voice, song, and calls are semiotic objects extensively used in the real world by human and nonhuman animals. In a technological world the production of sounds is used as an auditory display that returns non-speech sounds to warn, alarm, indicators of status, etc. (Walker and Kramer 2005).

There are several types of auditory display:

Alert and notification
The sound is used to indicate that an event is to happen; the long beep of a microwave is an example. This sound does not indicate if the food is cooked but simply that the time book is expired. The same is for the telephone ring that does not indicate who is calling, although actually we have ringtones in cell phones that can be personalized with the single number.
Cautions and warnings
It refers to sound that are linked with a specific event, such as fire alarm signals, or the sound that precedes announcements in airports.

There are two basic approach to increase the efficiency of signaling: auditory icons and earcons.

The *auditory icons* are equivalent to visual icons that are symbols which depict objects, such as the image of a deer along a road to alert about possible crossings by wild ungulates. Icons are extensively used in the computer language and in the Internet. So an auditory icon is a sound that copies the sound of the object represented, perhaps a typewriter.

Earcone is a more sophisticated representation of an event, so when a file is moved into the desktop trash the sound is that of crumpled paper. The file deletion may be represented by a plain beep that in a second is degraded in terms of loudness and pitch that metaphorically represent the vanishing of a file.

Audification is the direct translation of a data waveform into sounds. This process requires that the transformation is made along the human audible spectrum with possibilities to increase or decrease the time. One example is represented by seismic data: in this way is possible to distinguish the variations of strata complexity in oil exploration or in earthquake events. In one experiment listeners were able to separate more than 90 % of the underground explosions from earthquake events (Speeth 1961). Some waves when transformed in a sonic format can be easily distinguished from the background noise, and auditory expectation can be produced rapidly; for instance, 24 h of data can be represented in 5 min of acoustic output.

The auditory display is considered a form of applied auditory perception (Walker and Kramer 2005).

Between the different task of the auditory display perception is one of the first objectives. The listener must hear the sound and the changes of its parameters. Perception requires an integration between sound attributes, the environment, and the listener. Especially when we are in a source environment and where competitive sonic sources are available, the use of a *stream analysis* can improve the perceptual process. Spatial separation can help a listener to distinguish two pieces of similar acoustic information according to the position in space from which the two sources come.

After an auditory display is heard and streamed, it is absolutely necessary to attribute the meaning. For this, it is necessary to reduce the ambiguity; for example, the listener must recognize a fire from a depressurization or by the presence of toxic substances.

To optimize the design of an acoustic display it is important to fix some parameters such as loudness, pitch, masking, interacting dimensions, tempo and rhythm, timbre, pain and fatigue, spatial location, and streaming.

Loudness or perceived intensity is the first acoustic approach to a sound. A correct design should maintain this attribute in a way that can be strong enough to communicate but not too much to damage the auditory display.

Masking is the effect by which a sound is shaded by the background noise or by another sound. This is a well-known problem in bioacoustics where animals are often in competition for the sonic environment.

Pitch is one of the most used auditory display dimensions for the ease with which it can be manipulated and perceived. Pitch is less influenced by the sonic environment than loudness. The human hearing capacity ranges from 20 to 20,000 Hz, and the optimum ranges between 80 and 10,000 Hz. The sensitivity to pitch change is highest at 3,000 Hz, and for this reason the acoustic display is optimized for working between 200 and 5,000 Hz.

Interacting dimensions refers for instance to the relationships between pitch and loudness.

Tempo (from Latin, tempus: time) is the speed of music or any sequence of sounds. Traditional tempo indications are given in Italian, ranging (from slow to fast) as *largo, adagio, andante, moderato, allegro, presto, prestissimo.*

Tempo has been used for instance in the Geiger counter in which the change of tempo is related to the increase of radioactivity.

The discrimination between periodic and aperiodic events can be easily obtained by people with changes in temporal frequency of sounds.

Timbre refers to the properties of sound that cannot be expressed in loudness, pitch, or tempo. The capacity to distinguish sounds with different timbres is important for the auditory display. Timbre concurs to the completeness and efficiency of the auditory display.

Pain and fatigue have to be eliminated in every auditory display, of course, but often it is not easy to find the better threshold and connected trade-off. People in general reduce their acoustic attention after half an hour and this is important in auditory design.

Spatial location. The creation of artificial spatiality of sound is today a common technique in movies. The spatiality of sound source facilitates their identification or selection. The psychoacoustic research has established that spatial discrimination varies with angular resolution (1–2° of azimuth in the front, and 5–6° to the side). Differing from vision, sound perception does not requires that the listener be oriented in some direction. Thus, the auditory display can be used also when the eyes are used in another task; for this reason sound can be used to inform busy people for instance engaged in manual performances.

In the use of auditory display, the phase of training appears important; this is the case of blind people but also of sonar operators who must be trained to distinguish the different sound attributes produced by morphology, banks of fish, ships, and submarines.

Summary

Acoustic communication is the most powerful semiotic process with which organisms exchange information. Every phenomenon that reduces the acoustic communication can have severe consequences on the survivorship of many vocal species.

Sonic intrusion appears a growing threat to the maintenance of an high level of biological and ecological complexity. In particular noise of anthropogenic origin masks or degrades the acoustic signals between individuals and species. Urban noise, marine noise, habitat loss and isolation, land use transformation, chemical pollution, hunting activity, and tourism intrusion in remote areas are some of the principal menaces to the communication processes that connect individuals and groups sharing information on resources, refuges, and reproduction.

Long-term monitoring of the sonic environment appears a new and promising approach to understand the dynamic of natural and human-modified systems inside and outside protected areas. Monitoring schemes become important tools to create efficient guidelines to protect these areas on a long-term perspective.

Acoustics coupled with other data can produce a very efficient tool to try to respond to questions pertaining the effect of global warming, the change of land use, the effect of agriculture practice on natural systems or on pest spread.

Acoustic monitoring in marine ecosystems appears a new important methodology able to grasp the changes in the marine sonic environment and the level of noise intrusion. Long term-monitoring requires sonic observatories that represent a set of instruments and procedures that allow the recording, the storing, and the processing of a continuum flux of data scanned at different temporal scales according to the different purposes.

Soundscape design means to consider the sonic environment not simply a communication medium but also an important component of our environment rich with important information.

The research of quiet areas across regions and countries represents an important investigation, and their mapping allows exchange of information in different environmental and socioeconomic databases.

Noise is a process that creates a growing concern for national parks and protected areas managers. External nonrecreational noise intrusions, external recreational noise intrusions, onsite interactivity noise intrusion, and onsite intra-activity noise intrusions are the four main categories of noise sources that have an impact on protected areas.

To improve the recreational services of natural parks and protected areas, information collected from the visitors using interviews and "listening exercise" is important. Human voices and aircraft overflights are the major sources of visitor concern.

The explosive increase of environmental tourism has produced a great pressure of visitors on parks and protected areas also in remote regions, and human crowding is an important source of concern for planners and decision makers.

Perceived crowding (PC) is a measure of the level of acoustic confidence that tourists have after a visit. PC largely depends on the typology of park, the experience of individuals, and the context in which the evaluation is carried out. For instance, encounters at the border of a park were more tolerated than encounters in the core areas. To rating the quality of a park the category of tourists is a further element of difference and visitor behavior is more important than the real density of visitors.

Noise management is a difficult task in protected areas, largely because of the different nature of each protected area. Physical characters, ecosystem structure and dynamics, and fragility to human intrusion prevent the application of standardized management. At least three main actions can be common: the separation of conflicting activities, the reduction of noise effect, and the improvement of visitors expectations.

To visit mountain parks in wintertime the use of snowmobiles and snowcoaches that reduce the risk of individual trekker casualties is required. However, these vehicles are responsible for an altered soundscape that can decrease the rating of a protected area, as recently found by a survey conducted on a large set of visitors. The last generation of visitors were aware of the consequence of the use of vehicles during wintertime in the parks. The annoyance produced by snowmobiles on visitors is confirmed also for wild animals such as wolves and elk, which have an altered level of glucocorticoids (GCs) produced by the hypothalamic-pituitary-adrenal axis.

Aircraft noise is an important source of disturbance, especially in parks and protected areas, but the rating of this effect largely depends by the status of the interviewed persons and their experiences. Visitors who consider natural quiet very or extremely important are more sensitive to aircraft noise of 10–50 dB. Visitors that have experienced a past visit at the same site were 20–35 dB more sensitive to aircraft noise than other visitors. Visitors in adults-only groups were 10–20 dB more sensitive to aircraft noise than other visitors. Finally, visitors on short hikes were 35–70 dB more sensitive to aircraft noise.

Backcountry visitors were more sensitive to aircraft disturbance than frontcountry visitors.

A conservation agenda for natural parks and protected areas should be based on a reasonable trade-off between conservation of natural resources and limiting the impact of visitors.

The reduction of inappropriate sounds seems the major mission in conserving the wilderness of parks, but often this is not possible because of the large number of visitors and the growing intrusions from external sources.

Valuing the sonic heritage represents a new possibility for the holistic landscape evaluation. *The Careggi Landscape Declaration on Soundscape* – June 2012 is the first document that considers the soundscape as a primary component of local heritage and opens new perspectives for land assessment and landscape management.

Managing traffic noise is a priority in urban planning and design. This type of noise is very intrusive and produces annoyed reactions in people. Mitigation procedures to reduce traffic noise can utilize the masking effects of pleasant sounds such as waterfalls or bird choruses but this is possible where noise has low value of acoustic amplitude. A more traditional way to reduce noise is to build green barriers using trees that add further benefits in the urban context such as soil cover, habitat for animals, atmospheric pollutant absorption, and scenic value.

The urban sonic environment is peculiar for the presence of buildings, squares, and green areas that create very heterogeneous sources of sounds. The design of the soundscape in urban areas should be a priority goal for decision makers and urban planners.

Sonification is the use of nonspeech audio to convey information, representing the transformation of data relations into perceived relations through an acoustic signal. Sonification is an important source of anthrophonic sounds. Sonification is realized by auditory displays such as alert and notification, cautions, and warnings. An auditory display is particularly useful when the eyes are used in another task, and sound can be used to inform busy people who are engaged in manual performances.

References

Anderson GS, Rapoza AS, Fleming GG, Miller NP (2011) Aircraft noise dose-response relations for national parks. Noise Control Eng J 59(5):519–540

André M, der Schaar M, Zaugg S, Houégnigan L, Sànchez AM, Castell JV (2011) Listening to the deep: live monitoring of ocean noise and cetacean acoustic signals. Mar Pollut Bull 63:18–26

Barber JR, Burdett CL, Reed SE, Warner KA, Formichella C, Crooks KR, Theobald DM, Fristrup K (2011) Anthropogenic noise exposure in protected natural areas: estimating the scale of ecological consequences. Landsc Ecol 26:1281–1295

Bassett C, Polagye B, Holt M, Thomson J (2012) A vessel noise budget for Admiralty Inlet, Puget Sound, Washington (USA). J Acoust Soc Am 132(6):3706–3719

Benfield JA, Bell PA, Troup LJ, Soerstrom NC (2010) Aesthetic and affective effect of vocal and traffic noise on natural landscape assessment. J Environ Psychol 30:103–111

Boebel O, Kindermann L, Klinck H, Bornemann H, Plotz J, Steinhage D, Riedel S, Burkhardt E (2006) Real-time underwater sounds from the southern ocean. EOS 87(36):361–372

Booth KL, Jones NC, Devlin PJ (1999) Measuring the effects of aircraft overflights on recreationists in natural setting. Department of Conservation technical series, no. 18. New Zealand Department of Conservation, Wellington, 18, pp 1172–6873

Bowlby W, Harris RA, Cohn LF (1990) Seasonal measurements of aircraft noise in a national park. J Air Waste Manag Assoc 40:68–76

Calef G, DeBock EA, Lortie GM (1976) The reaction of barren-ground caribou to aircraft. Artic 29(4):210–212

Cato D, McCauley R, Rogers T, Noad M (2006) Passive acoustics for monitoring marine animals: progress and challenges. Proc Acoust 2006:453–460

Cessford GR (2000) Noise impact issues on the Great Walks of New Zealand. USDA Forest Service Proceedings RMRS-P-15, vol 4, pp 69–76

Codarin A, Wysocki LE, Ladich F, Picciulin P (2009) Effects of ambient and boat noise on hearing and communication in three fish species living in a marine protected area (Miramare, Italy). Mar Pollut Bull 58:1880–1887

Creel S, Fox JE, Hardy A, Sands J, Garrot B, Peterson RO (2002) Snowmobile activity and glucocorticoid stress responses in wolves and elk. Conserv Biol 16(3):809–814

De Coensel BD, Vanwetswinkel S, Botteldooren D (2011) Effects of natural sounds on the perception of road traffic noise. J Acoust Soc Am 129(4):el148–el153

Fan Y, Zhiyi B, Zhujun Z, Jiani L (2010) The investigation of noise attenuation by plants and the corresponding noise-reducing spectrum. J Environ Health 72(8):8–15

Fidell A, Silvati L, Howe R, Pearsons KS, Tabachnick B, Knopf RC, Gramann J, Buchanan T (1996) Effects of aircraft overflights in wilderness recreationists. J Acoust Soc Am 100 (5):2909–2918

Fowler MD (2012) Soundscape as a design strategy for landscape architectural praxis. Design Stud 34(1): 111–128 http://dx.doi.org/10.1016/j.destud.2012.06.001

Freimund W, Sacklin J, Patterson M, Bosak K, Saxen SW (2011) Soundscapes and the winter visitor experience. Yellowstone Sci 19(2):6–13

Gorissen L, Snoeijs T, Van Duyse E (2005) Heavy metal pollution affects dawn singing behaviour in a small passerine bird. Oecologia (Berl) 145:504–509

Gramman J (1999) The effect of mechanical noise and natural sound on visitor experience in units of the national park system. Soc Sci Res Rev 1(1):1–16

Grava T, Mathevon N, Place E, Balluet P (2008) Individual acoustic monitoring of the European Eagle Owl *Bubo bubo*. Ibis 150:279–287

Jensen M, Thompson H (2004) Natural soundscapes in the national park system. George Wright Forum 21(1):14–17

Jeon JY, Lee PJ, You J, Kang J (2010) Perceptual assessment of quality of urban soundscapes with combined noise sources and water sounds. J Acoust Soc Am 127(3):1357–1366

Kalish D, Klaphake A (2007) Visitors' satisfaction and perception of crowding in German National Park: a case study on the island of Hallig Hooge. For Snow Landsc Res 81:109–122

Kaplan S, Kaplan R (1989) The experience of nature: a psychological perspective. Cambridge University Press, Cambridge

Kim SO, Shelby B (2011) Effects of soundscapes on perceived crowding and encounter norms. Environ Manag 48:89–97

Kramer G, Walker BN, Bonebright T, Cook P, Flowers J, Miner N et al (1999) The sonification report: status of the field and research agenda. Report prepared for the National Science Foundation by members of the International Community for Auditory Display. International Community for Auditory Display (ICAD), Santa Fe

Krause B (1993) The niche hypothesis. Soundscape Newsl 6:6–10

Krause B (2002) An educational program guide to listening and recording. National Park Service (NPS) U.S. Department of the Interior

Laiolo P (2010) The emerging significance of bioacoustics in animal species conservation. Biol Conserv 143:1635–1645

Mace BL, Bell PA, Loomis RJ (1999) Aesthetic, affective and cognitive effects o noise on natural landscape assessment. Soc Nat Resour 12:225–242

Mace BL, Bell PA, Loomis RJ, Haas GE (2003) Source attribution of helicopter noise in pristine national park landscapes. J Park Recr Admin 21:97–119

Markman S, Leitner S, Catchpole C, Barnsley S, Müller CT, Pascoe D, Buchanan KL (2008) Pollutants increase song complexity and the volume of the brain area HVC in a songbird. PLoS One 3(2):e1674

Mason R, Roe P, Towsey M, Zhang J, Gibson J, Gage S (2008) Towards an acoustic environmental observatory. Fourth IEE International Conference on eScience. doi: 10.1109/eScience.2008.16
Mazaris AD, Kallimanis AS, Chatzigianidis G, Papadimitriou K, Pantis JD (2009) Spatiotemporal analysis of an acoustic environment: interactions between landscape features and sound. Landsc Ecol 24:817–831
McComb K, Moss C, Durant SM, Baker L, Sayialel S (2001) Matriarchs as repositories of social knowledge in African elephants. Science 292:491–494
Miller NP (2008) US National Parks and management of park soundscapes: a review. Appl Acoust 69:77–92
O'Connor P (2008) Sound of silence: valuing acoustics in heritage conservation. Geogr Res 46(3):361–373
Papadimitriou KD, Mazaris AD, Kallimanis AS, Pantis JD (2009) Cartographic representation of the sonic environment. Cartogr J 46(2):126–135
Pilcher EJ, Newman P, Manning RE (2009) Understanding and managing experiential aspects of soundscapes at Muir Woods National Monument. Environ Manag 43:425–435
Planitz B, Roe P, Sumitomo J, Towsey M, Williamson I, Wimmer J, Zhang J (2009) Listening to nature: acoustic monitoring of the environment. 2009 Microsoft eScience workshop, Pittsburgh
Ramis J, Alba J, Garcia D, Hernandez F (2003) Noise effects of reducing traffic flow through a Spanish city. Appl Acoust 64:343–364
Schafer RM (1977) The soundscape: our sonic environment and the tuning of the world. Destiny Books, Rochester
Schafer RM (1994) Soundscape: our sonic environment and the tuning of the world. Destiny Books, Rochester
Slobodchikoff CN, Ackers SH, Van Ert M (1998) Geographic variation in alarm calls of Gunnison's prairie dogs. J Mammal 79(4):1265–1272
Speeth SD (1961) Seismometer sounds. J Acoust Soc Am 33(7):909–916
Trenkel VM, Ressler PH, Jech M, Giannoulaki M, Taylor C (2011) Underwater acoustics for ecosystem-based management: state of the science and proposals for ecosystem indicators. Mar Ecol Prog Ser 442:285–301
Votsi N-E P, Drakou EG, Mazaris AD (2012) Distance-based assessment of open country quiet areas in Greece. Landsc Urban Plann 104(2):279–288
Walker BN, Kramer G (2005) Mappings and metaphors in auditory displays: an experimental assessment. ACM Trans Appl Percept 2(4):407–412
Watson D, Clark LA (1994) Manual for the positive and negative affect schedule (expanded form). Unpublished manuscript, University of Iowa, Iowa City
Yan HY, Anraku K, Babaran RP (2010) Hearing in marine fish and its application in fisheries. In: He P (ed) Behavior of marine fishes: capture processes and conservation challenges. Wiley-Blackwell, Ames, pp 45–64
Zanette L, Clinchy M, Sung H-C (2009) Food-supplementing parents reduces their sons' song repertoire size. Proc R Soc B 276:2855–2860

Index

A. Farina, *Soundscape Ecology: Principles, Patterns, Methods and Applications*,
DOI 10.1007/978-94-007-7374-5,

B

C

M

Q

R

T

U

The manufacturer's authorised representative in the EU is Springer Nature Customer Service Centre GmbH, Europaplatz 3, 69115 Heidelberg, Germany. If you have any concerns regarding our products, please contact ProductSafety@springernature.com

Printed and bound by CPI Group (UK) Ltd, Croydon, CR0 4YY
06/10/2026
02228539-0001